Photocatalysts and Electrocatalysts
in Water Remediation

Photocatalysts and Electrocatalysts in Water Remediation

From Fundamentals to Full Scale Applications

Edited by

Dr Prasenjit Bhunia
Silda Chandra Sekhar College, India

Dr Kingshuk Dutta
Central Institute of Petrochemicals Engineering and Technology (CIPET), India

Dr S. Vadivel
Saveetha Institute of Medical and Technical Sciences, India

WILEY

This edition first published 2023

Registered Office(s)
John Wiley & Sons, Inc., 111 River Street, Hoboken, NJ 07030, USA
John Wiley & Sons Ltd, The Atrium, Southern Gate, Chichester, West Sussex, PO19 8SQ, UK

Editorial Office
The Atrium, Southern Gate, Chichester, West Sussex, PO19 8SQ, UK
Boschstr. 12, 69469 Weinheim, Germany

For details of our global editorial offices, customer services, and more information about Wiley products visit us at www.wiley.com.

Wiley also publishes its books in a variety of electronic formats and by print-on-demand. Some content that appears in standard print versions of this book may not be available in other formats.

A catalogue record for this book is available from the Library of Congress

Hardback ISBN: 9781119855316; ePub ISBN: 9781119855330; ePDF ISBN: 9781119855323;
Obook ISBN: 9781119855347

Cover image: Courtesy of Kingshuk Dutta
Cover design by Wiley

Set in 9.5/12.5pt STIXTwoText by Integra Software Services Pvt. Ltd, Pondicherry, India

Contents

Preface

A wide array of wastewater treatment alternatives are being investigated nowadays, and this is owing to the increase in polluted wastewater generation because of the growth in population and industrial activities. Advanced oxidation processes (AOPs) have become, in the last few years, a selected alternative due to several advantages, such as their nonselective degradation of pollutants and their easy setup. Photo-based processes have always been one of the most preferred AOP options, due to the possibility of using solar radiation that may reduce the AOPs' high elevated costs. Photocatalysis processes fall under the AOP category, and it is studied worldwide for various applications. A photocatalyst is defined as a "catalyst able to produce, upon absorption of light, chemical transformations of the reaction partners" [1]. The excited state of a photocatalyst interacts repeatedly with the reaction partners, resulting in the formation of reaction intermediates, and regenerates itself after each cycle of such interactions. In effect, the photocatalyst is activated with radiation, which brings about the separation of electrons and holes from, respectively, the valence and conduction bands of semiconductor photocatalysts. This, in turn, starts a series of chain reactions that lead to the generation of oxidants and, ultimately, to pollutant degradation. However, even photocatalysis has limitations for future applications; for instance, the electrons and holes are usually recombined, long treatment times are required, etc. As a solution, the combination of photocatalysis with the application of an electrochemical field (photoelectrocatalysis) has been contemplated.

On the other hand, electrochemistry, an interdisciplinary field of interfacial charge transfer, has been introduced for the decontamination of both organic and inorganic pollutants, and therefore has developed significant worldwide interest toward water remediation. In this connection, electrocatalytic oxidation and reduction usually work together to decompose organic contaminants, and convert heavy metal ions from their toxic to nontoxic form in electrocatalytic advance oxidation processes (EAOPs). Although this technique was elucidated long back in the 1970s, it is only in recent years that the electrocatalysts have become highly encouraging materials toward water remediation. Electrocatalysts degrade or convert the contaminants (organic or inorganic) through profound collision of a very clean reagent – "the electron"; therefore, the technique is recognized as environmentally benign. In addition, this technique has been found to be highly versatile toward degradation of various contaminants, including dyes, pesticides and herbicides, phenolic compounds, pharmaceuticals, etc., and is also able to convert heavy metal ions from their toxic to nontoxic forms. Moreover, the most attractive features of EAOPs toward water treatment are high treatment efficiencies, accumulation of less toxic by-products, and environmental friendliness.

The subject of this book has been meticulously designed to cover both the photo and electrocatalytic water remediation aspects, starting from fundamentals to the applications. This book describes the major functions of catalysts (photo-, electro-, and electrophoto-) in the domain of water remediation, along with the involved chemical reactions, mechanisms, challenges, and up-to-date developments. In addition, the scope of further research on photo, electro, and electrophotocatalysts is also thoroughly discussed. Enriched with critically analyzed and expertly opined contributions from several well-known researchers around the world, this book is likely to serve as one of the most comprehensive and authoritative pieces of literature that has ever been published in this field, and will undoubtedly serve as a potent source of information for those interested in this field.

Prasenjit Bhunia
Kingshuk Dutta
S. Vadivel

Reference

1 IUPAC, (1997) *Compendium of Chemical Terminology: The Gold Book*. Oxford: Blackwell Scientific Publications.

About the Editors

Dr. Prasenjit Bhunia has obtained his Masters and Doctorate in Inorganic Chemistry from Jadavpur University, Kolkata, India. In addition, he has done Post-Doctoral research at National Taiwan University, Taiwan and Sungkyunkwan University, Suwon, South Korea, and has gained prolonged experience in Graphene/Graphene Oxide and their functionalization and applications. He has also served as an Institute Post-Doctoral Fellow in the Department of Chemical Engineering, Indian Institute of Technology Kharagpur, India, and has gathered significant experience in photocatalysis and electrocatalysis. Furthermore, he has noticeable industrial experience from working at Hindustan Unilever Limited, Bangalore and TCG Life Sciences, Kolkata, India. In addition, he has served as a Principal Researcher at TATA Steel Limited, Jamshedpur, India, where he has gained outstanding experience in the field of photocatalytic wastewater treatment. Now, he serves as an Assistant Professor in the Department of Chemistry, Silda Chandra Sekhar College, Jhargram, India (affiliated to Vidyasagar University, Paschim Medinipur, India). At this time, he has 23 international peer reviewed journal articles, 7 book chapters and 5 patents to his credit. His present research field is photocatalytic water treatment.

Dr. Kingshuk Dutta, *FICS*, is currently employed as a Scientist in the Advanced Polymer Design and Development Research Laboratory of the Central Institute of Petrochemicals Engineering and Technology, India. Prior to this appointment, he worked as an Indo–US Postdoctoral Fellow at Cornell University, USA (2018–19) and as a National Postdoctoral Fellow at the Indian Institute of Technology – Kharagpur, India (2016–17), both funded by the Science and Engineering Research Board, Govt. of India. Earlier, as a Senior Research Fellow funded by the Council of Scientific and Industrial Research, Govt. of India., he had carried out his doctoral study at the University of Calcutta, India (2013–16). He possesses degrees in both technology (B. Tech. and M. Tech.) and science (B. Sc.), all from the University of Calcutta. He was also a recipient of the prestigious Graduate Aptitude Test in Engineering (GATE) and National Scholarship, both from the Ministry of Human Resource Development, Govt. of India. His areas of research interest lie in the fields of fuel cells (including alcohol, bio/microbial, and hydrogen fuel cells), sensors, water purification, polymer blends and composites, and biodegradable polymers. At this time, he has contributed to 54 experimental and review papers in reputed international platforms, 25 book chapters, 1 patent application, and many national and international presentations. In addition, he has edited/co-edited four books published by Elsevier and two books published by the American Chemical Society. He has also served as a guest associate handling editor for *Frontiers in Chemistry* and a peer-reviewer for over 180 journal articles, conference papers, book chapters, and research project proposals. He is a life member and an elected fellow of the Indian Chemical Society, a life member of the International Exchange Alumni Network (US Department of State) and a member of the Science Advisory Board (USA). Earlier, he held memberships with the International Association for Hydrogen Energy (USA), the International Association of Advanced Materials (Sweden), the Institute for Engineering Research and Publication (India) and the Wiley Advisors Group (USA).

Dr. S. Vadivel is currently working as an Assistant Professor (Senior Grade) in the Department of Electrochemistry, Saveetha School of Engineering, Saveetha Institute of Medical and Technical Sciences, Tamil Nadu, India. He obtained his PhD in Chemistry (2015) from the AC Tech Campus, Anna University, Chennai, and then successfully completed his postdoctoral fellowship tenure at the Tokyo Institute of Technology, Tokyo, Japan, under the highly prestigious JSPS Fellowship. Furthermore, he received the highly prestigious PIFI fellowship from the Chinese Academy of Sciences in 2019. His current research involves the development of novel materials with graphene, graphitic carbon nitride, in combination with metals, metal oxides, polymers, and carbon nanotubes for energy conversion & storage, and removal of various toxic pollutants. His research results have been documented in 66 peer-reviewed journals, including 5 review articles and 10 book chapters. He also has more than 2200 citations with an h-index of 25. He served as a guest editor for a special issue of *Frontiers in Chemistry*, and is currently serving as an associate editor for Heliyon journal and editing two books for reputed publishers. He is also serving as a peer-reviewer for various high impact journals from Royal Society of Chemistry, American Chemical Society, and Elsevier journals. He has supervised one PhD Scholar under the DST-SERB (Early Career Research Award Scheme) and is currently guiding two PhD scholars.

List of Contributors

Ahmed M. Awad Abouelata Chemical Engineering & Pilot Plant Department, Engineering Division, National Research Centre (NRC), Dokki, Giza, Egypt

Mohsen Ahmadipour Institute of Microengineering and Nanoelectronics, Universiti Kebangsaan Malaysia, Bangi, Selangor, Malaysia

Baquir Mohammed Jaffar Ali Department of Green Energy Technology, Madanjeet School of Green Energy Technologies, Pondicherry University, Puducherry, India

Anamaria Baciu Politehnica University Timisoara, Faculty of Industrial Chemistry and Environmental Engineering, P-ta Victoriei, Timisoara, Romania

Prasenjit Bhunia Department of Chemistry, Silda Chandra Sekhar College, Jhargram, West Bengal, India

Dhruba Chakrabortty Department of Chemistry, B.N. College, Dhubri, Assam, India

Sin Ling Chiam School of Materials and Mineral Resources Engineering, Universiti Sains Malaysia, Nibong Tebal, Malaysia

Kinghshuk Dutta Advanced Polymer Design and Development Research Laboratory (APDDRL), School for Advanced Research in Petrochemicals (SARP), Central Institute of Petrochemicals Engineering and Technology (CIPET), Bengaluru, Karnataka, India

Baizeng Fang Department of Chemical and Biological engineering, University of British Columbia, Vancouver, BC, Canada

Peyman Gholami Department of Chemistry, University of Helsinki, Helsinki, Finland

Alamelu Kaliaperumal Department of Chemical Engineering, Indian Institute of Technology Madras, Chennai, India

Mehdi Al Kausor Department of Chemistry, Science College, Kokrajhar, BTR, Assam, India

Ashitha Kishore Department of Green Energy Technology, Madanjeet School of Green Energy Technologies, Pondicherry University, Puducherry, India

Vineet Kumar Chemistry and Bioprospecting Division, Forest Research Institute, Dehradun, India

Parteek Mandyal School of Advanced Chemical Sciences, Shoolini University, Solan (HP), India

Florica Manea Politehnica University Timisoara, Faculty of Industrial Chemistry and Environmental Engineering, P-ta Victoriei, Timisoara, Romania

Harshvardhan Mohan Department of Chemistry, Research Institute of Physics and Chemistry, Jeonbuk National University, Jeonju, Republic of Korea

Sorina Negrea National Institute of Research and Development for Industrial Ecology (INCD ECOIND), Romania; "Gheorghe Asachi" Technical University of Iasi, Department of Environmental Engineering and Management, Romania

Aashish Priye Department of Chemical Engineering, University of Cincinnati, Ohio, United States

Swee-Yong Pung School of Materials and Mineral Resources Engineering, Universiti Sains Malaysia, Nibong Tebal, Malaysia

Saravana Rajendran Laboratorio de Investigaciones Ambientales Zonas Áridas, Departamento de Ingeniería Mecánica, Facultad de Ingeniería, Universidad de Tarapacá, Avda. General Velásquez, Arica, Chile

Shabnam Sambyal School of Advanced Chemical Sciences, Shoolini University, Solan (HP), India

Pavithra Muthu Kumar Sathya Department of Microbiology, PSG College of Arts & Science, Tamilnadu, India

Albert Serrà Department of Materials Science and Physical Chemistry, Universitat de Barcelona, Barcelona, Catalonia, Spain; Institute of Nanoscience and Nanotechnology (IN2UB), Universitat de Barcelona, Barcelona, Catalonia, Spain

Pooja Shandilya School of Advanced Chemical Sciences, Shoolini University, Solan (HP), India

Rohit Sharma School of Advanced Chemical Sciences, Shoolini University, Solan (HP), India

Malavika Sunil Department of Green Energy Technology, Madanjeet School of Green Energy Technologies, Pondicherry University, Puducherry, India

R. Suresh Laboratorio de Investigaciones Ambientales Zonas Áridas, Departamento de Ingeniería Mecánica, Facultad de Ingeniería, Universidad de Tarapacá, Avda. General Velásquez, Arica, Chile

Sethumathavan Vadivel Department of Chemistry, Saveetha School of Engineering, Saveetha Institute of Medical and Technical Sciences, Chennai, India

Raja Viswanathan Department of Computational Physics, School of Physics, Madurai Kamaraj University, Madurai, India

Acknowledgments

First and foremost, we are thankful to Sarah Higginbotham (Senior Commissioning Editor), Sakeena Quraishi (Associate Commissioning Editor), Jenny Cossham (Associate Editorial Director), Stacey Woods (Managing Editor) and the entire team of Wiley for bringing out this unique book.

The reviewers of our book proposal have played a significant role in ensuring the quality of this book by providing their constructive suggestions and inputs, and we are grateful for that.

We would also like to extend our thankfulness to the subject/domain experts for contributing highly informative chapters.

Finally, we would like to express our love and gratitude towards our family members for their continuous support and unconditional patience.

Prasenjit Bhunia
Kingshuk Dutta
S. Vadivel

1

Fundamentals and Functional Mechanisms of Photocatalysis in Water Treatment

Rohit Sharma[1], Parteek Mandyal[1], Shabnam Sambyal[1], Baizeng Fang[2], Vineet Kumar[3], Peyman Gholami[4], Aashish Priye[5] and Pooja Shandilya[1,]*

[1] School of Advanced Chemical Sciences, Shoolini University, Solan (HP), India
[2] Department of Chemical and Biological engineering, University of British Columbia, 2360 East Mall, Vancouver, BC, V6P 1Z3, Canada
[3] Chemistry and Bioprospecting Division, Forest Research Institute, Dehradun, India
[4] Department of Chemistry, University of Helsinki, Helsinki, Finland
[5] Department of Chemical Engineering, University of Cincinnati, Ohio, United States
* Corresponding author

1.1 Introduction

The continuous hike in world population, industrialization, and agricultural development has increased the trend of organic pollutants in water sources. Fortunately, many biodegradable pollutants can be removed by sedimentation, biodegradation, filtration, adsorption processes, etc. But, the pollutants discharged in the textile, agricultural and pharmaceutical industries usually contain nonbiodegradable, toxic, and carcinogenic pollutants. The extensive usage of industrial chemicals and their partial removal not only pollutes the larger water bodies but also damages the quality of ground water [1, 2]. Water having a low ratio of biological oxygen demand to chemical oxygen demand carries pollutants with a complex composition that is difficult to remove. The use of an oxidizing agent for the water remediation process is one of the facile strategies that can be frequently applied. From the past few years, the realm of photocatalysis has been flourishing as an excellent and green approach toward wastewater remediation.

The term photocatalysis is generally adopted to explicate the activation of chemical reactions under light irradiation. Photocatalysis is based on advanced oxidation processes (AOPs) capable of decomposing the complex and less biodegradable contaminants. Furthermore, photocatalysis is a sustainable, economical, energy-saving, and green technology with zero generation of a secondary pollutant. Previously, various photocatalysts like metal oxides, metal phosphides, carbonaceous materials, layered double hydroxides, metal sulfide, etc. have been broadly investigated for photodegradation [3–8]. The reactive oxidation species (ROS) generated on irradiation are generally accountable for

Photocatalysts and Electrocatalysts in Water Remediation: From Fundamentals to Full Scale Applications, First Edition. Edited by Prasenjit Bhunia, Kingshuk Dutta, and S. Vadivel.

photodegradation. These reactive oxygen species have strong redox potential, which can degrade almost all types of organic pollutants from water. Thus, ideal photocatalysts must possess high light absorption ability, suitable bandgap value, low recombination process, and large surface area.

The construction of an ideal photocatalyst cannot be accomplished easily, however, and there are several infirmities such as wide bandgap, inappropriate separation, migration of charge carriers, high recombination rates of excitons, and low light absorption ability that hindered wider applicability. Especially for bare photocatalysts, these drawbacks are more prominent. For efficient solar light harvesting, semiconductors with a small bandgap are required; on the contrary, such semiconducting materials will speed up the recombination process. These two conditions contradict each other, thereby eventually reducing the overall quantum efficiency. Several modifications were thus done to overcome these drawbacks.

Heterojunction formation is one of the most common and efficient techniques where the different materials with suitable band edge potential and bandgap were fabricated together. Such heterojunction-based photocatalysts provide good charge transfer ability with high charge separation and better oxidation and reduction capability. This chapter, therefore, summarizes the characteristic properties and heterojunctions of various metal oxides, metal phosphides, carbonaceous materials, and other layered materials. Also, the fundamental and functional mechanisms of varied photocatalysts utilized in the water decontamination process are briefly described [5, 9–13]. The mechanism of charge transfer in different heterojunctions, specifically, type II, Z-scheme, and S-scheme are described in detail. Different characterization techniques including X-ray photoelectron spectroscopy (XPS), density functional theory (DFT) calculation, trapping experiments, and electron paramagnetic resonance (EPR) are also elucidated to confirm the successful fabrication of a heterojunction and charge transfer route. Finally, the applications of different metal oxides, metal phosphides, and carbonaceous material based heterojunctions are considered; a summary is then appended.

1.2 Different Photocatalytic Materials for Water Treatment

1.2.1 Metal Oxides Based Photocatalysts

Metal oxides having closely packed structures are generally formed by the coordination of metal ions and oxides. Such a configuration provides stoichiometric diversity and compositional simplicity to metal oxides; thus, they have been vastly investigated in several physical and chemical phenomena like magnetism, electron transport, gas sensing, and photocatalysis. Additionally, the physical, chemical, optical, and electronic characteristics of metal oxides can be enhanced by changing reaction conditions, doping with heteroatoms, and forming nanocomposites of metal oxides. Particularly, when the size of metal oxide particles is reduced to nanoscale dimensions, a large concentration of related atoms is present on their surface. This increases the surface area, which corresponds to its high efficiency, since a large number of active sites will be available for reaction. The absorption capacity of nanomaterials at nanoscale improves in comparison to their bulk counterpart as the reduction in particle size increases the surface-to-volume ratio and surface energy of adsorbent. Furthermore, good recyclability is an advantageous characteristic of metal oxides, which makes them persistent nanomaterials studied in applications from environmental remediation to energy conversion; this is true specifically for transition metal oxides owing to their good crystallinity, versatile surface characteristics, well-controlled structure, nontoxicity, high stability, cost-effectiveness, and nature as an earth-abundant material suitable for the formation of an ideal photocatalyst [14].

Typically, metal oxide semiconductors have a large bandgap value (E_g) of > 3 eV and are extensively employed in photocatalysis. They also demonstrate great potential for the generation of both reducing and oxidizing species on irradiation. The valance band maximum (VBM) and conduction band minimum (CBM) majorly consist of oxygen 2p and metal ns or nd orbitals, respectively. When photons of energy equal to or more than the bandgap value strike on the semiconducting surface, the photoexcited electrons accumulate on the CB while the holes remain on the VB. Once these charge species are created, they are typically trapped on metal oxide surfaces and can oxidize or reduce the pollutant. The basic photocatalytic mechanism is attained by following four steps: (i) light absorption and charge separation, (ii) electron migration to acceptor, (iii) oxidation of donor, and (iv) charge recombination. The dye photosensitization of metal oxide semiconductors occurs in the following steps: (i) visible light absorption, (ii) electron transport from the dye (excited state) to CB of the metal oxide semiconductor, (iii) electrons shifting to acceptor, (iv) charge recombination, and (v) regeneration of sensitizer. Similarly, the ligand-to-metal charge transfer (LMCT) photosensitization is accomplished by (i) visible light stimulated LMCT shifting, (ii) electron transport to acceptor, (iii) charge recombination, and finally (iv) regeneration of adsorbate or degradation of adsorbate in the presence and absence of electron donor. Despite several favorable features of metal oxide semiconductors, various properties should be precisely optimized for practical applications [14, 15].

Generally, wide bandgap and high recombination of excitons are the main drawbacks of metal oxide semiconductors, which reduces their visible light absorption response and also decreases the density of available charge carriers to carry photocatalysis. Wide bandgap allows the absorption of light radiation under the UV region, which consists of only 3–4% of the solar spectrum. Additionally, UV radiation possesses highly energetic photons that can directly activate the chemical bonds of some organic substances and cause the formation of radical intermediates, initiating the nonselective reactions. Furthermore, the study of UV light active metal oxide semiconductors in the laboratory needs specialized glassware and high-cost UV light sources. Several modifications, therefore, such as doping, tuning of surface, defects creation, and heterostructure formation, are commonly used to enhance charge separation and visible light response of semiconductors. Crystallinity, geometry, surface area, and conductivity are some other influencing factors that affect the photocatalytic performance of metal oxide-based photocatalysts. These semiconductors exhibit both n and p-type conductivity depending upon the interaction between oxides and metal orbitals. Usually, O 2p orbitals are more localized whereas oxide orbitals are highly dispersed, resulting in a smaller effective mass of electrons in comparison to holes. Since carrier mobility is inversely proportional to carrier effective mass, most metal oxide semiconductors show higher mobility of electrons than holes and predominately exhibit n-type conductivity.

To attain p-type conductivity, the effective mass of holes should be smaller, needing a more dispersive VBM. To acquire a dispersive VBM in metal oxides, the metal cations need to occupy s or d states close to the VBM. Such an arrangement causes p–s or p–d coupling between oxygen orbitals and metal cations, which consequently increases the VBM dispersion and reduces the effective mass of positive holes. It has been noticed, however, that examples of metal oxides with p-type conductivity are still limited in comparison with metal oxides having n-type conductivity. This is because in metal oxides the VB is mostly composed of localized O 2p orbitals, resulting in a deep acceptor level with a large effective mass of positive charge holes. Such an electronic band structure of both p-type or n-type metal oxide semiconductors allows transport of charge carriers beneficial to photocatalysis [16–18].

1.2.2 Transition Metal Phosphides

Phosphorus can react with most of the elements to form a broad range of compounds called phosphides. Bonding in these compounds is also diverse, ranging from covalent for transition metals to ionic for alkaline earth metals and alkali metals. In general, metal phosphides are the solid substances produced by the incorporation of phosphorus with semimetallic or metallic elements. This class of compounds is defined by the general formula of M_xP_y where, x and y represent metal and phosphorus content, respectively. On the basis of this concentration of metal and phosphorus, metal phosphides can be classified into metal-rich phosphides (for $x > y$), stoichiometric metal phosphides (for $x = y = 1$), and phosphorus-rich metal phosphides ($x < y$). Interestingly, different types of metal phosphide exhibit distinct properties. For example, due to a large number of strong metal–metal bonds, metal-rich and stoichiometric metal phosphides possess superconducting and semiconducting behavior. Metal phosphides also generally form clusters or oligomers as they generally have rich phosphorus–phosphorus bonding.

The phosphorus-rich metal phosphides generally demonstrate high reactivity, low thermal stability, and softer material characteristics that are different from their stoichiometric counterparts. Most phosphorus-rich metal phosphides also display low thermal stability and disintegrate into more-rich metal phases and elemental phosphorus [19, 20]. Most physical and chemical characteristics of metal phosphides are similar to nitride and carbide, which form relatively simple lattices where N and C atoms are situated in interstitial spaces between metal atoms. But, in the case of phosphides, due to the larger atomic radius of the phosphorus atom (0.109 nm) compared to N (0.065 nm) and C (0.071 nm), phosphorus atoms resist fitting into the simple octahedral holes created by closely packed metal atoms. For the same explanation, metal atoms can form a triangular prism where nonmetal atoms are surrounded by metal atoms, similar to boride and sulfide. However, this structure changes in the case of metal-rich phosphides, where an increased number of metal atoms form 9-fold tetrakaidecahedral coordination in which the extra metal atoms occupy the space near the center of vertical faces of the triangular prism (Figure 1.1a).

The different arrangement of metals and phosphorus signifies different structures (Figure 1.1b). For instance, MoP is isostructural with WC, VP adopts the NiAs structure, and TaP and NbP have a closely related NbAs structure. The monophosphides of group 6–10 elements mostly acquire NiP and MnP structures, which resemble a distorted NiAs structure. In contrast to sulfide, phosphide has no layered structure and therefore avails of

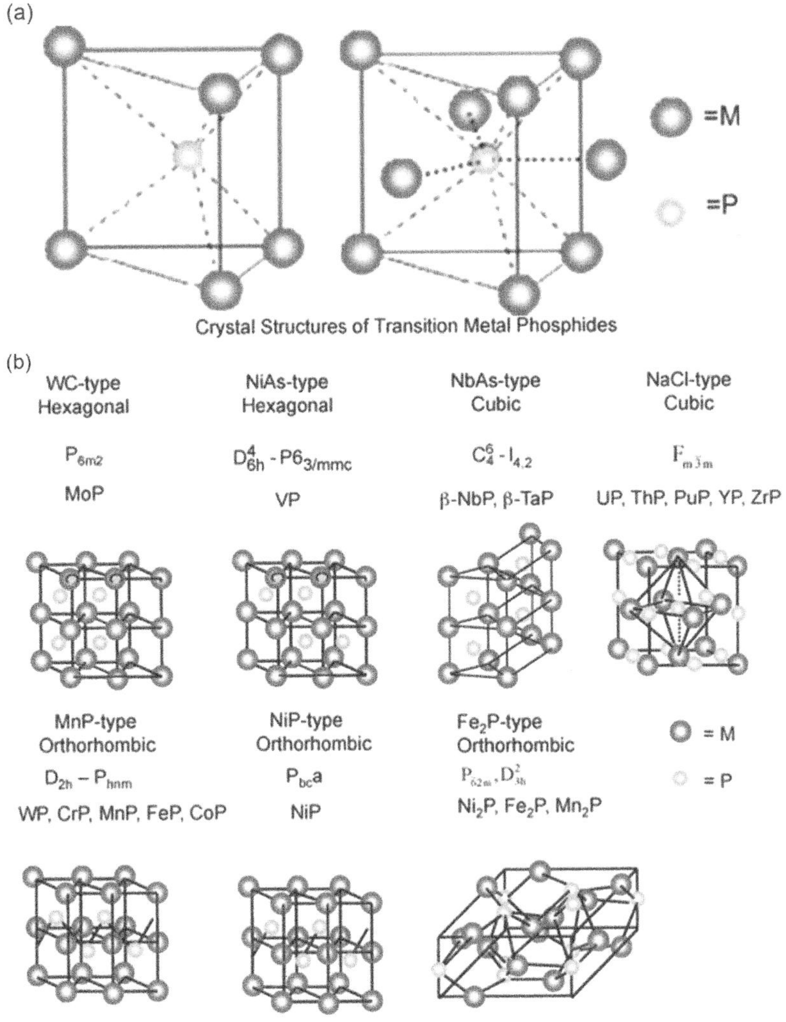

Figure 1.1 (a) Triangular prism and tetrakaidecahedral structures in phosphides (b) Crystal structures of metal-rich phosphides. Reprinted with permission from Ref [20].

greater access to edge sites and active corners. Among the diverse class of metal phosphides, transition metal phosphides are recognized as one of the most important types, composed of transition metals and phosphorus. Nearly all transition metals can be phosphorized to produce transition metal phosphides. In general, transition metal phosphides are regarded as a dopant of phosphorus atoms into the crystal lattice of transition metals. The high electronegativity of the P atom attracts the electrons from metal atoms, and the negatively charged phosphorus atoms, therefore, function as a base to attract positive species during catalysis. By increasing the concentration of P atoms, transition metal phosphides are highly beneficial especially for the electrocatalytic hydrogen evolution reaction (HER), where negatively charged phosphorus atoms can capture protons. Previous reports showed that among $Ni_{12}P_5$, Ni_5P_4, and Ni_2P, Ni_5P_4 with the highest concentration of phosphorus (44%) revealed the highest activity [21, 22].

Moreover, the conductivity of metal phosphides is also influenced by the content of P atoms. Generally, with increased concentration of phosphorus atoms, metal sulfide displays semiconducting or insulating characteristics. This is for the reason that the more electronegative nature of phosphorus atoms inhibits the delocalization of electrons in metals, which consequently decreases electrical conductivity with a large number of phosphorus atoms. However, this nature of metal phosphide can be minimized and electrical conductivity improved by incorporating a large ratio of metal ions. The corrosion resistance in acidic media of transition metal phosphides is also related to the number of phosphorus atoms. It was suggested that metal dissolution became thermodynamically less favorable when alloyed with phosphorus atoms [23]. The research stated that less soluble phosphate is produced by the course of oxidation on the surface that prevents transition metal phosphides from dissolution. Additionally, a few metal phosphides are susceptible to showing magnetic behavior. Brock et al. investigated how the magnetic characteristics of MnP and FeP depended upon their crystallite size [24]. In bulk phases, both types of phosphide express antiferromagnetic interaction with complex low-temperature magnetic structures (helimagnetic). The transition in magnetic behavior for FeP takes place above 125 K where it becomes paramagnetic, while MnP has ferromagnetic characteristics above 50 K. When the crystal size of both FeP and MnP is at nanoscale dimensions, the helimagnetic nature was completely diminished. The small nanoparticle size destabilizes the helimagnetic configuration state, due to which nanoparticles of FeP showed paramagnetic behavior over the entire range of temperature (5–300 K). For the same reason, MnP nanoparticles have ferromagnetic properties below the Curie temperature (291 K). Also, the temperature-dependent transition to superparamagnetic state was investigated for MnP nanoparticles. Therefore, owing to versatility in chemical composition imparting different structural, optical, magnetic, and physiochemical properties, metal phosphides have emerged as an earth-abundant and highly active photocatalytic material.

1.2.3 Carbonaceous Materials

Carbon-based nanomaterials have a proclaimed huge versatility and can be chemically combined with a large number of different nanomaterials. Carbonaceous materials have been designed successfully for various applications due to their large surface area, high conductivity, facile synthesis, and magnificent physicochemical properties. By regulating the preparation conditions carbonaceous materials with desirable properties can be obtained. So far several carbonaceous materials like carbon nanotubes, carbon dots, graphene, fullerene, and activated carbon have been developed to utilize in various photocatalytic applications. The photocatalytic activity is generally governed by various parameters like number of layers, size of materials, interfacial contact surface, type of functionality present on the surface, density, and type of defects. The large surface area of carbonaceous materials increases the number of active sites that affect the catalytic activity [25, 26]. The specific surface area can be determined by nitrogen adsorption–desorption isotherm.

Among different allotropes of carbon, one-dimensional carbon nanotubes have a potential utilization in several applications. The sp^2 bonds between the carbon atoms enhance the tensile strength of carbon nanotubes compared to steel. At high-pressure conditions, carbon nanotubes can form a bond together and interchange some of the bonds from sp^3 to

sp^2 hybridization, with characteristic features imparted such as excellent electronic and thermal properties. Depending upon their surface structure, carbon nanotubes can behave as metallic or semiconducting materials, displaying high electronic mobility. These excellent properties of carbon nanotubes allow them to serve as a fantastic material in photocatalysis that can increase the lifetime of charge carriers by acting as an effective electron sink, enlarging the surface area, and improving the electron transfer rate [27–29].

Graphene is a sheet-like material consisting of sp^2 hybridized carbon atoms arranged in a honeycomb-like crystal matrix. The sheet-like morphology of graphene has unique mechanical, electronic, thermal, and photonic properties with high thermal conductivity (5000 $W\ m^{-1}\ k^{-1}$), fast charge mobility (250 000 $cm^2\ V^{-1}\ s^{-1}$), excellent conductivity (up to 6000 $S\ cm^{-1}$), and large surface area (2630 $m^2\ g^{-1}$). Additionally, the sheet-like structural characteristics of graphene make it highly transparent with high visible light absorption capacity ($< 2.3\%$) and high strength (130 Gpa) [30–32]. Graphene provides a new passage for an electron to enhance its migration during photocatalysis [33]. It acts as good support material to anchor the metal nanoparticles and prevent their leaching into the system. Due to its unique properties, two-dimensional (2D) graphene has potential applications in photonics, optoelectronics, biomedicine, sensors, and environmental remediation. Among diverse applications, a combination of graphene with a metal oxide semiconductor-based photocatalyst is considered a green and effective approach for photodegradation. The graphene nanoparticles act as an absorbent for various organic pollutants, metals (Mg, Fe, Ca), and gases such as CO_2 from the aquatic environment. Graphene nanosheets facilitate the charge transfer from the semiconductor surface, which reduces charge recombination. Lanthanide orthovanadate nanoparticles such as $EuVO_4$, $SmVO_4$, and $GdVO_4$ can be loaded over fluorine-doped graphene nanosheets that have high photocatalytic degradation and bacterial disinfection efficiency [34–36]. Singh et al., synthesized an AgBr/BiOBr/graphene heterojunction for phenol photodegradation under visible light [37]. Further, an Eu^{3+} doped ZnO/Bi_2O_3 heterojunction was fabricated with a graphene oxide sheet to assess the photocatalytic efficiency [10]. Fortunately, the construction of graphene-based nanomaterials provides a facile method to accelerate their activities and stabilities.

Graphitic carbon nitride, $g\text{-}C_3N_4$, is a polymeric form of carbonaceous material containing N and C atoms with a minute amount of H atoms linked through a tris-triazine-based unit. Compared to other carbon materials, $g\text{-}C_3N_4$ displays primitive surface functionality and electron-rich properties due to the presence of H and N atoms. Perfect $g\text{-}C_3N_4$ ideally contains C–N bonds causing electron delocalization in π-stage. The synthesis of $g\text{-}C_3N_4$ through polycondensation of cyanamide gives an insufficient quantity of H in the form of 1° and 2° amine groups at the edges. The existence of H atoms reveals that the actual $g\text{-}C_3N_4$ is partially reduced and thus induces surface defects worthwhile to improve delocalization. Further, high thermal stability permits this material to be used in either gaseous or liquid surroundings at a high temperature, which is potentially advantageous in heterogeneous catalysis. The metal-free nature, low synthesis cost, and suitable bandgap value of nearly 2.7 eV make $g\text{-}C_3N_4$ highly promising for photodegradation, CO_2 reduction, and water splitting [27, 38].

Carbon quantum dots (C-dots, CDs, or CQDs) are novel carbon nanomaterials having a size less than 10 nm. This class of carbon materials exhibits high resistance against photobleaching, strong chemical inertness, good solubility, and facile modification. CQDs possess electron donating and accepting properties that endow chemical and electrochemical luminescence, having immense use in catalysis, sensors, and optronics. The novel chemiluminescence properties for CQDs were obtained in the presence of a strong basic solution of NaOH and KOH. The CQDs show good electron donating ability toward dissolved O_2 to form $O_2^{\bullet-}$ in basic solution. Additionally, the radiative recombination of electrons generated by chemical reduction of CQDs and thermally excited holes was responsible for chemiluminescence properties under alkaline conditions [39–41]. Activated carbon is another beneficial form of carbon widely used due to its high surface area, simple synthetic process, and robust physiochemical stability. It is not only used as an effective support material for various metal-based catalysts, but its synergistic effects with metal also enhance its photoefficiency. The ultrahigh surface area of activated carbon enhances its capacity to arrest organic pollutants from air or water and also increases the number of active sites. The intermediated species formed after completion of photocatalysis are absorbed by activated carbon, and can be further used in the next degradation cycle. Furthermore, the incorporation of activated carbon with metal nanoparticles reduces the charge recombination rate [42, 43].

1.2.4 Other Photocatalysts

Apart from the nanomaterials discussed in Sections 1.2.1–1.2.3, metal–organic framework (MOF), layered double hydroxides (LDH), and metal sulfide are also extensively explored for photocatalysis [6, 44, 45]. In recent years, MOF has been extensively explored in the field of photocatalysis [46, 47]. MOF particularly consists of metal nodes interconnected with organic linkers. The metal nodes serve as a connecter, and linker refers to different organic ligands; thus, it is a mingling of inorganic and organic chemistry that builds an open structure. In detail, MOF is a class of porous polymer crystals that are theoretically connected in an infinite lattice between two secondary building units (SBUs), organic linkers, and metal clusters through coordination bonds. The different types of organic linkers, ditopic (having two coordination functionalities), tritopic (three coordination functionalities), and polytopic (more than three coordination functionalities) can be utilized during MOF construction. Similarly, a finite polyatomic inorganic cluster (with two or more metal atoms) or infinite inorganic units can be incorporated in MOF structure.

On the basis of point of extension or connectivity of organic linker with metal cluster, MOF appears in different shapes like polygon, infinite rod, or polyhedron. As an organic substance, MOFs are highly flexible; therefore, it is crucial to properly identify all the edges, vertices, and points of extension present in their structure. The SBUs of MOFs are majorly interconnected through coordination bonds along with other weak interactions like H-bonds, van der Waals, or pi-bonds, providing more flexibility to their structure. In addition to this, MOFs possess ultrahigh surface area (1000–10 000 $m^2\ g^{-1}$), tunable metals/organic linker cluster, well-ordered porosity (micro/mesopores), high pore volume (go up to 4.4 $m^3\ g^{-1}$), and high thermal stability (200°C), which enhances photophysical and photochemical characteristics. Further, the topology, surface area, and pore size of MOF can be altered by choosing suitable SBUs. The coordination networks between the components of MOFs play a key role in regulating their band gaps, and depending upon this network, MOFs can act as semiconductors or insulators with a wide range of bandgap (1.0–5.5 eV).

The photocatalytic process of MOFs is very similar to inorganic semiconductors initiated by the absorption of photons having sufficient energy. Unlike inorganic semiconductors, however, the higher and lower energy levels in MOFs are identified as LUMO (lowest unoccupied molecular orbitals) and HOMO (highest occupied molecular orbitals) instead of the conduction band (CB) and valence band (VB). Typically, modified MOFs mainly have three active sites: i) reactive functional groups in organic ligands, ii) active metal sites, and iii) any guest species installed in porous MOFs during functionalization. These active sites are responsible for deriving photocatalytic mechanisms on MOFs [44, 46, 47].

LDHs are the class of 2D layered materials having structural similarities to hydrotalcite-like compounds, and are also known as host–guest materials. These compounds are derived by the general formula $[M^{II}_{1-x} M^{III}_{x} (OH)_2]^{x+} A^{n-}_{x/n}.yH_2O$, where M^{III} and M^{II} are the trivalent and divalent metal cations, A^{n-} are the intercalated anions, and y is the count of water molecules. The host structure of LDHs is composed of a brucite $Mg(OH)_2$ like layer where central metal ions are coordinated octahedrally by six OH^- ions. In host layers of LDHs, some of the bivalent metal cations are replaced by trivalent or very rarely by tetravalent metal cations, which induce net positive charge into these layers. To provide electric neutrality and stability to the structure of LDHs, negatively charged anions occupy the space between brucite layers, known as interlayer galleries. Interestingly, either bi, tri, or tetravalent metal cations are uniformly distributed within the hydroxide layers without forming any cluster materials. Most metal cations used to form LDHs generally belong to the third and fourth periods of the periodic table; therefore, the high dispersibility of transition metal octahedrons allows the charge migration and suppresses the recombination rate beneficial for photocatalysis.

Furthermore, the hydroxide ions present on the positively charged brucite layers of LDHs react with VB holes and produce $^{\bullet}OH$ species, which are a powerful intermediate generated during photocatalysis that possess strong oxidation potential. Thus, the presence of hydroxide ions on an LDH surface increases its photoactivity. The role of doping is well-known in modifying the photocatalytic properties of photocatalysts. Interestingly, it has been noticed that the existence of M–O bonds and more metal cations in LDHs allow them to behave as a doped semiconductor. However, if the concentration of these metal cations is optimized, then metal cations with the highest oxidation state will behave as a dopant. Other than these properties, high surface area, thermal stability, variability in the bandgap, compositional flexibility, and structure memory effect are unique characteristics of LDHs. Compared with bulk nonlayered photocatalysts, the layered structure of LDHs supplies a more favorable environment for the separation and diffusion of charge species by MO_6 octahedrons [6, 48, 49]. Similarly, metal sulfide is another important class of nanomaterials significantly used in several potential applications due to their interesting optical, electrical, field emission, thermoelectrical, magnetic, and photoelectrical properties, and photocatalytic activity.

1.3 In-depth Mechanisms of Photocatalysis

Photocatalysts are semiconducting materials that carry the ability to produce electron–hole pairs on illumination as illustrated in Figure 1.2a. Based on the bandgap value, which is determined by calculating the energy difference between the VB and CB, materials can be

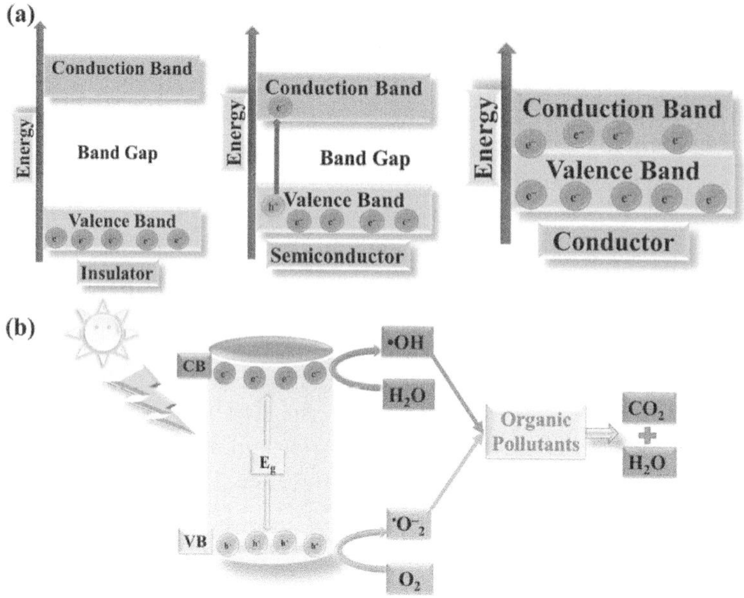

Figure 1.2 (a) Band gap distribution between different types of materials. Redrawn with permission from Ref [13] (b) General mechanism of photocatalysis.

categorized basically into three types: insulators, semiconductors, and conductors, having band gap values > 5.0 eV, 1.5–3.0 eV, and < 1.0 eV, respectively. From these values, semiconducting materials seem most suitable to be employed as photocatalysts. During the excitation process on irradiation, positively charged holes are generated in the VB, and the process leads to the formation of charge carriers. These photogenerated electrons and holes are used for reducing an acceptor and oxidation of a donor, respectively, which reveals the importance of photocatalysts providing both oxidation and reduction environments.

The future of these photogenerated charge carrier species strongly depends upon the redox level of the substrate and relative positions of the VB and CB. Based on different redox potentials and VB and CB alignment, four types of interaction can take place between a substrate and a photocatalyst: i) Reduction of the substrate occurs if the CB of the photocatalyst is aligned at a higher position than the redox level of the substrate. ii) Oxidation of the substrate occurs when the VB of the photocatalyst is positioned at a lower energy level than the redox level of the substrate. iii) Both oxidation and reduction occur when the positions of VB and CB respectively lie at lower and higher energy levels as compared to the redox level of the substrate. iv) Neither reduction nor oxidation occurs when the CB and VB of the photocatalyst are positioned at lower and higher levels than the redox level of the substrate (Figure 1.2b) [13]. In general, the photocatalytic principle can be defined by the generation of electrons and holes having a lifetime in the femtosecond range. These charge carriers undergo separation and migration from bulk to the surface where they interact with the substrate through any of the conditions mentioned here. These photogenerated charge carriers react with water or surface absorbed oxygen to form reactive oxygen species (ROS), such as hydroxyl radicals and superoxide radical anions possessing strong redox abilities. These ROS are the major species accountable for the degradation of various kinds of pollutants in wastewater. These photogenerated charge pairs, however, show a tendency to recombine both on the surface and in the bulk of a photocatalyst. During the process, the photoexcited electrons come back to their nonexcited states by dissipating energy in the form of heat and bringing the whole process to the initial stage, which suppresses the photocatalytic efficiency [50]. Thus, the progress of some simple strategies to overcome these drawbacks is gaining paramount significance for enhancing photocatalytic activity.

Till now, several approaches have been followed to conquer these challenges. Amid various strategies, heterostructure formation becomes much more promising due to its advantages in separating charge carriers [51]. Several kinds of heterojunctions have been evolved to solve the aforementioned issues, among which type II, Schottky, Z-scheme, and S-scheme draw considerable attention. Type II heterojunctions are the combination of two semiconductors aligned in a staggered band configuration followed by the type II route for charge carrier transport. When the surface is exposed to light of appropriate energy, both photocatalysts can be activated to produce electron–hole pairs. In a detailed mechanism, the electrons present in the CB of PC-1 aligned at a higher position are migrated to the CB of PC-II, whereas the holes transfered between two semiconductors follow reverse migration (Figure 1.3a). This migration route provides the efficient separation of excitons and suppresses the recombination rate. For example, Shen and coworkers synthesized type II heterojunctions of black TiO_2/g-C_3N_4 that degraded almost 95% of methyl orange and produced 555.8 μmol h^{-1} g^{-1} H_2 [52]. The photoexcited electrons from the CB of g-C_3N_4 transferred to the CB of TiO_2, and holes migrated from the VB of TiO_2 to the VB of g-C_3N_4 thus following the type II route. They confirmed the type II heterojunction formation through X-ray photoelectron spectroscopy. The N 1s spectra of g-C_3N_4/TiO_2 showed a shift in XPS peaks by the amount of 0.5 eV toward higher energy as compared to pure g-C_3N_4. This result indicates that when two photocatalysts are brought close to each other, electron density decreases due to the shifting of photoelectrons from g-C_3N_4 to TiO_2.

Similarly, the C 1s XPS peak of TiO_2/g-C_3N_4 has a 0.5 eV positive shift compared to bare g-C_3N_4 (Figures 1.3b and c). The shifting in C 1s and N 1s XPS peaks signifies the decrease in electron density on the g-C_3N_4 surface when combined with TiO_2, which implies the transfer of electrons from the CB of g-C_3N_4 to the CB of TiO_2. Type II heterojunctions seem ideal owing to their charge separation ability, hence triggering innumerable investigations, but on closer inspection numbers of inadequacies are observed. For example, during this transfer photoexcited electrons accumulate on the photocatalyst, which has low CB potential, whereas holes accumulate on that VB position which has low oxidation potential.

Additionally, the charge carriers feel an electrostatic repulsion from the similar charge as electrons move from the CB of PC 1 to the CB of PC 2 and holes migrate from the VB of PC 2 to the VB of PC 1, which makes this transfer somehow difficult. Although this transfer mechanism provides the charge separation, it correspondingly weakens the redox ability. To improve the redox ability of the photosystem, therefore, a new type of charge transfer mode was proposed in 1979 by Bard [52]. Initially, the traditional Z-scheme was involved where suitable redox ion mediators such as I_3^-/I^-, Fe^{3+}/Fe^{2+}, and IO_3^-/I^- are coupled between two photocatalysts. In this heterojunction two photocatalysts align in a staggered band alignment; the holes from the VB of PC 1 react with electron donor species to yield an electron acceptor, while the photoexcited electrons from the CB of PC 2 react with the

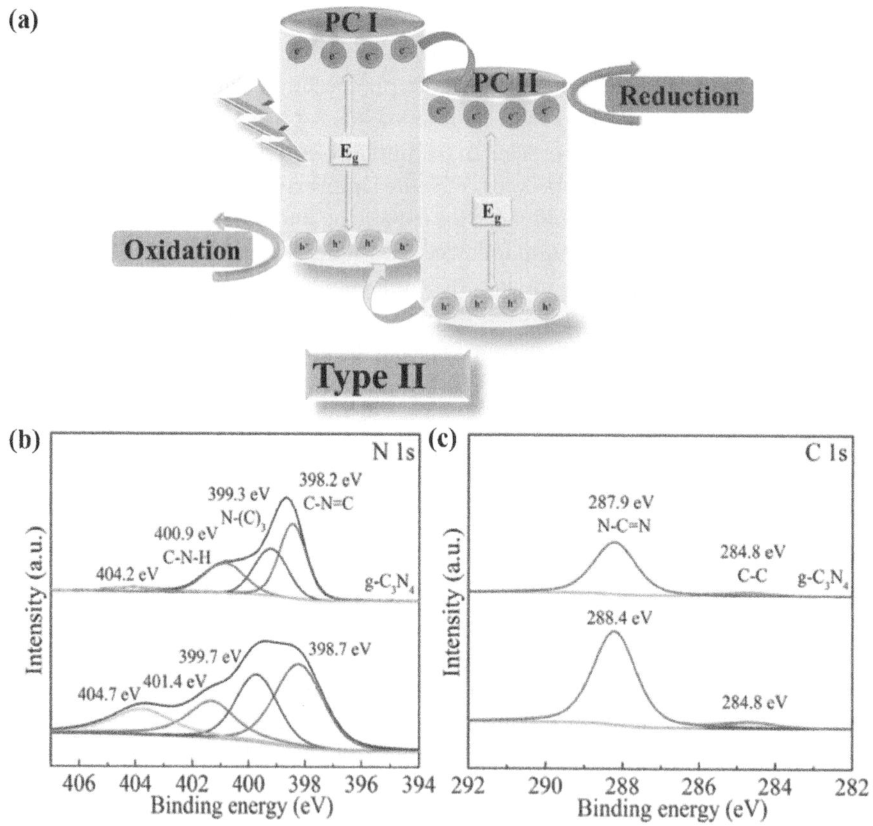

Figure 1.3 (a) General mechanism of Type II heterojunction (b) XPS N 1s spectra of g-C_3N_4 and b-TiO_2/g-C_3N_4 (c) C 1s spectra of g-C_3N_4 and b-TiO_2/g-C_3N_4. Reprinted with permission from Ref [52].

electron acceptor to yield an electron donor (Figure 1.4a). The rest of the electrons and holes reserved in the CB and VB of PC 1 and PC 2 ultimately participate in photoreduction and photo-oxidation reactions.

The traditional Z-scheme, therefore, provides high charge separation efficiency together with high redox ability. This system, however, possesses many drawbacks: such a system can be operated only in the solution phase, chances of side reactions are higher, and redox mediators are colored substances that therefore induce the light-shielding effect, reducing photocatalytic performance. In addition to these disadvantages, such a system is also pH sensitive; for example, while using Fe^{3+}/Fe^{2+} as a redox ions pair, strong acidic conditions are required, because under alkaline or weakly acidic conditions precipitation of iron may occur. To broaden the scope of Z-scheme heterojunctions, therefore, the second generation of Z-scheme was developed by Tada and groups in 2006, deemed the all-solid-state Z-scheme [53]. As its name suggested this system can be used in a solid phase where instead of using the shuttle redox ions pair, solid conductors are utilized. When this system is activated by light irradiation, the electrons from the CB of PC II are shifted to the solid conductor and from there to the VB of PC I (Figure 1.4b).

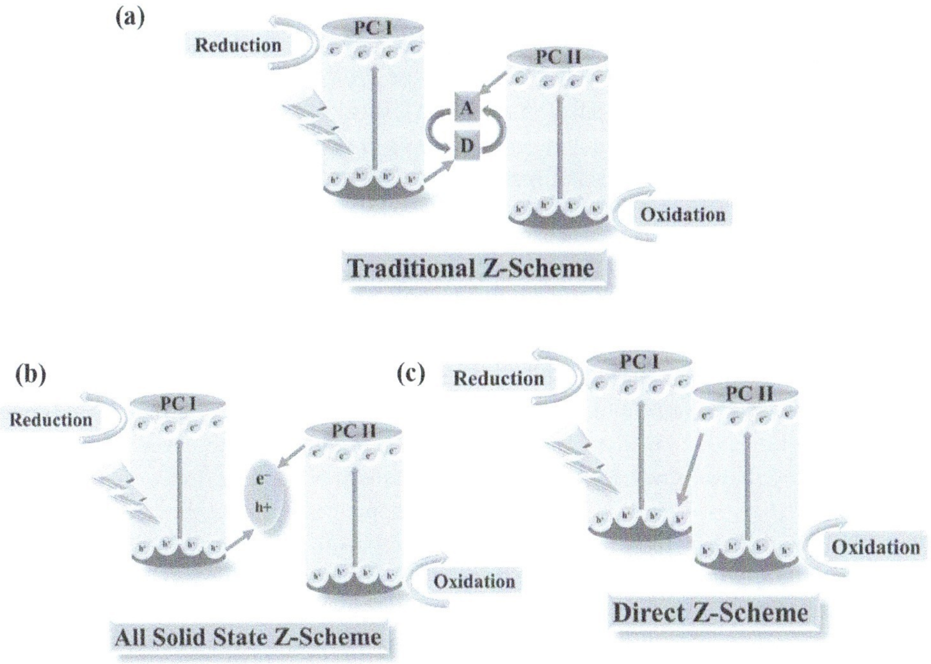

Figure 1.4 Charge transfer route via (a) Traditional Z-scheme (b) All-solid-state Z-scheme (c) Direct Z-scheme.

Such a system, therefore, accelerates charge transfer by shortening the charge transport length and enhancing photoactivity. The desired electron transfer route is difficult to attain, however, since in the all-solid-state Z-scheme, electrons are transferred through solid conductors. For example, the metal conductor will possibly accept the electrons from both the photocatalysts and can even drag the electrons having strong oxidation potential into the recombination process. Secondly, it is difficult to precisely control or guarantee the installation of metal conductors exactly between two photocatalysts. It may be located randomly on the surface of photocatalysts and may act as a cocatalyst instead of a charge transfer shuttle. Also, most of the metal conductors exhibit a light absorption property that induces the light-shielding effect resulting in a poor light absorption tendency. The incorporation of metal conductors also boosts the cost of production; therefore, the much superior third generation of the Z-scheme series was developed in 2013 by Yu et al. [54]. This latest make of Z-scheme is known as the direct Z-scheme, where no intermediates, conductor, or redox mediator are incorporated (Figure 1.4c). This transfer mode thus reduces the backward reaction tendency and shielding effect caused by redox mediators or metal conductors. The direct Z-scheme efficiently separates the charge species, suppresses recombination, and provides the electrons and holes stronger redox ability for the photocatalytic reactions. There are many techniques used to support the charge transfer mechanism via the Z-scheme pathway. For example, Zheng and co-workers synthesized an all tungsten-derived Z-scheme photocatalyst between lead tungstate ($PbWO_4$) and hydrated tungsten trioxide ($WO_3 \cdot 0.33\ H_2O$) [55].

A platinum (Pt) photodeposition experiment was performed to identify the type of charge transfer mechanism. From transmission electron microscopy (TEM) analysis, the lattice spacing of small particles of 0.225 nm was related to Pt (111) facets. The results demonstrate that the reduction of Pt occurs on the $PbWO_4$ surface, illustrating that photoexcited electrons hold on $PbWO_4$ rather than on $WO_3{\cdot}0.33\ H_2O$. These findings were also supported by the band positions of lead tungstate and hydrated tungsten trioxides having VBs at 1.77 eV and 2.95 eV and CBs at −2.11 eV and −0.2 eV, respectively. The radical trapping experiment is another superb technique that can be used to confirm the charge transfer. By investigating the type of generated reactive species and band potential, the migration route of photoexcited electrons can be considered. Li and groups constructed a BiOCl-Au-CdS heterojunction via stepwise deposition of Au and CdS [56]. They proved the Z-scheme transfer mechanism in a BiOCl-Au-CdS heterojunction via radical trapping experiments. If BiOCl-Au-CdS were a conventional heterojunction, the photoexcited electrons would confidently move to the CB of BiOCl from the CB of CdS. This migration route did not generate any kind of $^{\bullet}O_2^-$ radical species, as the reduction potential of BiOCl (0.14 eV) is more positive than $E_0(O_2/^{\bullet}O_2^-)$ (−0.046 eV). But, the radical trapping experiment showed the generation of $^{\bullet}O_2^-$ radicals in BiOCl-Au-CdS during photocatalysis, indicating the Z-scheme charge migration where electrons were transferred to the CB of CdS (−0.74) to produce $^{\bullet}O_2^-$ radicals (Figure 1.5a, b and c). Similarly, Xu et al. used XPS and DFT calculation to confirm the Z-scheme shifting in $TiO_2/CuInS_2$ [57]. The in situ XPS analysis is done under light illumination, and binding energy of O 1s and Ti 2p in $TiO_2/CuInS_2$ was shifted in a positive direction by the amount of (0.2 eV) compared to the value in dark (Figure 1.5d and e). Similarly, the binding energy of In 3d and Cu 2p of $TiO_2/CuInS_2$ displayed a negative shift compared with bare $CuInS_2$ under light irradiation (Figure 1.6a). The positive and negative shift in binding energy, therefore, confirmed that electron transfer from TiO_2 to $CuInS_2$ agreed well with the Z-scheme transfer route. DFT is another approach they used to support the construction of Z-scheme transfer between $CuInS_2$ and TiO_2. These calculations were based upon the well-known fact that electrons can transport from higher Fermi levels to lower Fermi levels of involved semiconductors. The Fermi levels in this work calculated by DFT were found to be lower for TiO_2 than $CuInS_2$ indicating the electron transfer from $CuInS_2$ to TiO_2 (Figure 1.6b). These electron transfer processes rearranged the Fermi levels and generated a built-in electric field responsible for photoexcited electron transport from TiO_2 to $CuInS_2$ as confirmed by XPS analysis (Figure 1.6c). Although the Z-scheme solved many shortcomings of type II heterojunctions, still there is a lot of hesitancy over it, since its first and second generations are problematic, premature, and do not explain the reaction at the interface.

To overcome grievous fundamental issues of type II and Z-scheme heterojunctions, then, a brand new concept known as the step scheme (S-scheme) was recently introduced. The S-scheme mechanism describes charge transfer in photocatalytic heterojunctions clearly and vividly. According to this concept, the photocatalysts used in heterojunction formation are divided into reduction photocatalyst (RP) and oxidation photocatalyst (OP), depending on their band structure. One with a high CB potential is known as the RP, while a photocatalyst with lower VB potential is considered as OP. The S-scheme heterojunction is formed between the OP and RP aligned in a staggered band configuration wherein CB electrons of the RP and VB holes of the OP are only effective to carry out photocatalysis.

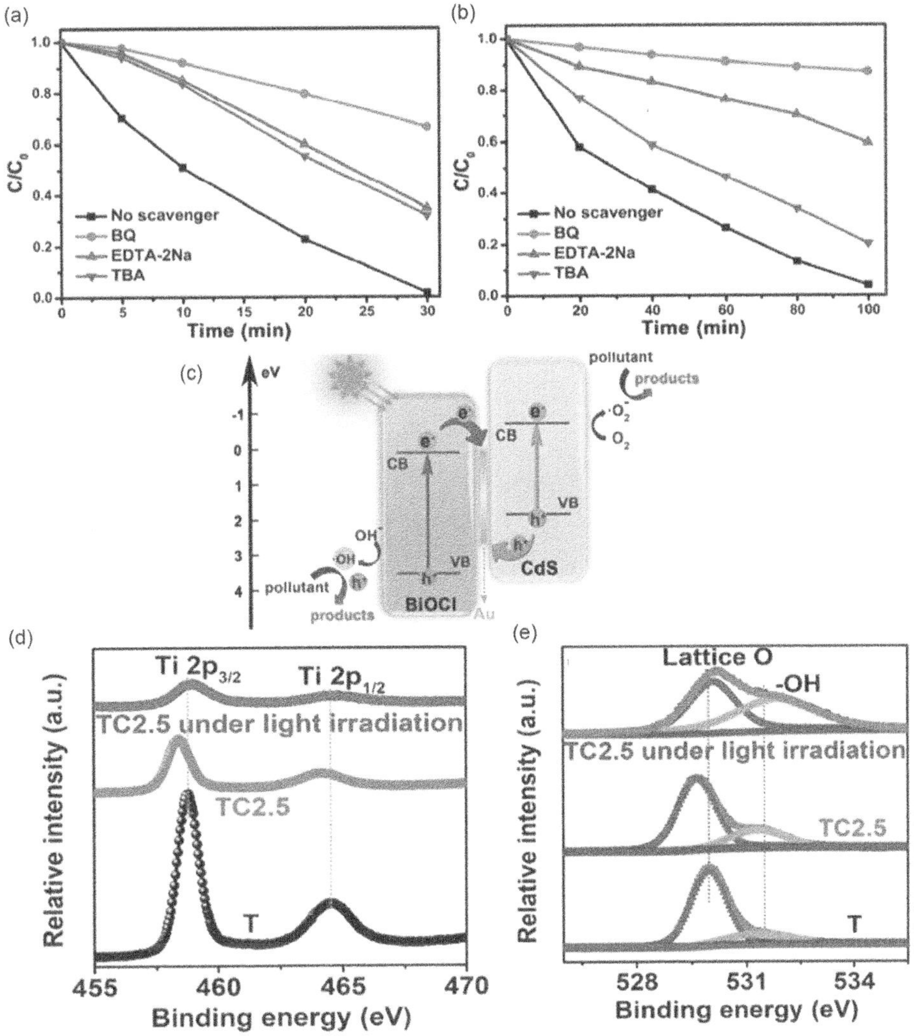

Figure 1.5 Photocatalytic degradation curves of RhB (a) and phenol (b) over BiOCl-Au-CdS in the presence of different radical scavengers including BQ (1 mM), EDTA-2Na (1 mM), and TBA (2 mM) under simulated solar light irradiation (c) Proposed photocatalytic mechanism of Z-scheme BiOCl-Au-CdS. Reprinted with permission from Ref [56], XPS spectra of (d) Ti 2p and (e) O 1s of T. Reprinted with permission from Ref [57].

The arrangement of an S-scheme heterojunction is almost analogous to type II but follows a nonidentical charge migration mechanism. In the complete mechanism, the RP is that photocatalyst that possesses VB and CB at a higher position and smaller work function comparative to the OP. When both OP and RP are put in close contact, electrons diffuse spontaneously from RP to OP. This diffusion creates electron accumulation and depletion layers close to the interface in the OP and RP, respectively. Thus, the OP becomes negatively charged whereas the RP becomes positively charged at the interface. This charge

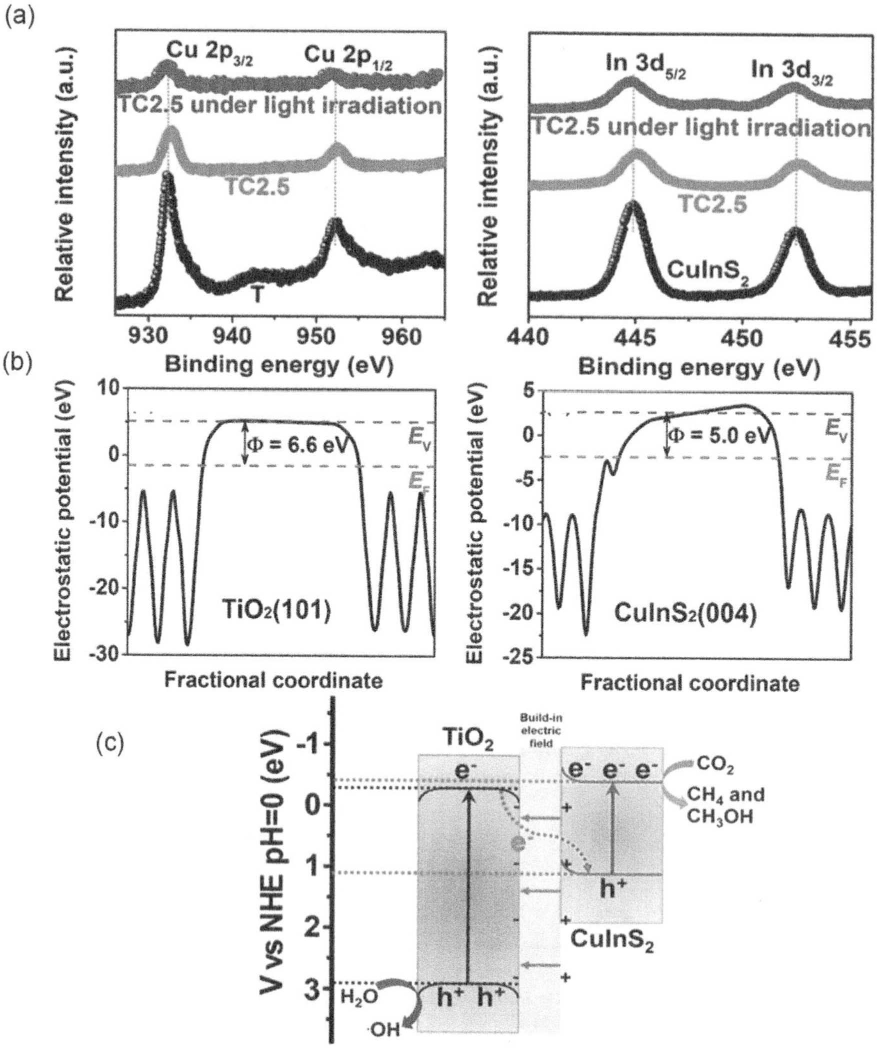

Figure 1.6 XPS spectrum of (a) In 3d and Cu 2p (b) Calculated electrostatic potentials for $CuInS_2$ (004) face and TiO_2 (101) face (c) Schematic representation of charge migration and separation in TC 2.5. Reused with permission from Ref [57].

establishment creates an electric field directing from RP to OP which accelerates electron transfer from OP to RP rather than the other way round (Figure 1.7a). In the meantime, when two photocatalysts come in close contact, their Fermi levels also get aligned at similar energy levels by shifting in a downward and upward direction in RP and OP, respectively. The rearrangement of the Fermi levels cause the bands to bend in the direction opposite to their Fermi level movement. This band bending also facilitates the recombination of useless photoexcited electrons and holes in the CB of OP and VB of RP at the interface. The coulombic interactions between oppositely charged electrons and holes also promote the recombination of excitons. The S-scheme is therefore defined by three factors, namely internal electric field, band bending, and coulombic interactions, which help in the

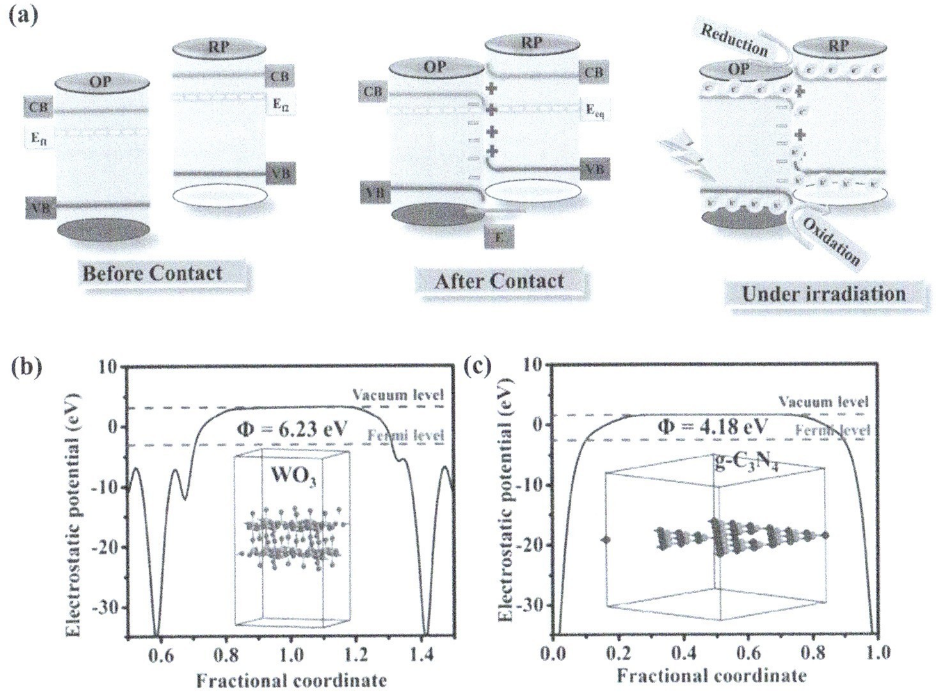

Figure 1.7 (a) General mechanism of S-scheme; (b) Electrostatic potentials of WO_3 (001) surface and (c) g-C_3N_4 (001) surface. Reprinted with permission from Ref [61].

recombination of useless electrons and holes and separate powerful charge species to efficiently take part in photocatalysis [57–60].

The construction of S-scheme heterojunctions can be assured by DFT and XPS analysis. For instance, work functions of g-C_3N_4 and WO_3 were calculated to be 4.18 and 6.23 eV, respectively by DFT analysis (Figures 1.7b and c) [61]. This difference in work function indicates a sign of charge migration as well as direction of migration. The larger work function of WO_3 leads to charge shifting from g-C_3N_4 to WO_3 until Fermi levels align at equal energy levels. This migration process thus produces band bending and a built-in electric field beneficial for the separation of photogenerated electrons and holes having high oxidation and reduction potential values, respectively. The direction of charge shifting via the S-scheme mechanism is also supported by XPS analysis. The XPS results show that C 1s and N 1s peaks of WO_3/g-C_3N_4 are shifted in the positive direction of binding energy by 0.2 eV as compared to C 1s and N 1s peaks of bare g-C_3N_4.

Contrary to this, the binding energy peaks of W^{6+} display a negative shift by −0.2 eV in WO_3/g-C_3N_4 compared with pristine WO_3 (Figure 1.8a). The decrease in electron density in g-C_3N_4 increases the effective nuclear charge, thereby a positive shift in binding energy is obtained. Conversely, a negative shift in WO_3 affirms the increase in electron density. The electronic characteristics of WO_3/g-C_3N_4 were characterized via different analysis techniques. Mott–Schottky plots show a positive slope of both WO_3 and g-C_3N_4 that explains their n-type behavior (Figure 1.8b). The CB positions of g-C_3N_4 and WO_3 were found to be −0.61 and +0.30 eV, respectively, while the VB potentials of g-C_3N_4 and WO_3 are situated at

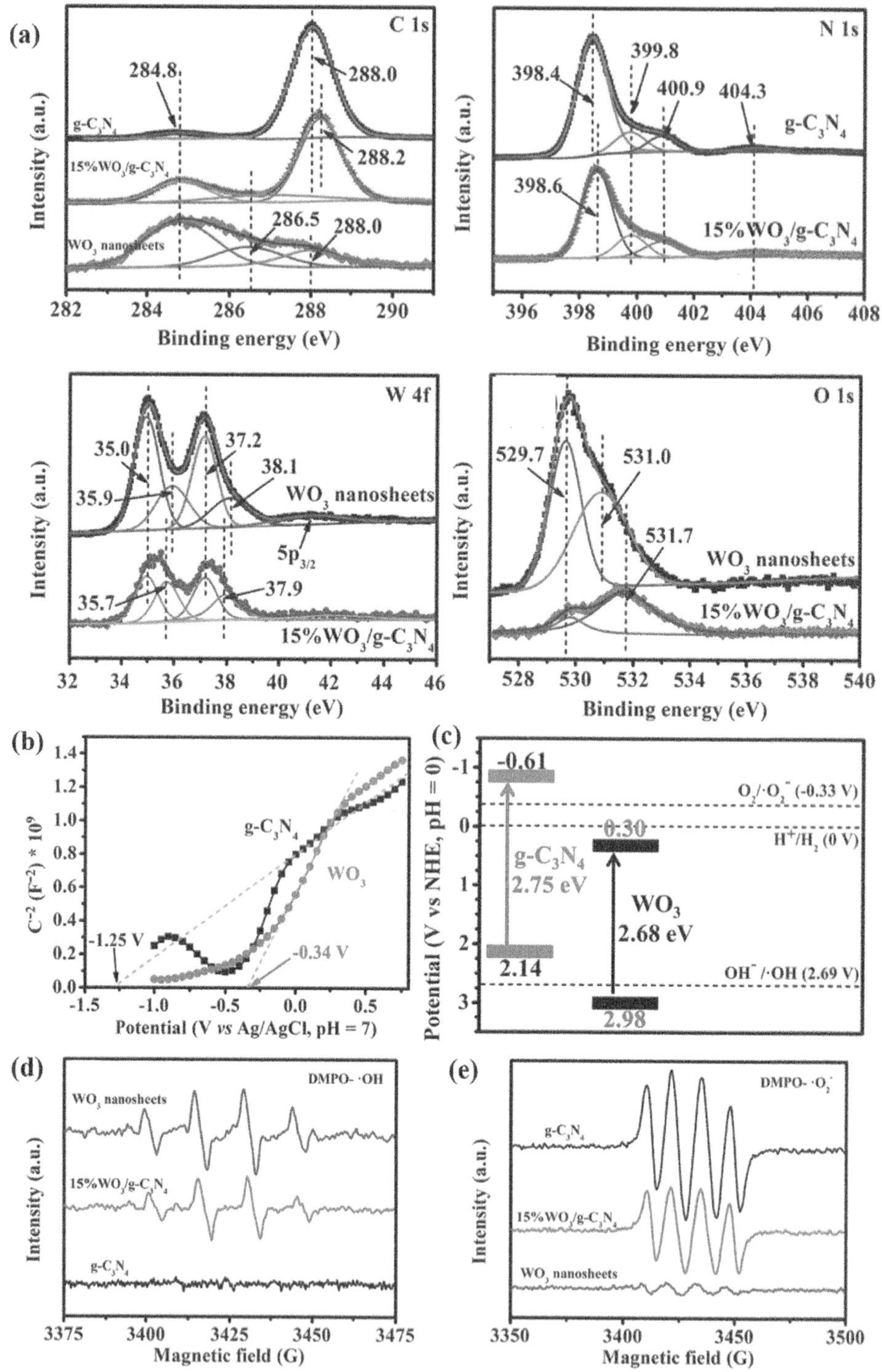

Figure 1.8 (a) Comparison of C 1s, N 1s, W 4f and O 1s XPS spectra of g-C_3N_4, 15%WO_3/g-C_3N_4 and WO_3 nanosheets; (b) Mott–Schottky plots and (c) Band structures of g-C_3N_4 and WO_3 nanosheets. EPR spectra of (d) DMPO–$^{\bullet}OH$ in aqueous and (e) DMPO–$^{\bullet}O_2{}^-$ in methanol dispersion in the presence of g-C_3N_4, WO_3 nanosheets, and 15%WO_3/g-C_3N_4. Reprinted with permission from Ref [61].

+2.14 and +2.98 eV, respectively (Figure 1.8c). Furthermore, electron paramagnetic resonance (EPR) was utilized to identify the presence of electrons and holes. 5,5-dimethyl-1-pyrroline N-oxide (DMPO)–$^{\bullet}$OH signals were found for bare WO_3 and WO_3/g-C_3N_4 while no such signals were observed for pure g-C_3N_4 owing to the weak oxidation potential of its photogenerated VB holes.

The presence of a DMPO–$^{\bullet}$OH signal in WO_3/g-C_3N_4, however, demonstrates that photogenerated holes still hold in the VB of WO_3 and are not transferred to the VB of g-C_3N_4. Similarly, strong DMPO–$^{\bullet}O_2^-$ signals were found for pure g-C_3N_4 and WO_3/g-C_3N_4, whereas negligible signals were obtained for bare WO_3. This implies that photoexcited electrons in the CB of g-C_3N_4 possess enough reduction potential to generate superoxide radicals. This outcome also indicates that when a heterojunction forms between g-C_3N_4 and WO_3, photoexcited electrons stay in the CB of g-C_3N_4 instead of moving to the CB of WO_3 (Figure 1.8d and e). The EPR results suggest that the charge migration pathway in WO_3/g-C_3N_4 follows the S-scheme transfer mechanism rather than type II (Figure 1.9a). The S-scheme heterojunction possesses strong redox ability, reduced recombination rate, and efficiently separated charge carriers because of its unique charge migration mode. The two semiconductors

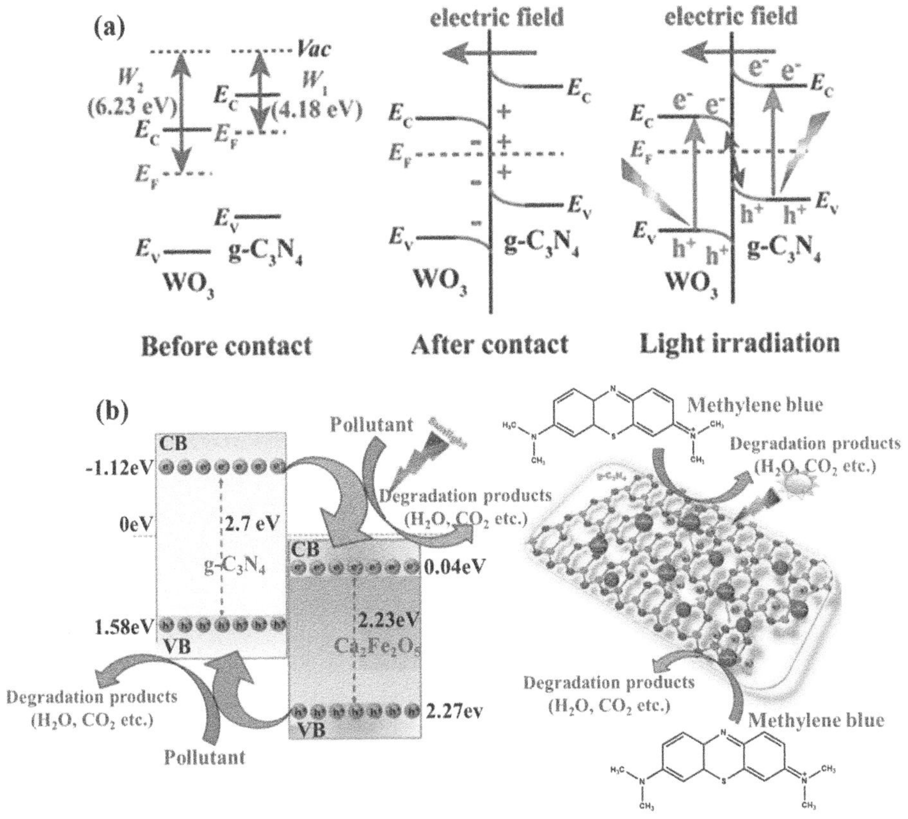

Figure 1.9 (a) S-scheme migration between WO_3 and g-C_3N_4. Reprinted with permission from Ref [61] (b) Type-II migration in g-C_3N_4/CFO (c) Photodegradation of MB via g-C_3N_4/CFO heterojunction. Reused with permission from Ref [68].

used to construct an S-scheme heterojunction, however, are mostly composed of n-type semiconductors that should possess appropriate band positions with suitable Fermi level differences.

1.4 Visible Light Driven Photocatalysts for Water Decontamination

Photocatalysis has been accepted as a simple, effective, and promising approach for eliminating the various kinds of pollutants from water sources. Different kinds of semiconductor photocatalysts, carbonaceous materials based photocatalysts, metal phosphide, and chalcogenides photocatalysts have been utilized extensively to degrade the harmful pollutants from water in presence of light irradiation. The process of degradation of pollutants by photocatalysts is generally termed photodegradation and involves the following steps: 1) When the photocatalyst system is irradiated with photons having sufficient energy, electrons get excited from the VB to CB and leave holes in the VB. 2) The VB holes can oxidize the donor molecules or can react with H_2O to produce hydroxyl radicals, which have strong oxidizing ability to degrade pollutants. 3) Photoexcited electrons react with the absorbed O_2 to produce superoxide radicals having a strong reduction potential and facilitate the redox reactions [62–66]. It must be noted, however, that only those electrons or holes that have more than −0.33 eV and +2.73 eV, respectively, can produce superoxide and hydroxyl radicals. To construct significant heterojunctions, therefore, it is important to choose the semiconductor photocatalyst whose VB potential is more than +2.73 or whose CB potential is less than −0.33 eV, to carry out the photodegradation.

1.4.1 Type II Heterojunctions

Kshirsagar et al. synthesized $CuSbSe_2/TiO_2$ heterojunctions, via thermal and microwave methods, that exhibit high photodegradation efficiency toward methyl orange (MO), Rhodamine B (RhB), and methylene blue (MB) [67]. The obtained efficiency was ascribed to the high surface area for absorption of dye molecules as well as the presence of hydroxyl and superoxide radicals produced during photocatalysis. $CuSbSe_2/TiO_2$ has a combination of wide and small bandgap photocatalysts that aid in suppressing the recombination process and improve the generation of charge pairs. The type II heterostructure provides a large density of reactive oxygen species involved in the photodegradation of organic pollutants.

Recently, Vavilapalli and coworkers fabricated a g-C_3N_4/$Ca_2Fe_2O_5$ heterostructure for photodegradation of MB and Bisphenol-A (BPA) under visible light irradiation [68]. The 50% g-C_3N_4 content displayed the highest activity. At high loading of g-C_3N_4 content, the $Ca_2Fe_2O_5$ nanoparticles get fully covered thus decreasing the photon absorption, whereas at low g-C_3N_4 the recombination process is more pronounced. In the reported work, therefore, 50% g-C_3N_4 loading onto $Ca_2Fe_2O_5$ is the ideal photocatalyst for the photodegradation process and degraded 96% of MB and 63.1% of BPA. The CB of g-C_3N_4 and $Ca_2Fe_2O_5$ was measured at −1.12 and 0.04 eV and the VB at 1.58 and 2.27 eV, respectively, as displayed in (Figures 1.9b and c).

These band positions are suitable to form type II heterostructures where electrons migrate from the CB of g-C_3N_4 to the CB of $Ca_2Fe_2O_5$, and holes move from the VB of $Ca_2Fe_2O_5$ to the VB of g-C_3N_4. This kind of migration consequently enhances the charge separation and improves the photoelectrochemical characteristic of composites. Linear sweep voltammetry and chronoamperometry analysis reveals that photocurrent response studied at different potentials is higher (1.9 mA) as compared to pristine $Ca_2Fe_2O_5$ electrodes (67.4 μA at 0.9 V) (Figure 1.10a–f). These results demonstrate type II heterojunction formation having a good photoresponse. Jo et al. constructed a black Cu doped TiO_2 (BCT)/NiAl LDH photosystem for photodegradation of MO and isoniazid under visible light irradiation [69]. The high activity of composites is based on various factors like the surface area (42.2 m^2/g^{-1}) and their mesoporous nature. The large density of active centers accelerates the transport of pollutants within the interconnected porous structure promoting absorption ability and thus high photodegradation ability.

Transient photocurrent response studied over four on–off cycles showed remarkable electron–hole separation efficiency of the nanocomposite. The transient photocurrent response was much stronger for BCT/LDH samples, which confirmed the electron/hole transfer rate between BCT and LDH. Trapping experiments explored whether $^{\bullet}O_2^-$ is the major species responsible for the photodegradation process rather than h^+ and $^{\bullet}OH$. During this experiment, TBA, AO, and BZQ were used as a trapping agent for $^{\bullet}OH$, h^+, and $^{\bullet}O_2^-$, respectively. It was observed that the photodegradation of MO was affected most when BZQ was used as a scavenger. The BZQ used trapped the generated $^{\bullet}O_2^-$, which was no longer available for photoreactions (Figure 1.10g).

1.4.2 Z-scheme Heterojunctions

Zhang et al. synthesized a Z-scheme-based MoS_2/Bi_2WO_6 heterojunction, which almost completely degraded RhB in 90 min under visible light irradiation [70]. The kinetic study of photocatalysis proved that the relation between photocatalytic reaction and the initial concentration of pollutants followed pseudo-first-order kinetics. The rate constant thus observed was found to be larger for all MoS_2/Bi_2WO_6 samples, specifically 0.029 87 min^{-1} for MB-9 (9 wt% of Bi_2WO_6) compared with pristine Bi_2WO_6 (0.006 92 min^{-1}) and MoS_2 (0.009 81 min^{-1}) indicating improved photocatalytic activity of the heterostructure (Figure 1.10h).

The enhanced activity of the MoS_2/Bi_2WO_6 heterojunction was explained by charge separation due to the presence of the Z-scheme pathway. The transfer of charge carriers by Z-scheme was confirmed by the relative band positions of MoS_2 and Bi_2WO_6 obtained using Mulliken electronegativity theory. From this theory, the CB and VB were observed for MoS_2 at 0.02 and 1.64 eV and for Bi_2WO_6 at 0.5 and 3.22 eV, respectively. The charge separation, therefore, was highly promoted when two photocatalysts are in close contact owing to the appropriate potential difference between Bi_2WO_6 and MoS_2. The high separation of charge carriers due to such a pathway resulted in high photoactivity. The high density of available charge carriers for photocatalysis was also explained by photoelectrochemistry analysis. Transient photocurrent response showed that the highest photocurrent density was obtained for the MoS_2/Bi_2WO_6 heterostructure which displayed the formation of a large number of charge species, in the presence of light,. Furthermore, electrochemical

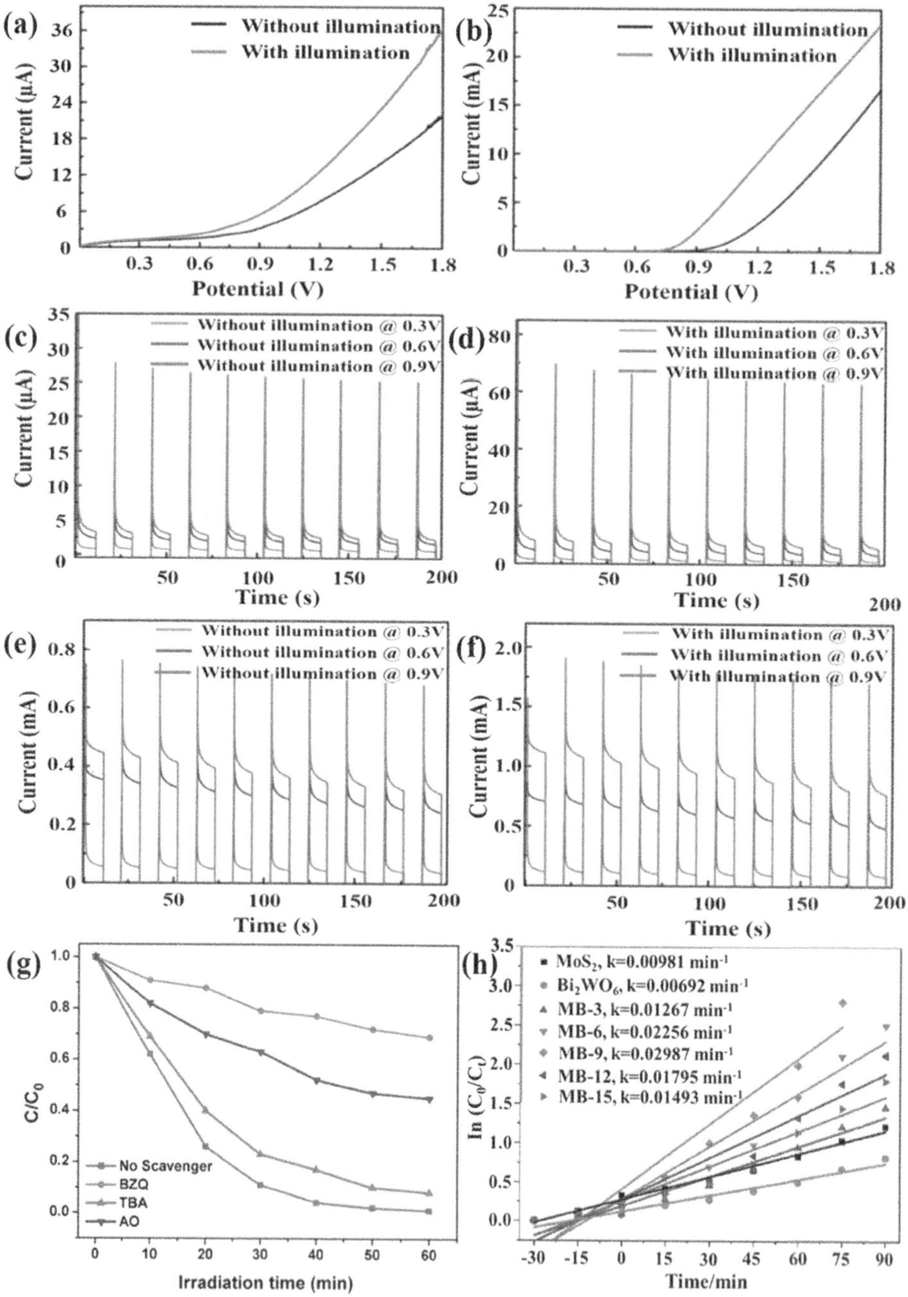

Figure 1.10 (a, b) PEC characteristics of pure CFO and g-C_3N_4/CFO (CCN50) in 0.5 M KOH electrolyte under 100 mW/cm^2 light intensity (c, d) LSV curves of CFO and CCN50 with and without light (e, f) CA plots of CFO with and without light. CA plots of CCN50 with and without light. Reprinted with permission from Ref [68] (g) Effects of various trapping reagents on MO decomposition via BCT/LDH-30. Reprinted with permission from Ref [69] (h) The dynamics of RhB degradation reaction. Reprinted with permission from Ref [70].

impedance spectroscopy (EIS) used to determine the electron transfer resistance showed less small arc radius for the heterostructure than bare Bi_2WO_6 and MoS_2 confirming less charge transfer resistance in the heterostructure (Figure 1.11a, b).

In 2020, He and co-workers fabricated a 6.7% $Au/Ni_2P/g\text{-}C_3N_4$ ternary heterojunction via the hydrothermal method for photodecomposition of levofloxacin hydrochloride antibiotics under visible light irradiation [71]. The Z-scheme-based ternary heterojunction degraded nearly 88.2 % of antibiotics due to the strong optical properties of Ni_2P and the surface plasmon resonance (SPR) effect of Au particles resulting in high light absorption capacity and charge generation. Additionally, Au and Ni_2P act as electron sinks, improving charge migration and separation. The incorporation of three also increases the overall surface area responsible for a large number of active sites. Figure 1.11c reveals that the CB potential of $g\text{-}C_3N_4$ is more positive than the CB of Ni_2P and more negative than the VB of Ni_2P; therefore, the Z-scheme transfer mechanism was exclusively proposed. Thereby, the CB electrons of $g\text{-}C_3N_4$ effectively migrated to the VB of Ni_2P through the Z-scheme transfer route.

The impressive Au cocatalyst endowed SPR effects that increased light absorption ability and induced a hot electron transfer process, owing to which the photoexcited electrons

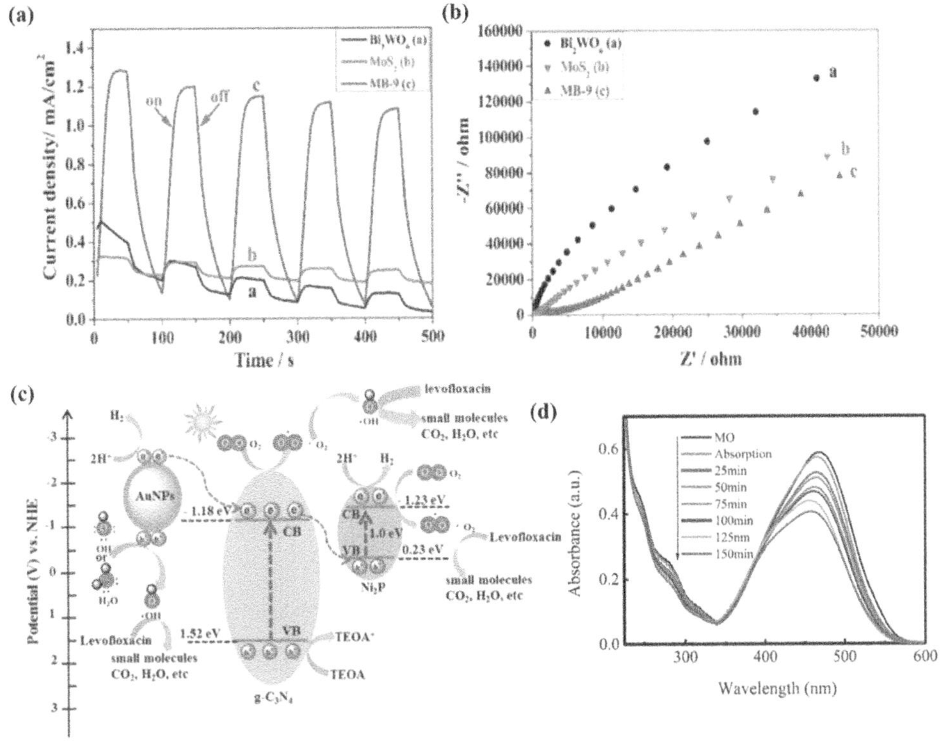

Figure 1.11 Transient photocurrent response (a) EIS Nyquist plots (b) for MoS_2, Bi_2WO_6, and MB-9. Reprinted with permission from Ref [70] (c) Schematic illustration of the proposed mechanism for levofloxacin hydrochloride degradation and H_2 evolution. Reprinted with permission from Ref [71] (d) UV–Vis spectrum changes of MO (10 mg/L) over 4% under simulated sunlight irradiation. Reprinted with permission from Ref [73].

stored in its empty orbitals can move efficiently toward the CB of g-C_3N_4. Due to the strong oxidative and reductive properties of photoexcited electrons and holes, powerful $•O_2^-$ and •OH radicals were produced, which easily decomposed the levofloxacin hydrochloride molecules into harmless CO_2 and H_2O molecules. In another work, a ternary Z-scheme-based $CaTiO_3$/g-C_3N_4/AgBr photocatalyst was designed by Yan et al., which displayed 99.6% of RhB degradation in 30 min [72]. The obtained photoactivity was almost 2, 20, and 28 times higher than bare AgBr, g-C_3N_4, and $CaTiO_3$, respectively. The high photoactivity was obtained by the double action of the Z-scheme, due to which the photogenerated electrons moved from the CB of AgBr and $CaTiO_3$ into the VB of g-C_3N_4. This transfer mechanism successfully suppresses the recombination of charge carrier species, resulting in a greater number of high potential charge species available to participate in the photodegradation process.

1.4.3 S-scheme Heterojunctions

An ionic liquid assisted BiOI/Bi_2WO_6 S-scheme heterojunction was constructed by Lei et al., which degraded 90.1% RhB and 72.1 % MO. The efficiency was nearly three times greater than pure Bi_2WO_6 [73]. To support the mineralization ability of BiOI/Bi_2WO_6 toward MO degradation, the UV–Vis spectrum alteration of MO over BiOI/Bi_2WO_6 was observed. Figure 1.11d suggests that the absorption peak of 464 nm was an induced blue shift due to the generation of intermediates, and the absorption intensity was decreased on decreasing the illumination time. This analysis firmly detected the remarkable activity of BiOI/Bi_2WO_6 for MO degradation into small molecules. In addition to the high photoactivity, BiOI/Bi_2WO_6 also possesses excellent photostability: no noticeable changes were observed in the degradation efficiency up to four cycles. In another work, an S-scheme heterojunction was prepared between black phosphorus (BP) and BiOBr displaying seven times higher photocatalytic efficiency than their bare counterparts [74].

The understanding of the band structure of photocatalysts is crucial for exploring the photocatalytic mechanism. The band potential of BP and BiOBr was measured with the help of Mott–Schottky and VB-XPS. To examine the charge shifting between BP and BiOBr, the work functions of BP/BiOBr, BP, and BiOBr were determined using a small range scan of VB-XPS plots. During the analysis, a solid sample had fine electrical proximity with a metal sample holder of the analyzer, and electron transport was balanced, where Fermi levels of both attained the same level. Due to work function difference, the kinetic energy of free electrons is altered by contact potential difference, $\Delta V = \Phi - \varphi$ (where Φ is the work function of samples, and φ is the work function of the XPS analyzer with value 4.55 eV), which causes a change in binding energy. By calculating a change in binding energy, ΔV could be obtained from the distance between two inflection points, which gave Φ values of 5.95, 5.65, and 6.25 eV for BP/BiOBr, BP, and BiOBr, respectively. The band structure and work functions results thus showed that BP is a reduction type photocatalyst with higher Fermi level and smaller work function, whereas BiOBr is an oxidation type photocatalyst with low Fermi level and large work function. When BiOBr and BP are in close contact with each other, electrons would shift from BP to BiOBr until their Fermi level equilibrated. This movement of electrons generated negatively and positively charged layers at the interface along with the formation of an electric field directed from BP to BiOBr. The rearrangement of the Fermi levels also causes downward band bending in BiOBr and upward band bending in BP. When this system was irradiated, the electrons were excited from the VB to the CB in both BP and BiOBr. In the meantime, the useless photoexcited electrons at the CB of BiOBr were effectively recombined with the holes at the VB of BP under the influence of the coulombic interaction due to the internal electric field, causing band bending at the interface. The electrons and holes with high redox potential thus easily participate in the formation of reactive oxygen species responsible for the degradation process.

Hu et al. synthesized an S-scheme-based Bi_2S_3/porous g-C_3N_4 that photodegraded 90.9% of MB [75]. The S-scheme formation improved the light absorption capacity of Bi_2S_3/porous g-C_3N_4 of the heterostructure. UV–Vis DRS spectra of Bi_2S_3/porous g-C_3N_4 indicated a red-shift in the absorption peak as compared to pure Bi_2S_3 and porous g-C_3N_4, which helps to generate a large number of charge carriers. A $BiVO_4/MoS_2$ heterojunction was formed by Xu and co-workers that completely degraded RhB dye in 20 min without using any additional agent [71]. EPR analysis was done to confirm the type of heterojunction formed by finding the charge transfer route. EPR spectra of DMPO–$^{\bullet}$OH and DMPO–$^{\bullet}O_2^-$ were analyzed for $BiVO_4$, MoS_2, and $MoS_2/BiVO_4$. In the case of pure $BiVO_4$, strong DMPO–$^{\bullet}$OH signals were observed while weak DMPO–$^{\bullet}O_2^-$ signals were obtained, suggesting the strong ability of $BiVO_4$ to produce hydroxyl radicals rather than superoxide radicals due to strong oxidation ability. On the other hand, EPR spectra of pure MoS_2 gave a clear DMPO–$^{\bullet}O_2^-$ signal and weak DMPO–$^{\bullet}$OH signal due to the strong reduction ability. When EPR spectra were taken for $MoS_2/BiVO_4$, however, both DMPO–$^{\bullet}O_2^-$ and DMPO–$^{\bullet}$OH signals were remarkably improved as compared to bare $BiVO_4$ and MoS_2. These findings suggested the strong reduction and oxidation ability of $MoS_2/BiVO_4$ and the formation of abundant reactive sites providing reduction sites (on MoS_2) and oxidation sites (on $BiVO_4$) via the S-scheme transfer mechanism. From the above discussion, we can thus generalize that heterojunction formation is a more promising strategy to design excellent photocatalysts for wastewater remediation. Table 1.1 shows that different types of heterojunction reported have efficiently removed various kinds of organic pollutants under mild reaction conditions; in particular, the S-scheme type of heterojunction displayed far superior degradation ability.

1.5 Summary

With the continuous hike in world population, usage of pharmaceuticals, pesticides, and dyes, etc. has considerably increased and is expected to continue increasing. Much-deliberated efforts are therefore needed to develop an effective, inexpensive, and environmentally friendly approach. Photocatalysis is one of the most widely used areas of green chemistry as it utilizes renewable solar light as an energy source, making it clean and economical. Although various techniques are employed for wastewater treatment, photocatalysis has proved one of the most efficient, green, and cost-effective techniques. Other techniques of wastewater treatment either generate secondary pollutants or are not capable of decomposing the pollutant fully, having a complex and stable structure. Photocatalysis

Table 1.1 Photodegradation efficiency of different heterojunctions toward various organic and pharmaceutical pollutants.

Nanocomposite	Type of heterojunction	Active species	Reaction conditions	Pollutants	Degradation	Recycle ability	References
CuS/TiO_2	Type II	$^{\bullet}O_2^-$ and $^{\bullet}OH$	Dye Conc. = 20 mg/L Hg lamp $\lambda = 365$ nm	Methylene blue and 4-chlorophenol	60% of methylene blue and 87% of 4-chlorophenol degraded in 90 min and 150 min	Up to 3 cycles	[76]
$TiO_2–ZnFe_2O_4$	Type II	e^- and h^+	450 W Xe arc lamp 350 to 550 nm	Methyl-red and thymol blue	90% of methyl-red and thymol blue degraded in 12 h	Up to 4 cycles	[77]
$CoFe_2O_4$-PANI	Type II	$^{\bullet}OH$	Conc. = 20 mg/L 500 W Hg and Xe lamp,	Methyl orange and methyl blue	Methyl orange and methyl blue degraded in 120 min	Up to 3 cycles	[78]
$NiFe_2O_4/Bi_2O_3$	Type II	h^+	Conc. = 10mg/L 150 W Xe lamp 420 nm	Tetracycline	90.78 % of tetracycline degraded in 80 min.	Up to 3 cycles	[79]
$TiO_2@CuFe_2O_4$	Type II	SO_4^{2-}, $^{\bullet}OH$ and $^{\bullet}O_2^-$	Conc. = 10 to 60 mg/L, 6W low pressure mercury UV-C lamp ($\lambda = 254$ nm pH = 6.5 ± 0.2	2,4- Dichlorophen-oxyacetic acid	97.2% of 2,4- dichloro-phenoxyacetic acid degraded in 60 min	Up to 5 cycles	[80]
Ag_2S/Bi_2WO_6	Type II	h^+	Conc. = 4 .8 mg/L 500W Xe lamp with 400 nm	Rhodamine B	85% Rhodamine B degraded in 135 min	Up to 5 cycles	[81]
$CQDs/Bi_2WO_6$	Type II	$^{\bullet}O_2^-$ and h^+	Conc. = 10 mg/L A 300 W Xe lamp with a 400 nm	Rhodamine B, ciprofloxacin	100% of Rhodamine B and 85% of ciprofloxacin degraded in 120 min	Up to 5 cycles	[82]
$AgBr/Bi_2WO_6$	Type II		Conc. = 4 .8 mg/L Xe lamp	Rhodamine B	99.83% of Rhodamine B degraded in 440 min	Up to 5 cycles	[83]

UiO-66@ZnO/GO	Z-scheme	$^{\bullet}O_2^-$	Conc. = 20 mg/L Hg lamp (500 W, wavelength > 400 nm	Tetracycline and malathion	81% of tetracycline and 100% of malathion degraded in 90 min	Up to 4 cycles	[84]
CdS@LDHs	Z-scheme	$^{\bullet}OH$ and $^{\bullet}O_2^-$	Conc. = 50 mg/L 500 W Xe lamp	Tetracycline	93.04% of tetracycline degraded in 300 min	Up to 5 cycles	[85]
$Co_3O_4TiO_2$	Z-scheme	$^{\bullet}O_2^-$, $^{\bullet}OH$ and h^+	Conc. = 20 mg/L 500 W Xe lamp	Enrofloxacin	95% of enrofloxacin degraded in 100 min	Up to 4 cycles	[86]
$Cs_2O - Bi_2O_3 - ZnO$	Z-scheme	$^{\bullet}OH$ and $^{\bullet}O_2^-$	Conc. = 10 mg/L A 300 W Xe lamp	4-chlorophenol	82% of 4-chlorophenol degraded in 80 min	Up to 5 cycles	[87]
$BiOBr/Bi_2MoO_6$	Z-scheme	$^{\bullet}O_2^-$, h^+ and $^{\bullet}OH$	Conc. = 50 mg/L 300 W Xe Lamp ($\lambda \geq 420$ nm)	Organic carbon and Rhodamine B	66.7% of total organic carbon and 95% of Rhodamine B degraded in 20–25 min	Up to 5 cycles	[88]
$BiOI/g-C_3N_4$	Z-scheme	$^{\bullet}O_2^-$, h^+ and $^{\bullet}OH$	Conc. = 100 mg/L 300W Xe Lamp ($\lambda > 400$ nm)	Phenol	60% of phenol degraded in	-	[89]
$TCPP/rGO/Bi_2WO_6$	Z-scheme	$^{\bullet}O_2^-$ and h^+	Conc. = 15 mg/L 300 W Xe lamp ($\lambda > 420$ nm)	Tetracycline	83.60% of tetracycline degraded in 60 min	Up to 5 cycles	[90]
MoO_3/Bi_2O_4	Z-scheme	$^{\bullet}O_2^-$ and h^+	Conc. = 10 mg/L 100 W LED lamp ($\lambda = 420$ nm)	Rhodamine B	99.6% Rhodamine B degraded in 40 min	Up to 5 cycles	[91]

(Continued)

Table 1.1 (Continued)

Nanocomposite	Type of heterojunction	Active species	Reaction conditions	Pollutants	Degradation	Recycle ability	References
Bi_2MoO_6/TiO_2	Z-scheme	$^{\bullet}OH$ and h^+	Conc. = 50 mg/L 800 W Xe lamp	4-Nitrophenol	95.3% of 4-nitrophenol degraded in 180 min	Up to 4 cycles	[92]
Ag/AgI/BiOI	Z-scheme	$^{\bullet}O_2^-$ and h^+	Conc. = 10 mg/L 500 W Xe lamp 420 nm	Methylene orange	93% of methylene orange degraded in 3 h	-	[93]
$Ag_3PO_4/TiO_2@MoS_2$	Z-scheme	h^+ and $^{\bullet}OH$	Conc. = 5 mg/L 800 W Xe arc lamp	Oxytetracycline and enrofloxacin	75% and 92% of oxytetracycline and enrofloxacin in 25 min	Up to 10 cycles	[94]
$g-C_3N_4/TiO_2$	S-scheme	$^{\bullet}O_2^-$ and h^+	Conc. = 10 and 15 mg/L LED lamp with wavelength 460 nm	Methylene blue and Rhodamine B	94.92% methylene blue and 93.07% Rhodamine B degraded in 80 min	-	[95]
$BiVO_3/SnO_2$	S-scheme	$^{\bullet}O_2^-$ and $^{\bullet}OH$	300 W Xe lamp	Amaranth dye	90% amaranth degraded in 120 min	Up to 5 cycles	[96]
$Bi_2WO_6/Fe_2O_3/WO_3$	S-scheme	$^{\bullet}O_2^-$ and $^{\bullet}OH$	Conc. 20 mg/L 300 W Xe lamp	Bisphenol-A	99% of bisphenol-A degraded in 30 min	Up to 7 cycles	[5]
$Bi_2MoO_6/g-C_3N_4/Au$	S-scheme	h^+, $^{\bullet}O_2^-$ and $^{\bullet}OH$	Conc. = 5 mg/L 300 W Xe lamp 60 for 8 h	Rhodamine B	97.6% of Rhodamine B degraded in 40 min	Up to 5 cycles	[97]
BP/BiOBr	S-scheme	H_2O_2, $^{\bullet}O_2^-$ and $^{\bullet}OH$	Visible light	Total organic carbon	79% of total organic carbon degraded in 90 min	Up to 4 cycles	[74]

$TiO_2/W_{18}O_{49}$	S-scheme	$^{\bullet}O^{2-}$ and h^+	Conc. of Rhodamine B = 10 mg/L Conc. of tetracycline = 30 mg/L	Rhodamine B and tetracycline	82.1% of Rhodamine B degraded in 60 min and 80.3% of tetracycline in 75 min	-	[98]
$BiVO_4@MoS_2$	S-scheme	$^{\bullet}O_2^-$ and $^{\bullet}OH$	Conc. = 20 mg/L Xe lamp of 500 W	Rhodamine B	Rhodamine B degraded in 20 min	Up to 3 cycles	[99]
$BiOI/Bi_2WO_6$	S-scheme	$^{\bullet}O_2^-$ and h^+	Conc. = 10 mg/L 500 W Xe lamp	Rhodamine B and methylene orange	90.1% of Rhodamine B and 72.1% of methylene orange in 60 min	Up to 4 cycles	[73]
$Bi_2S_3/MoO_3/C_3N_4$	S-scheme	h^+, $^{\bullet}O_2^-$ and $^{\bullet}OH$	Conc. = 10 mg/L 500 W Xe lamp	Methylene orange	75.02 % of methylene orange degraded in 120 min	Up to 3 cycles	[100]
Bi_2S_3/porous g-C_3N_4	S-scheme	h^+ and $^{\bullet}O_2^-$	Conc. = 20 mg/L LED 50 W	Methylene blue	90.9% methylene blue degraded in 90 min	Up to 3 cycles	[75]
OVs-Bi_2O_3/Bi_2SiO_5	S-scheme	$^{\bullet}O_2^-$ and h^+	Conc. = 10 mg/L Xe lamp	Methylene orange and phenol	Methylene orange and phenol degraded in 7 h	Up to 5 cycles	[101]

based on an advanced oxidation process provides a promising alternative for the degradation of various pollutants. Depending upon the band edge potential, ROS with strong redox potential can be generated, which are capable of carrying out the degradation process.

Metal oxides and metal phosphides provide more choices for construction of heterostructures with desirable properties. The physicochemical and electronic properties of semiconductors can also be modified via doping, reducing size to a nanoscale dimension, or inducing n or p-type semiconducting behavior. Similarly, carbonaceous materials, due to the high surface area, high conductivity, fast charge mobility, and compatibility with other materials, were significantly employed in photocatalysis. Other materials like MOF and LDH have also been explored vastly in photocatalysis due to their high porosity, pore volume, and large surface area. The presence of hydroxide ions in the interlayer spaces of LDHs additionally generates ample hydroxyl radicals. Furthermore, we have examined that the most popular S-scheme heterojunction has high productivity in comparison to conventional Type II and Z-scheme heterojunctions. The different characterization techniques were used to confirm the transfer routes in this heterojunction. Finally, the photodegradation application of these heterojunctions based on metal oxides, metal phosphides, and carbonaceous materials was reported. Thus, before performing the photocatalytic application, careful selection of the most suitable combination of heterojunction with acceptable bandgap and band edge potential must be made.

References

1 Zhu, S. and Wang, D. (2017). Photocatalysis: basic principles, diverse forms of implementations and emerging scientific opportunities. *Advanced Energy Materials* 7: 1700841.

2 Rueda-Marquez, J.J., Levchuk, I., Ibañez, P.F., and Sillanpää, M. (2020). A critical review on application of photocatalysis for toxicity reduction of real wastewaters. *Journal of Cleaner Production* 258: 120694.

3 Koe, W.S., Lee, J.W., Chong, W.C., Pang, Y.L., and Sim, L.C. (2020). An overview of photocatalytic degradation: photocatalysts, mechanisms, and development of photocatalytic membrane. *Environmental Science and Pollution Research* 27: 2522–2565.

4 Zhou, H., Xiao, C., Yang, Z., and Du, Y. (2020). 3D structured materials and devices for artificial photosynthesis. *Nanotechnology* 31: 282001.

5 Shandilya, P., Guleria, A., and Fang, B. (2021). A magnetically recyclable dual step-scheme $Bi_2WO_6/Fe_2O_3/WO_3$ heterojunction for photodegradation of bisphenol-A from aqueous solution. *Journal of Environmental Chemical Engineering* 9: 106461.

6 Sharma, R., Arizaga, G.G.C., Saini, A.K., and Shandilya, P. (2021). Layered double hydroxide as multifunctional materials for environmental remediation: from chemical pollutants to microorganisms. *Sustainable Materials and Technologies* 29: e00319.

7 Shandilya, P., Sharma, R., Arya, R.K., Kumar, A., Vo, D.-V.N., and Sharma, G. (2021). Recent progress and challenges in photocatalytic water splitting using layered double hydroxides (LDH) based nanocomposites. *International Journal of Hydrogen Energy* 16: 162614.

8 Shandilya, P., Mandyal, P., Kumar, V., and Sillanpaa, M. (2021). Properties, synthesis, and recent advancement in photocatalytic applications of graphdiyne: a review. *Separation and Purification Technology* 281: 119825.

9 Guleria, A., Sharma, R., Singh, A., Upadhyay, N.K., and Shandilya, P. (2021). Direct dual-Z-scheme PANI/Ag_2O/Cu_2O heterojunction with broad absorption range for photocatalytic degradation of methylene blue. *Journal of Water Process Engineering* 43: 102305.

10 Shandilya, P., Sudhaik, A., Raizada, P., Hosseini-Bandegharaei, A., Singh, P., Rahmani-Sani, A. et al. (2020). Synthesis of Eu^{3+}– doped ZnO/Bi_2O_3 heterojunction photocatalyst on graphene oxide sheets for visible light-assisted degradation of 2, 4-dimethyl phenol and bacteria killing. *Solid State Sciences* 102: 106164.

11 Serpone, N. and Emeline, A. (2012). Semiconductor photocatalysis past, present, and future outlook. *ACS Publications* 3: 673–677.

12 Nguyen, V.H., Nguyen, T.P., Le, T.H., Vo, D.V.N., Nguyen, D.L., Trinh, Q.T. et al. (2020). Recent advances in two-dimensional transition metal dichalcogenides as photoelectrocatalyst for hydrogen evolution reaction. *Journal of Chemical Technology & Biotechnology* 95: 2597–2607.

13 Ameta, R., Solanki, M.S., Benjamin, S., and Ameta, S.C. (2018). Photocatalysis. In: *Advanced Oxidation Processes for Waste Water Treatment*, 135–175. Elsevier.

14 Riente, P. and Noël, T. (2019). Application of metal oxide semiconductors in light-driven organic transformations. *Catalysis Science & Technology* 9: 5186–5232.

15 Danish, M.S.S., Estrella, L.L., Alemaida, I.M.A., Lisin, A., Moiseev, N., Ahmadi, M. et al. (2021). Photocatalytic applications of metal oxides for sustainable environmental remediation. *Metals* 11: 80.

16 Sato, H., Minami, T., Takata, S., and Yamada, T. (1993). Transparent conducting p-type NiO thin films prepared by magnetron sputtering. *Thin Solid Films* 236: 27–31.

17 Kawazoe, H., Yasukawa, M., Hyodo, H., Kurita, M., Yanagi, H., and Hosono, H. (1997). P-type electrical conduction in transparent thin films of $CuAlO_2$. *Nature* 389: 939–942.

18 Pan, Y., Abazari, R., Yao, J., and Gao, J. (2021). Recent progress in 2D metal-organic framework photocatalysts: synthesis, photocatalytic mechanism and applications. *Journal of Physics: Energy* 3: 032010.

19 Callejas, J.F., Read, C.G., Roske, C.W., Lewis, N.S., and Schaak, R.E. (2016). Synthesis, characterization, and properties of metal phosphide catalysts for the hydrogen-evolution reaction. *Chemistry of Materials* 28: 6017–6044.

20 Oyama, S.T., Gott, T., Zhao, H., and Lee, Y.-K. (2009). Transition metal phosphide hydroprocessing catalysts: a review. *Catalysis Today* 143: 94–107.

21 Shi, Y. and Zhang, B. (2016). Recent advances in transition metal phosphide nanomaterials: synthesis and applications in hydrogen evolution reaction. *Chemical Society Reviews* 45: 1529–1541.

22 Pan, Y., Liu, Y., Zhao, J., Yang, K., Liang, J., Liu, D. et al. (2015). Monodispersed nickel phosphide nanocrystals with different phases: synthesis, characterization and electrocatalytic properties for hydrogen evolution. *Journal of Materials Chemistry A* 3: 1656–1665.

23 Kucernak, A.R. and Sundaram, V.N.N. (2014). Nickel phosphide: the effect of phosphorus content on hydrogen evolution activity and corrosion resistance in acidic medium. *Journal of Materials Chemistry A* 2: 17435–17445.

24 Brock, S.L., Perera, S.C., and Stamm, K.L. (2004). Chemical routes for production of transition-metal phosphides on the nanoscale: implications for advanced magnetic and catalytic materials. *Chemistry–A European Journal* 10: 3364–3371.

25 Goodman, P.A., Li, H., Gao, Y., Lu, Y., Stenger-Smith, J., and Redepenning, J. (2013). Preparation and characterization of high surface area, high porosity carbon monoliths from pyrolyzed bovine bone and their performance as supercapacitor electrodes. *Carbon* 55: 291–298.

26 Pinheiro, V.S., Paz, E.C., Aveiro, L.R., Parreira, L.S., Souza, F.M., Camargo, P.H. et al. (2019). Mineralization of paracetamol using a gas diffusion electrode modified with ceria high aspect ratio nanostructures. *Electrochimica Acta* 295: 39–49.

27 Khan, M.E. (2021). State-of-the-art developments in carbon-based metal nanocomposites as a catalyst: Photocatalysis. *Nanoscale Advances* 3: 1887–1900.

28 Attia, A.K., Abo-Talib, N.F., and Tammam, M.H. (2017). Voltammetric determination of ivabradine hydrochloride using multiwalled carbon nanotubes modified electrode in presence of sodium dodecyl sulfate. *Advanced Pharmaceutical Bulletin* 7: 151.

29 Yao, Y., Li, G., Ciston, S., Lueptow, R.M., and Gray, K.A. (2008). Photoreactive TiO_2/carbon nanotube composites: synthesis and reactivity. *Environmental Science & Technology* 42: 4952–4957.

30 Tong, Z., Yang, D., Li, Z., Nan, Y., Ding, F., Shen, Y. et al. (2017). Thylakoid-inspired multishell g-C_3N_4 nanocapsules with enhanced visible-light harvesting and electron transfer properties for high-efficiency photocatalysis. *ACS nano* 11: 1103–1112.

31 Zheng, Y., Jiao, Y., Zhu, Y., Li, L.H., Han, Y., Chen, Y. et al. (2014). Hydrogen evolution by a metal-free electrocatalyst. *Nature Communications* 5: 1–8.

32 Khan, M.E., Khan, M.M., and Cho, M.H. (2018). Recent progress of metal–graphene nanostructures in photocatalysis. *Nanoscale* 10: 9427–9440.

33 Di Paola, A., García-López, E., Marcì, G., and Palmisano, L. (2012). A survey of photocatalytic materials for environmental remediation. *Journal of Hazardous Materials* 211: 3–29.

34 Shandilya, P., Mittal, D., Soni, M., Raizada, P., Lim, J.H., Jeong, D.Y., and Singh, P. (2018). Islanding of $EuVO_4$ on high-dispersed fluorine doped few layered graphene sheets for efficient photocatalytic mineralization of phenolic compounds and bacterial disinfection. *Journal of the Taiwan Institute of Chemical Engineers* 93: 528–542.

35 Shandilya, P., Mittal, D., Soni, M., Raizada, P., Hosseini-Bandegharaei, A., Saini, A.K., and Singh, P. (2018). Fabrication of fluorine doped graphene and $SmVO_4$ based dispersed and adsorptive photocatalyst for abatement of phenolic compounds from water and bacterial disinfection. *Journal of Cleaner Production* 203: 386–399.

36 Shandilya, P., Divya, M., Sudhaik, A., Soni, M., Raizada, P., Saini, A.K., and Singh, P. (2019). $GdVO_4$ modified fluorine doped graphene nanosheets as dispersed photocatalyst for mitigation of phenolic compounds in aqueous environment and bacterial disinfection. *Separation and Purification Technology* 210: 804–816.

37 Singh, P., Raizada, P., Sudhaik, A., Shandilya, P., Thakur, P., Agarwal, S., and Gupta, V.K. (2019). Enhanced photocatalytic activity and stability of AgBr/BiOBr/graphene

heterojunction for phenol degradation under visible light. *Journal of Saudi Chemical Society* 23: 586–599.
38 Cao, S., Low, J., Yu, J., and Jaroniec, M. (2015). Polymeric photocatalysts based on graphitic carbon nitride. *Advanced Materials* 27: 2150–2176.
39 Dou, X., Lin, Z., Chen, H., Zheng, Y., Lu, C., and Lin, J.-M. (2013). Production of superoxide anion radicals as evidence for carbon nanodots acting as electron donors by the chemiluminescence method. *Chemical Communications* 49: 5871–5873.
40 Zhao, L., Di, F., Wang, D., Guo, L.-H., Yang, Y., Wan, B. et al. (2013). Chemiluminescence of carbon dots under strong alkaline solutions: a novel insight into carbon dot optical properties. *Nanoscale* 5: 2655–2658.
41 Wang, Y. and Hu, A. (2014). Carbon quantum dots: synthesis, properties and applications. *Journal of Materials Chemistry C* 2: 6921–6939.
42 Dąbrowski, A., Podkościelny, P., Hubicki, Z., and Barczak, M. (2005). Adsorption of phenolic compounds by activated carbon—a critical review. *Chemosphere* 58: 1049–1070.
43 Jiang, Z., Huang, B., Lou, Z., Wang, Z., Meng, X., Liu, Y. et al. (2014). Immobilization of BiOX (X= Cl, Br) on activated carbon fibers as recycled photocatalysts. *Dalton Transactions* 43: 8170–8173.
44 Lee, S.L. and Chang, C.-J. (2019). Recent progress on metal sulfide composite nanomaterials for photocatalytic hydrogen production. *Catalysts* 9: 457.
45 Li, Y., Xu, H., Ouyang, S., and Ye, J. (2016). Metal–organic frameworks for photocatalysis. *Physical Chemistry Chemical Physics* 18: 7563–7572.
46 Baumann, A.E., Burns, D.A., Liu, B., and Thoi, V.S. (2019). Metal-organic framework functionalization and design strategies for advanced electrochemical energy storage devices. *Communications Chemistry* 2: 1–14.
47 Younis, S.A., Kwon, E.E., Qasim, M., Kim, K.-H., Kim, T., Kukkar, D. et al. (2020). Metal-organic framework as a photocatalyst: Progress in modulation strategies and environmental/energy applications. *Progress in Energy and Combustion Science* 81: 100870.
48 Duan, X. and Evans, D.G. (2006). *Layered Double Hydroxides*, 119, 1–87. Berlin: Springer Science & Business Media.
49 Mohapatra, L. and Parida, K. (2016). A review on the recent progress, challenges and perspective of layered double hydroxides as promising photocatalysts. *Journal of Materials Chemistry A* 4: 10744–10766.
50 Cardoza-Contreras, M.N., Sánchez-Serrano, S., and Contreras, O.E. (2020). Highly efficient photocatalytic and antimicrobial AgGaCl Tri-Doped ZnO nanorods for water treatment under visible light irradiation. *Catalysts* 10: 752.
51 Xu, Q., Zhang, L., Yu, J., Wageh, S., Al-Ghamdi, A.A., and Jaroniec, M. (2018). Direct Z-scheme photocatalysts: principles, synthesis, and applications. *Materials Today* 21: 1042–1063.
52 Shen, L., Xing, Z., Zou, J., Li, Z., Wu, X., Zhang, Y. et al. (2017). Black TiO_2 nanobelts/ gC_3N_4 nanosheets laminated heterojunctions with efficient visible-light-driven photocatalytic performance. *Scientific Reports* 7: 1–11.
53 Tada, H., Mitsui, T., Kiyonaga, T., Akita, T., and Tanaka, K. (2006). All-solid-state Z-scheme in CdS–Au–TiO_2 three-component nanojunction system. *Nature Materials* 5: 782–786.

54 Yu, J., Wang, S., Low, J., and Xiao, W. (2013). Enhanced photocatalytic performance of direct Z-scheme $gC_3 N_4$–TiO 2 photocatalysts for the decomposition of formaldehyde in air. *Physical Chemistry Chemical Physics* 15: 16883–16890.

55 Zheng, Y., Chen, G., Yu, Y., Sun, J., Zhou, Y., and He, F. (2015). Ion exchange synthesis of an all tungsten based Z-scheme photocatalytic system with highly enhanced photocatalytic activity. *RSC Advances* 5: 46897–46903.

56 Li, Q., Guan, Z., Wu, D., Zhao, X., Bao, S., Tian, B. et al. (2017). Z-scheme BiOCl-Au-CdS heterostructure with enhanced sunlight-driven photocatalytic activity in degrading water dyes and antibiotics. *ACS Sustainable Chemistry & Engineering* 5: 6958–6968.

57 Xu, F., Zhang, J., Zhu, B., Yu, J., and Xu, J. (2018). $CuInS_2$ sensitized TiO_2 hybrid nanofibers for improved photocatalytic CO2 reduction. *Applied Catalysis B: Environmental* 230: 194–202.

58 Wang, J., Li, Q., Cheng, Y., Chen, L., Sun, Q., Zhao, J. et al. (2021). Construction of a novel direct Z-scheme heterostructure consisting of ReS_2 nanoflowers and In_2S_3 nanohoneycombs for improving photoelectrochemical performance. *Journal of Physics D: Applied Physics* 54: 175111.

59 Xu, Q., Zhang, L., Cheng, B., Fan, J., and Yu, J. (2020). S-scheme heterojunction photocatalyst. *Chem* 6: 1543–1559.

60 Cheng, C., He, B., Fan, J., Cheng, B., Cao, S., and Yu, J. (2021). An inorganic/organic S-scheme heterojunction H_2-production photocatalyst and its charge transfer mechanism. *Advanced Materials* 33: 2100317.

61 Fu, J., Xu, Q., Low, J., Jiang, C., and Yu, J. (2019). Ultrathin 2D/2D WO_3/g-C_3N_4 step-scheme H2-production photocatalyst. *Applied Catalysis B: Environmental* 243: 556–565.

62 Kaur, K., Badru, R., Singh, P.P., and Kaushal, S. (2020). Photodegradation of organic pollutants using heterojunctions: a review. *Journal of Environmental Chemical Engineering* 8: 103666.

63 Wang, Y., Wang, Q., Zhan, X., Wang, F., Safdar, M., and He, J. (2013). Visible light driven type II heterostructures and their enhanced photocatalysis properties: a review. *Nanoscale* 5: 8326–8339.

64 Huu, H.T., Thi, M.D.N., Nguyen, V.P., Thi, L.N., Phan, T.T.T., Hoang, Q.D. et al. (2021). One-pot synthesis of S-scheme MoS_2/g-C_3N_4 heterojunction as effective visible light photocatalyst. *Scientific Reports* 11: 1–12.

65 Wang, J., Zhang, Q., Deng, F., Luo, X., and Dionysiou, D.D. (2020). Rapid toxicity elimination of organic pollutants by the photocatalysis of environment-friendly and magnetically recoverable step-scheme $SnFe_2O_4$/$ZnFe_2O_4$ nano-heterojunctions. *Chemical Engineering Journal* 379: 122264.

66 Lai, C.-H., Lu, M.-Y., and Chen, L.-J. (2012). Metal sulfide nanostructures: synthesis, properties and applications in energy conversion and storage. *Journal of Materials Chemistry* 22: 19–30.

67 Kshirsagar, A.S. and Khanna, P.K. (2019). $CuSbSe_2$/TiO_2: novel type-II heterojunction nano-photocatalyst. *Materials Chemistry Frontiers* 3: 437–449.

68 Vavilapalli, D.S., Peri, R.G., Sharma, R., Goutam, U., Muthuraaman, B., Rao, R. et al. (2021). g-C_3N_4/$Ca_2Fe_2O_5$ heterostructures for enhanced photocatalytic degradation of organic effluents under sunlight. *Scientific Reports* 11: 1–11.

69 Jo, W.-K., Kim, Y.-G., and S, T. (2018). Hierarchical flower-like NiAl-layered double hydroxide microspheres encapsulated with black Cu-doped TiO_2 nanoparticles: highly efficient visible-light-driven composite photocatalysts for environmental remediation. *Journal of Hazardous Materials* 357: 19–29.

70 Zhang, Y., Ju, P., Hao, L., Zhai, X., Jiang, F., and Sun, C. (2021). Novel Z-scheme MoS_2/Bi_2WO_6 heterojunction with highly enhanced photocatalytic activity under visible light irradiation. *Journal of Alloys and Compounds* 854: 157224.

71 He, Y., Zhang, F., Ma, B., Xu, N., Junior, L.B., Yao, B. et al. (2020). Remarkably enhanced visible-light photocatalytic hydrogen evolution and antibiotic degradation over g-C_3N_4 nanosheets decorated by using nickel phosphide and gold nanoparticles as cocatalysts. *Applied Surface Science* 517: 146187.

72 Yan, Y., Yang, H., Yi, Z., Li, R., and Xian, T. (2020). Design of ternary $CaTiO_3$/g-C_3N_4/AgBr Z-scheme heterostructured photocatalysts and their application for dye photodegradation. *Solid State Sciences* 100: 106102.

73 Lei, S., Luo, R., Li, H., Chen, J., Zhong, J., and Li, J. (2020). Ionic liquid assisted in-situ construction of S-scheme BiOI/Bi_2WO_6 heterojunctions with improved sunlight-driven photocatalytic performance. *Inorganic Chemistry Communications* 121: 108192.

74 Li, X., Xiong, J., Gao, X., Ma, J., Chen, Z., Kang, B. et al. (2020). Novel BP/BiOBr S-scheme nano-heterojunction for enhanced visible-light photocatalytic tetracycline removal and oxygen evolution activity. *Journal of Hazardous Materials* 387: 121690.

75 Hu, T., Dai, K., Zhang, J., Zhu, G., and Liang, C. (2019). One-pot synthesis of step-scheme Bi_2S_3/porous g-C_3N_4 heterostructure for enhanced photocatalytic performance. *Materials Letters* 257: 126740.

76 Lu, Y., Zhang, Y., Zhang, J., Shi, Y., Li, Z., Feng, Z. et al. (2016). In situ loading of CuS nanoflowers on rutile TiO_2 surface and their improved photocatalytic performance. *Applied Surface Science* 370: 312–319.

77 Hankare, P., Patil, R., Jadhav, A., Garadkar, K., and Sasikala, R. (2011). Enhanced photocatalytic degradation of methyl red and thymol blue using titania–alumina–zinc ferrite nanocomposite. *Applied Catalysis B: Environmental* 107: 333–339.

78 Xiong, P., Chen, Q., He, M., Sun, X., and Wang, X. (2012). Cobalt ferrite–polyaniline heteroarchitecture: a magnetically recyclable photocatalyst with highly enhanced performances. *Journal of Materials Chemistry* 22: 17485–17493.

79 Ren, A., Liu, C., Hong, Y., Shi, W., Lin, S., and Li, P. (2014). Enhanced visible-light-driven photocatalytic activity for antibiotic degradation using magnetic $NiFe_2O_4$/Bi_2O_3 heterostructures. *Chemical Engineering Journal* 258: 301–308.

80 Golshan, M., Kakavandi, B., Ahmadi, M., and Azizi, M. (2018). Photocatalytic activation of peroxymonosulfate by TiO_2 anchored on cupper ferrite (TiO_2@ $CuFe_2O_4$) into 2, 4-D degradation: process feasibility, mechanism and pathway. *Journal of Hazardous Materials* 359: 325–337.

81 Tang, R., Su, H., Sun, Y., Zhang, X., Li, L., Liu, C. et al. (2016). Facile fabrication of Bi_2WO_6/Ag_2S heterostructure with enhanced visible-light-driven photocatalytic performances. *Nanoscale Research Letters* 11: 1–12.

82 Di, J., Xia, J., Ge, Y., Li, H., Ji, H., Xu, H. et al. (2015). Novel visible-light-driven CQDs/Bi_2WO_6 hybrid materials with enhanced photocatalytic activity toward organic pollutants degradation and mechanism insight. *Applied Catalysis B: Environmental* 168: 51–61.

83 Jonjana, S., Phuruangrat, A., Thongtem, S., and Thongtem, T. (2018). Synthesis, characterization and photocatalysis of heterostructure AgBr/Bi_2WO_6 nanocomposites. *Materials Letters* 216: 92–96.

84 Fakhri, H. and Bagheri, H. (2020). Two novel sets of UiO-66@ metal oxide/graphene oxide Z-scheme heterojunction: insight into tetracycline and malathion photodegradation. *Journal of Environmental Sciences* 91: 222–236.

85 Dai, T., Yuan, Z., Meng, Y., Xie, B., Ni, Z., and Xia, S. (2021). Performance and mechanism of photocatalytic degradation of tetracycline by Z–scheme heterojunction of CdS@ LDHs. *Applied Clay Science* 212: 106210.

86 Wang, Y., Zhu, C., Zuo, G., Guo, Y., Xiao, W., Dai, Y. et al. (2020). 0D/2D Co_3O_4/TiO_2 Z-Scheme heterojunction for boosted photocatalytic degradation and mechanism investigation. *Applied Catalysis B: Environmental* 278: 119298.

87 Hezam, A., Namratha, K., Ponnamma, D., Drmosh, Q., Saeed, A.M.N., Cheng, C. et al. (2018). Direct Z-scheme Cs_2O–Bi_2O_3–ZnO heterostructures as efficient sunlight-driven photocatalysts. *ACS omega* 3: 12260–12269.

88 Wang, S., Yang, X., Zhang, X., Ding, X., Yang, Z., Dai, K. et al. (2017). A plate-on-plate sandwiched Z-scheme heterojunction photocatalyst: BiOBr-Bi_2MoO_6 with enhanced photocatalytic performance. *Applied Surface Science* 391: 194–201.

89 He, R., Zhou, J., Fu, H., Zhang, S., and Jiang, C. (2018). Room-temperature in situ fabrication of Bi_2O_3/g-C_3N_4 direct Z-scheme photocatalyst with enhanced photocatalytic activity. *Applied Surface Science* 430: 273–282.

90 Hu, K., Chen, C., Zhu, Y., Zeng, G., Huang, B., Chen, W. et al. (2019). Ternary Z-scheme heterojunction of Bi_2WO_6 with reduced graphene oxide (rGO) and meso-tetra (4-carboxyphenyl) porphyrin (TCPP) for enhanced visible-light photocatalysis. *Journal of Colloid and Interface Science* 540: 115–125.

91 Jiang, T., Wang, K., Guo, T., Wu, X., and Zhang, G. (2020). Fabrication of Z-scheme MoO_3/Bi_2O_4 heterojunction photocatalyst with enhanced photocatalytic performance under visible light irradiation. *Chinese Journal of Catalysis* 41: 161–169.

92 Yang, Q., Guo, E., Liu, H., and Lu, Q. (2019). Engineering of Z-scheme 2D/3D architectures with Bi_2MoO_6 on TiO_2 nanosphere for enhanced photocatalytic 4-nitrophenol degradation. *Journal of the Taiwan Institute of Chemical Engineers* 105: 65–74.

93 Cao, J., Zhao, Y., Lin, H., Xu, B., and Chen, S. (2013). Facile synthesis of novel Ag/AgI/BiOI composites with highly enhanced visible light photocatalytic performances. *Journal of Solid State Chemistry* 206: 38–44.

94 Shao, N., Wang, J., Wang, D., and Corvini, P. (2017). Preparation of three-dimensional Ag_3PO_4/TiO_2@ MoS_2 for enhanced visible-light photocatalytic activity and anti-photocorrosion. *Applied Catalysis B: Environmental* 203: 964–978.

95 Barzegar, M.H., Sabzehmeidani, M.M., Ghaedi, M., Avargani, V.M., Moradi, Z., Roy, V.A. et al. (2021). S-scheme heterojunction g-C_3N_4/TiO_2 with enhanced photocatalytic activity for degradation of a binary mixture of cationic dyes using solar parabolic trough reactor. *Chemical Engineering Research and Design* 174: 307–318.

96 Abd-Rabboh, H.S., Galal, A., Aziz, R.A., and Ahmed, M. (2021). A novel $BiVO_3$/SnO_2 step S-scheme nano-heterojunction for an enhanced visible light photocatalytic degradation of amaranth dye and hydrogen production. *RSC Advances* 11: 29507–29518.

97 Li, Q., Zhao, W., Zhai, Z., Ren, K., Wang, T., Guan, H. et al. (2020). 2D/2D Bi_2MoO_6/g-C_3N_4 S-scheme heterojunction photocatalyst with enhanced visible-light activity by Au loading. *Journal of Materials Science & Technology* 56: 216–226.

98 Wang, R., Shen, J., Zhang, W., Liu, Q., Zhang, M., and Tang, H. (2020). Build-in electric field induced step-scheme $TiO_2/W_{18}O_{49}$ heterojunction for enhanced photocatalytic activity under visible-light irradiation. *Ceramics International* 46: 23–30.

99 Xu, A., Tu, W., Shen, S., Lin, Z., Gao, N., and Zhong, W. (2020). $BiVO_4$@ MoS_2 core-shell heterojunction with improved photocatalytic activity for discoloration of Rhodamine B. *Applied Surface Science* 528: 146949.

100 Chen, J., Liu, T., Zhang, H., Wang, B., Zheng, W., Wang, X. et al. (2020). One-pot preparation of double S-scheme $Bi_2S_3/MoO_3/C_3N_4$ heterojunctions with enhanced photocatalytic activity originated from the effective charge pairs partition and migration. *Applied Surface Science* 527: 146788.

101 Dou, L., Jin, X., Chen, J., Zhong, J., Li, J., Zeng, Y. et al. (2020). One-pot solvothermal fabrication of S-scheme OVs-Bi_2O_3/Bi_2SiO_5 microsphere heterojunctions with enhanced photocatalytic performance toward decontamination of organic pollutants. *Applied Surface Science* 527: 146775.

2

Different Synthetic Routes and Band Gap Engineering of Photocatalysts

Ashitha Kishore[1], Malavika Sunil S.[1], Raja Viswanathan[2], Alamelu Kaliaperumal[3], and Baquir Mohammed Jaffar Ali[1,]*

[1] *Department of Green Energy Technology, Madanjeet School of Green Energy Technologies, Pondicherry University, Puducherry, India*
[2] *Department of Computational Physics, School of Physics, Madurai Kamaraj University, Madurai, India*
[3] *Department of Chemical Engineering, Indian Institute of Technology Madras, Chennai India*
[*] *Corresponding author*

2.1 Introduction

Photocatalysis, as the word depicts it, is the catalytic action (change in rate of reaction) of materials in the presence of light. Thermodynamically, photocatalysis is an energy uphill process, and photosynthesis is an energy downhill process. The photocatalyst material, which serves as a platform to mediate the reaction, is a semiconductor. Photocatalysis is a phenomenon whereby electron–hole pairs are generated upon light interaction and then participate in the chemical reactions. It is broadly classified into two categories based on the nature of the state of the catalyst and the medium, namely, *Homogeneous catalysis*, wherein semiconductor catalyst and reactant are in the same phase, and *Heterogeneous catalysis*, wherein semiconductor catalyst and reactant are in different phases.

Typical photocatalytic processes include photodecomposition of water, photocatalytic fuel formation, and photocatalytic pollutant removal. Solar energy assisted photocatalytic water splitting, considered as artificial photosynthesis, consists of simultaneous oxidation and reduction to produce solar fuel. The general equation can be represented as

$$2H_2O \rightarrow 2H_2 + O_2\left(\Delta G^{\circ} = 237\,\mathrm{kJmol^{-1}}\right) \tag{2.1}$$

Recent developments in the technology of photocatalysts have focused on visible-light active and cost-effective materials optimized for high photoconversion efficiency. This enables the harnessing of sunlight for the application of the material, thereby rendering it an

Photocatalysts and Electrocatalysts in Water Remediation: From Fundamentals to Full Scale Applications, First Edition. Edited by Prasenjit Bhunia, Kingshuk Dutta, and S. Vadivel.

environmentally benign process. Depending on the nature of materials chosen and the desired bandgap to be engineered, the synthesis methodologies and process protocol may vary. In this chapter, we shall survey different methodologies adopted in the synthesis of photocatalysts and summarize how a photocatalyst band gap is engineered to impart optimal properties to the photocatalysts.

2.2 Synthesis of Photocatalysts

There are several nanomaterial-based methods reported so far for synthesizing photocatalysts for applications. To engineer the desired properties in the photocatalysts, the nanomaterial-based synthesis pursues realization of controlled size, shape, structure, composition, and purity of constituents. The methodologies are broadly classified into two approaches, namely, *top-down* and *bottom-up* approaches, as depicted in Figure 2.1.

In the top-down approach, the bulk material is taken as the starting material, followed by breaking it down into nanosized structures in the range of 1–100 nm. In the bottom-up approach, build-up of the desired material is achieved atom by atom or molecule by molecule, which subsequently leads to agglomeration of the particles, finally acquiring the nano dimension. These agglomerated clusters come together to form a self-assembled monolayer.

Conventional nanomaterial synthesis involves high temperatures, expensive equipment, and high energy consumption. By overcoming the limitations of the conventional approach, however, and by keeping in mind the need to synthesize various types of photocatalyst, several other methods of synthesis have been reported and are discussed below.

2.2.1 Sol-gel Method

The sol-gel technique belongs to the bottom-up approach. It is based on chemical processing in which the chemical substances used are in wet conditions. Glassy/ceramic solid porous structures are obtained at relatively low temperatures. This method involves evolution of a three-dimensional (3D) network by the dispersion of solid particles in a colloidal suspension (sol) followed by gelation of the sol. The final product obtained can be in different forms: powders as micro/nanocrystalline or amorphous, monoliths, aerogels (porous structure), glasses, ceramics, hybrid material coatings, fibre, and film [1]. The chosen precursor should have the ability to form the gel. Generally used precursors are metal alkoxides $M(ROH)_n$, where M = Al and Si, and ROH is alcohol dissolved in water. Usually, the metal alkoxides are insoluble in water; thus, to obtain a homogeneous solution, the alkoxides are first dissolved in alcohol, followed by the addition of water, which acts as a catalyst. The preliminary reaction mixture is hence comprised of three components, namely, metal alkoxides, alcohol, and water. Consequently, the

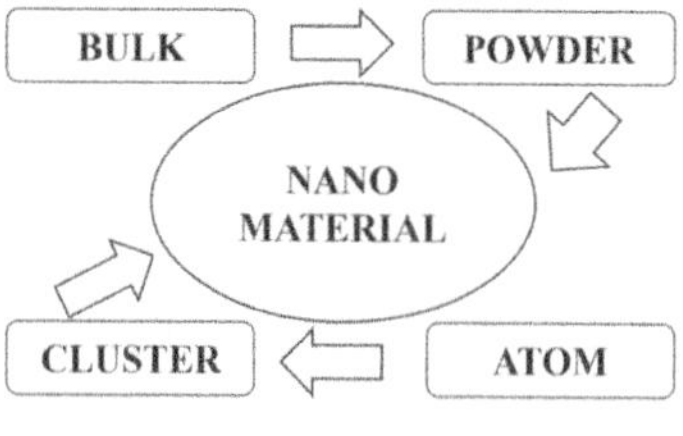

Figure 2.1 Schematic representation of bottom-up and top-down approach of nanomaterial synthesis.

production of ceramic materials is carried out by preparation of sol followed by its reaction, and finally, the removal of the solvent. Sol-gel transition is carried out to prepare nanosized metal oxide as the end product. The reaction involves the following steps: hydrolysis of precursors, condensation followed by polycondensation, gelation, aging, drying, and identification [2]. Figure 2.2 illustrates the various steps of the sol-gel technique.

The metal alkoxides react with the water molecule in the hydrolysis step as described in Eq. (2.2).

$$M-OR + H_2O \rightarrow M-OH + ROH \tag{2.2}$$

Here, M–OR is the metal alkoxide, and M–OH is the metal hydroxide/sol.The reaction leads to splitting of large molecules into smaller ones, resulting in metal hydroxide or sol, which is facilitated by protons or hydroxide present in the aquatic environment. Since the metal hydroxide is insoluble in water, the resultant heterogeneous mixture of solid in liquid forms a colloidal suspension with a modest dispersed phase in the range of 1–100 nm. For instance, the amount of hydrolysis water used during the sol-gel procedure influences the structure of nanocrystalline titania powders [3].

Hydrolysis of precursors takes place in both the acidic and basic media, having two different reaction mechanisms. In the acidic medium, the reaction initiates with the transfer of a proton from water to the metal alkoxides resulting in the removal of the –ROH group followed by the addition of a hydroxide ion to the metal, removing metal hydroxide (as shown in Eq. (2.3)). The rate of hydrolysis in TiO_2 thin films produced by the sol-gel method was accelerated in an acidic environment [4, 5].

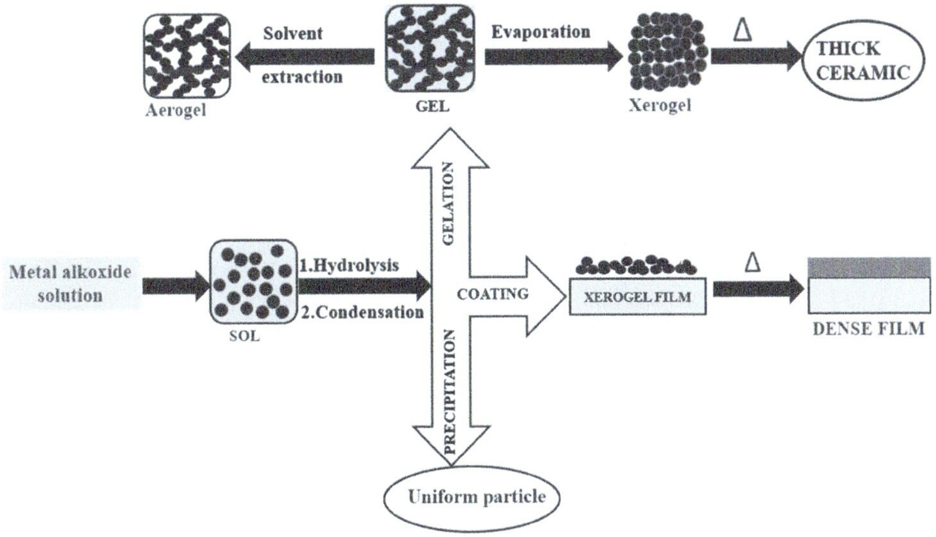

Figure 2.2 Schematic representation of stages involved in sol-gel method.

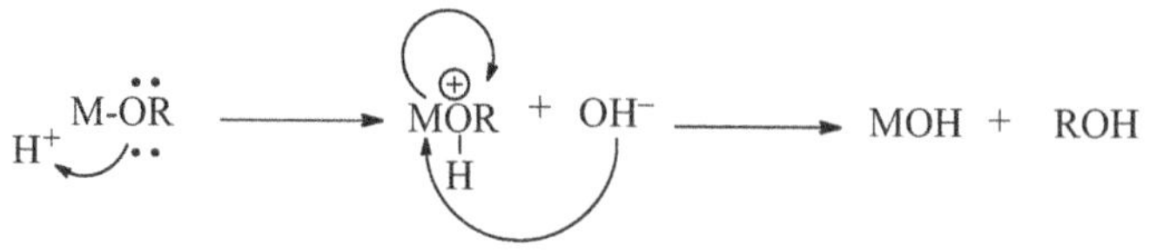

(2.3)

Nucleophilic substitution ($S_N{}^2$) takes place in the basic medium with the inversion of the configuration. The hydroxide ion acts as a nucleophile and attacks the metal center from the opposite side of the leaving group (–OR), leading to the formation of MOH and ROH (as shown in Eq. (2.4)). At basic pH, smaller crystalline particle formation of mixed oxide ZnO-Fe_2O_3 is favored at a reduced temperature [6].

$$MOR + H_2O \rightarrow [HO - M - OR] \rightarrow MOH + ROH \quad (2.4)$$

The sol particle aggregation occurs due to the weak van der Waals force to decrease total surface energy. To reduce the weak attractive forces ranging a few nanometres, repulsive counter forces must be established. This strategy can be achieved by introducing electrostatic repulsion and steric hindrance. Introducing a charge on the surface of the particle results in electrostatic repulsion, thereby reducing the tendency toward agglomeration amongst similar charges. Stabilization by electrostatic repulsion is due to the formation of a double layer. The proton concentration shall be very high at lower pH; therefore, they would be adsorbed more than other counter cations. Similarly, when the pH is high, hydroxyl ion concentration shall be high, and these adsorb on the surface rather than any other counter anion.

Successfully coating a surface with a bulky organic layer prevents the attraction between the particles, thus reducing the weak van der Waals attraction. Usually, heavy branched chains are used for nanomaterial synthesis. Bulky ligands often slow down the rate of hydrolysis. In the TiO_2 thin film preparation, the incorporation of secondary and tertiary alkyl groups in the alkoxy chain control the rate of hydrolysis, resulting in a much more homogeneous gel with less cross-linking [7]. The pH at which the net charge of the whole particle surface is zero is called the point of zero charges (PZC). When the pH is below the PZC value, a positive charge is formed on the adsorbent surface, which adsorbs anions; however, if the pH is above the PZC threshold, cations will be adsorbed.

The metal hydroxide obtained after the hydrolysis process combines with another molecule of the sol or unreacted metal alkoxides, accompanied with the elimination of water and small molecules like alcohol in the condensation process. This procedure is usually performed in the presence of a catalyst (as shown in Eqs. (2.5) and (2.6)).

$$MOH + MOH \rightarrow MOM + H_2O \quad (2.5)$$

$$MOH + MOR \rightarrow MOM + ROH \quad (2.6)$$

This dynamic process can be controlled in the desired direction by modifying the relevant parameters: precursor type, alkoxides/water ratio, the catalyst employed, solvent, temperature, pH, and relative and absolute reactant concentrations [8]. As the sol aggregation continues, the viscosity increases, resulting in gel formation. The sol-gel transition or the gel point is attained when a continuous network is established. The gel formation can be confirmed when the liquid is fully resistant to flow. At gel point, many sol particles and clusters will remain unreacted, and thus aging of the gel is a crucial stage. During the aging process, the solvent molecules will be trapped inside the gel. Usually, the groups on neighboring branches will condense, making the gel even more viscous. It will squeeze out the liquid from the interior part of the gel, which increases the gel's flexibility. When the liquid in the gel is replaced by air, significant changes in the interior structure may occur. The main aim of the drying process is to remove the trapped solvent molecule and the alcohol from the gel. During standard heating, the already formed framework-like structure may collapse, leading to the formation of xerogel. This happens mainly due to the capillary forces from the walls of the pores, thus reducing the pore size. Cracking also may happen due to the increased tension in the gel, which may lead to shrinkage.

Further heating of the xerogel results in the formation of thick glass or ceramic materials. Thus, aerogel formation occurs by carefully eliminating the solvent, yielding nanomaterials with tremendous porous surface area. This can be achieved in three ways: heating till supercritical conditions, thermal evaporation, and calcination.

When the gel is heated above the supercritical point to remove the solvent, it retains the morphology and 3D network of the gel. This method is most preferable; however, it is not economical, and maintaining the pressure and temperature beyond supercritical conditions is complicated. In this context, the aerogel was in the anatase phase with a homogeneous and uniform particle dispersion upon high-temperature supercritical TiO_2 drying, and the xerogel sample was a mixture of anatase, brookite, and amorphous TiO_2. Aerogel demonstrated more excellent photocatalytic activity than xerogel [9]. The gel is heated at highly economical room temperatures. The only limitation of this process is the change in morphology after heat treatment.

Calcination of the gel in the environment of oxygen at high temperatures around 600–800 °C is widely used for producing nanomaterials with extremely porous and large surface area architectures. In the case of $SrTiO_3$ it was shown to improve the grain size and crystallinity of the material [10].

As a whole, we can say that the sol-gel process is crucial in obtaining nanostructures with good morphology and high surface area. The reaction proceeds through several stages, and each step is dependent on various parameters, which include pH, temperature, solvent, and water. Very recently, Portia et al. studied the influence of temperature on morphological and optical properties of Eu-doped $CaTiO_3$ catalysts [11]. It was detected that as the annealing temperature increased, it displayed a uniform spherical shape with a bandgap lowered from 2.99 to 2.45 eV, indicating better photocatalytic efficiency in degrading organic dyes. Similarly, [12] developed a composite of CdO/La-doped $SrTiO_3$ (LNTO) with a 732-fold increased photodegradation rate of antibiotic waste material in water compared to pristine LNTO.

The main advantages of the sol-gel process are high versatility, highly economical procedure, extended compositional ranges, better homogeneity, less energy consumption, and no use of expensive instruments. The only disadvantages of this method are the cost of the precursor and shrinkage of the wet gel upon drying.

2.2.2 Solid-state Method

The solid-state method is a typical solvent-free process for preparing ceramic powders and polycrystalline material from simple metal salts, oxides, carbonates, nitrates, oxalates, alkoxides, and hydroxides. Usually, the reaction is performed at a very high temperature to facilitate the interdiffusion of reactants driven by the chemical potential gradient due to concentration difference [13]. Solid-state reactions are simple to perform and cost-effective. Also, it leaves limited side product formation, and is compatible for large-scale production. Figure 2.3 illustrates the schematic depiction of the product formation in solid-state reactions. Fick's law governing diffusion flux J due to a concentration gradient dC/dX is shown in Eq. (2.7):

$$J = -DdC / dXJ = -DdC / dX \tag{2.7}$$

where D is the diffusion coefficient, dC is the change in concentration of the particles, and dX is the change in position of the diffusing particle. For proper diffusion to take place, D should be around 10–12 cm^2s^{-1}. Proper grinding of reactants in stoichiometric ratios results in thermodynamically stable products at high temperatures. To set up efficient solid-state reactions, chemical and morphological properties affecting reaction conditions need to be assessed; these include temperature, pressure, surface area, and free energy change [14], and are discussed below.

Surface area: As the principal step involved in the reaction is the interdiffusion of starting materials, the contact area between them should be maximized. To achieve this condition, the surface area of reacting species should be very high. This can be accomplished by several milling and annealing steps to reduce the particle size, and increase homogeneity and the surface area. In addition to milling, high-pressure pelletization effectively decreases grain porosity and improves the contact between the starting materials. For example, a powdered $BaBiO_3$ photocatalyst was synthesized using a stoichiometric mixture of Bi_2O_3 and $Ba(NO_3)_2$ as precursors, followed by a two-step heating solid-phase procedure. A highly photocatalytic semiconductor was obtained after thorough grinding, pressing at high pressure, and annealing at high temperature [15].

Reactivity: Reactive starting materials are usually used instead of inert materials. The reactant molecules have restricted contact between them in solid-state reactions compared to the gas and liquid molecules. In that, the lattice structure and diffusion are the prime factors. The reaction undergoes two steps, namely, formation of nuclei of the product, followed by growth of the nuclei, to obtain a thermodynamically stable solid-state product.

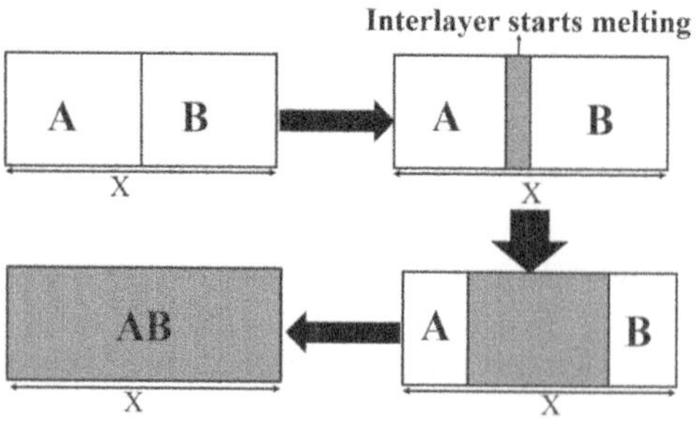

Figure 2.3 Schematic representation of solid-state reaction via interdiffusion of ions.

Thus, after the formation of nuclei, subsequent transfer of diffusion of ions is required for the formation of a thick product layer. It must be noted that a massive difference in the structure between the reactants and products requires a huge structural reorganization. Precursors that are incredibly reactive, therefore, should always be considered for the reaction to take place effectively.

Temperature: Solid-state reactions are usually done under high temperatures where the ion diffusion rate increases rapidly until the melting point is reached. According to Tamman's rule, the diffusion coefficient increases rapidly with temperature, and decomposition of the starting material occurs only when the temperature of two-thirds of the low melting reactant is achieved. Generally, using volatile precursors in the reactions should be avoided as there is a high chance of escaping of the material during high-temperature synthesis, and it is challenging to incorporate an ion that is prone to forming volatile species. Figure 2.4 exemplifies the formation of $BaTiO_3$ perovskite by solid-state reaction at high temperatures [16].

Further, a visible-light active direct Z-scheme Cu_2O/WO_3 nanocomposite photocatalyst was obtained by the solid-state reaction to avoid the inevitable production of the $CuWO_4$ phase in wet syntheses. The reaction is carried out under high temperature, indicating the superiority of solid-state synthesis over usual wet chemical methods [17]. A homogenous NH_4TiOF_3 nanoplate of approximately 100 nm thickness was synthesized by utilizing a green, solvent-free low-temperature solid-state chemical method [18]. The anatase TiO_2 obtained in this research had excellent heat stability, tolerating high temperatures up to 1000°C without any phase change. Due to the exposed (001) facet and high crystalline nature, the anatase TiO_2 nanoplates had a high rate of H_2 production.

Similarly, a simple solid-state method was demonstrated producing a $BiOCl/Bi_2O_2CO_3$ p-n heterojunction photocatalyst [19]. The RhB degradation efficiency was found to be more significant in the heterojunction. Increased photoactivity was connected with heterojunction formation, inhibited electron–hole recombination, and improved charge separation efficiency [19].

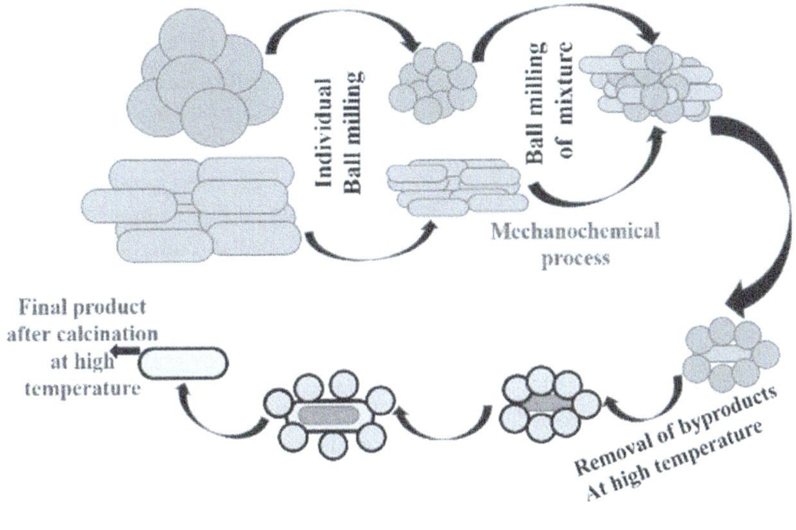

Figure 2.4 Synthesis of $BaTiO_3$ by modified solid-state reaction.

In recent times, room temperature solid-state synthesis has been developed in place of high-temperature processes. Solid-state synthesis of tin-doped ZnO has been demonstrated at room temperature [20], and the formation of wurtzitic ZnO was confirmed by analyzing the XRD patterns. No additional phases were found. Doped and undoped ZnO nanocrystals were synthesized to compare the photocatalytic and gas sensing characteristics. It was observed that the photocatalytic performance against methyl orange was highest. Another example of this method is the Co_3O_4/ZnO heterostructure synthesized by a simple solid-state reaction approach involving crushing zinc acetate, Co_3O_4, and sodium hydroxide at room temperature at a fixed molar ratio without the use of a surfactant or template [21].

2.2.3 Ball Milling

Ball milling, also known as mechanical milling, is an extensively used synthesis route for metallic and ceramic nanomaterials due to its simplicity, ease, solvent-free synthesis, cost-effectiveness, and high yield. Carbon nanotubes, boron nitride nanotubes, and metal oxide nanoparticles are commonly synthesized using this approach. The process involves milling followed by annealing to obtain nanoscale particles. To avoid undesirable oxidation of the precursor, both processes are performed in an inert environment. Several factors, namely, plastic deformation, cold welding, and fracturing determine the particle size after milling [22]. Ball milling improves the stability and photocatalytic efficiency of composite catalysts by generating imperfections and additional active edge sites. In this context, synergistic effects such as change of size and shape after ball milling, faster interfacial charge separation, increased specific surface area, more vital redox ability, and excellent visible light absorption, lead to superior photocatalytic efficiency of g-C_3N_4/$BiPO_4$ and CeO_2/g-C_3N_4heterojunctions [23, 24]. A MoS_2/BiOBr photocatalyst was made by using an ionic liquid-assisted mechanical ball milling process at room temperature and pressure. The produced material was identified according to FT-IR, Raman, X-ray photoelectron spectroscopy (XPS), and HRTEM findings. Under visible light illumination, the composite had better photocatalytic efficiency than pure BiOBr, with one percent MoS_2/BiOBr hybrid nanosheets having the greatest photocatalytic annihilation activity [25].

The ball milling apparatus consists of mills equipped with grinding media made up of stainless steel, tungsten carbide, or silicon carbide. Usually, a powder mixture is alloyed by placing it in the equipment, namely, a vibration ball mill, jet mill, bead mill, horizontal rotary ball mill, attritor, or a planetary mill, and is subjected to high-energy collision from the balls to obtain a nanoparticle in the range of 2–20 nm. Several factors determine the size and shape of the nanoparticle: temperature, nature of the balls, rotation speed, and size and number of balls. Depending on the requirement of the final product, these parameters can be altered. Transformation of Bi_2O_3 from monoclinic to cubic to tetragonal phase happens over a while with ball milling of the appropriate mixture of Bi_2O_3 and WO_3 powders to obtain the orthorhombic γ-Bi_2WO_6 phase. The dependence of milling time on the nanoparticle size has been studied extensively in the literature. The morphology and band-gap energies of the obtained nanosheets decreased considerably with the increment in milling time [26]. A dry mechanical alloying approach performed in a horizontal ball mill was explored to synthesize tungsten carbide, WC, 10 wt% Co powder. A rise in average domain size was detected with a fivefold increase in milling duration, whereas a drop in mean square strain was reported, which could be attributed to dynamic recovery occurring during extended milling. As a result, extended milling increased the number of nanosized particles [27]. The unique Z-scheme heterojunction CeO_2/g-C_3N_4 nanocomposites were synthesized by ball milling three different mass ratios of CeO_2 and g-C_3N_4 [23]. SEM analysis indicated powdered CeO_2 with uneven nanoparticles and partially agglomerated structures. It showed a better charge carrier separation and stronger UV light sensitivity, while 70% CeO_2/g-C_3N_4 mass ratio photocatalyst exhibited a higher potential to remove methylene blue (MB) [23].

When a small amount of sample is required, the planetary ball mill is a commonly used technology for mechanical alloying. The ball mill system consists of a single-turn disc rotating on its own axis with a velocity (v_1) and two or four bowls rotating in opposite directions with another velocity (v_2), creating a centrifugal force which aids to the fracture and cold welding of the powder mixture under high-energy impact [28]. This unique design takes advantage of opposite movements with different rotational velocities, using friction, impact force, the Coriolis effect, and centrifugal force to grind the materials finely. Cao et al. recently synthesized a ternary Z-type photocatalytic material, g-C_3N_4/TiO_2@Ti_3C_2, starting from a two-dimensional (2D) multi-layered Ti_3C_2 material known as MXenes (2D materials of a-few-atoms-thick layers of transition metal carbides, nitrides, or carbonitrides) as the carrier and g-C_3N_4 as raw material by a one-step calcination process, assisted by high-energy ball milling. The impact of Ti_3C_2 and calcination temperature on the composite was studied to determine the photocatalytic degradation efficiency [29].

There are several stages of the ball milling process in which powder particles are pressed due to the collision of the balls in the initial stage. Subtle morphological changes are observed without any net change of mass in this stage. In the subsequent stages, the particle remains heterogeneous even though slight dissolution happens. Cold welding and fracture become more significant in this stage. During the final stage, a reduction in particle size become more evident with the appearance of a highly deformed homogeneous metastable structure.

Contamination on the surface and interface of particles is a significant disadvantage of this mechanical alloying process. Careful maintenance of inert atmosphere, and rotating speed and time can prevent the contamination to an extent. There are several other strategies to reduce contamination; coating the surfaces of grinding media or balls with a ductile material is an effective method. Also, introduction of inert gases in ball-mill-based synthesis is one of the ways to reduce the trapped atmospheric gases [30].

2.2.4 Hydrothermal Method

Water represents a very good solvent at elevated temperature and pressure. Exploiting the solvent property of water, a heterogeneous chemical reaction enabled under these conditions is referred to as a hydrothermal reaction. Since this method of synthesis is purely based on water, avoiding the use of hazardous organic solvents, it is considered a greener and sustainable way of nanoceramic synthesis. In this method, the temperature of the solvent is brought to a critical point by continuous heating with autogenous pressure. The reaction chamber consists of a Teflon container in a stainless steel casing to maintain the temperature and pressure. When choosing an autoclave, the most significant aspects to consider are temperature and pressure range, and the corrosion resistance of the material in the given solvent/hydrothermal medium at that pressure–temperature range. The precursors are thoroughly mixed and transferred to the autoclave, where they are heated over the water boiling point, substantially increasing the pressure inside the reaction mixture above atmospheric pressure. Without any postannealing treatments, this synergistic action of high temperature and pressure favors the generation of highly crystalline material. Due to the excellent solvent property of water at the supercritical temperature, hydrothermal synthesis possesses the following advantageous characteristics: (1) the production of ultrafine particles, (2) the possibility of morphological changes through controlled manipulation of pressure or temperature, and (3) the application of a reducing or oxidizing atmosphere by introducing oxygen, hydrogen, or other gases [31]. The formation of a nanoparticle can take place at any point above room temperature. This method is widely used for the insoluble precursors in any solvent at room temperature or low temperature. To control the morphology of the product, either low or high-pressure conditions can be employed depending on the vapor pressure of the main reaction composition. This synthesis route is employed to produce nanometal oxides, layered oxides, nanotubes, nanowires, elemental nanostructures, and carbon nanotubes.

Many scientists have explored the hydrothermal technique to prepare photoactive materials for various energy applications. For example, TiO_2 based nanotubes produced by the hydrothermal method can exchange ions, allowing further modification by doping them with additional ions. As a result, the exceptional physicochemical qualities, such as large surface area and nanotubular characteristics with multilayer walls, open up many catalytic possibilities [32]. Other research showed that using 10% graphene–TiO_2 nanotube composites, maximal photocatalytic breakdown of malachite green was accomplished in both the UV and visible wavelength ranges. Thus, a threefold increase in photocatalytic efficiency was reported compared to pure TNTs [33]. SnS_2 nanoparticles of varying sizes were produced at 120–180°C for 12 hours using a hydrothermal process involving $SnCl_4.5H_2O$ and variable quantities of thioacetamide in a 5 vol % acetic acid aqueous solution, and their plausible formation mechanism was postulated in this research [34]. Physical and chemical properties of $BiOBr/Bi_2WO_6$ composites manufactured through the ionic–liquid hydrothermal technique with varying amounts of Bi_2WO_6 showed improved optoelectronic ability and photocatalytic performance for heavy metal ion removal, and drug and organic dye degradation [35].

Doping via hydrothermal synthesis is successful at improving the photocatalytic activity as metal/codoped nanostructured hybrids can absorb visible light by minimizing the recombination of electron–hole pairs and altering the bandgap by forming an interstitial level. Nitrogen-doped TiO_2 was synthesized by hydrothermally treating isopropyl titanate and tetramethylammonium hydroxide (TMAOH), which is a nitrogen and a carbon source. Simple post-thermal treatment was performed to increase the crystal nature and photocatalytic efficiency of the as-prepared precursor. The end products after post-thermal treatment at temperatures ranging from 200 to 700°C showed diverse characteristics [36]. The bandgap of TiO_2 was lowered to 2.65 eV via low-temperature hydrothermal sulphur doping, resulting in better visible photocatalytic activity than sulfur-TiO_2 oxidized at high temperatures. The higher photocatalytic activity was observed to be due to the greater surface area versus sintered sulfur-TiO_2 via various chemical reaction pathways [36]. Sulfur-doped reduced titania photocatalysts synthesized via the hydrothermal method using thiourea dioxide as both the sulphur source and reductant showed excellent photodegradation rates of RhB about 31, 2.5, and 3.6 times higher than those of pure TiO_2, pristine TiO_{2-x}, and sulfur-doped TiO_2 [37]. $Fe_3O_4@SiO_2@C_3N_4/TiO_2$ nanocomposite prepared via combination of sol-gel and hydrothermal methods showed greater utilization of photons from sunlight, thereby enhancing photocatalytic degradability of organic dyes [38]. Similarly, rGO-TiO_2 wt% below 1 prepared by the hydrothermal method showed excellent degradation of MB in 15 min under 50 mW/cm^2 UV light intensity [39]. 25% TiO_2/Bi_2WO_6 nanocomposite prepared via the hydrothermal process efficiently degraded starch [40].

$FeVO_4$ nanoparticles ($FeVO_4$-NPs) have recently been effectively synthesized using the hydrothermal technique. To remove MB dye from the solution, $FeVO_4$-NPs were employed as a Fenton-like photocatalyst [41]. SnS_2/SnO_2/rGO nanocomposites with a one-step in situ hydrothermal procedure by altering the molar ratio of L-cysteine to $SnCl_4.5H_2O$, exhibiting flowerlike SnO_2/SnS_2 sheet heterostructure evenly anchored on rGO sheets in the nanocomposites [42], offered a new technique for increasing the activity and selectivity of CO_2 to reduce methane preferentially. A two-step hydrothermal approach was used to successfully synthesize a trimodal porous silica (TPS)/g-C_3N_4 nanotube composite photocatalyst for converting CO_2 to ethanol. The crystallinity and chemical composition of g-C_3N_4 nanotubes were not altered; however, the addition of TPS increased the composite's specific surface area and reaction sites. The process of enhancing the photocatalytic performance of the composites was confirmed based on UV–visible diffuse reflectance spectroscopy (UV–Vis DRS), photoluminescence spectroscopy (PL), photocurrent responses, and electrochemical impedance spectroscopic (EIS) investigations [43].

2.2.5 Chemical Vapor Deposition

Chemical vapor deposition (CVD) is another chemical route used to synthesize high-performance, high-purity solid materials without the need for post-heat treatment to enhance crystallinity. This technique is also used to manufacture thin-film coatings, powders, single crystal, fibres, and monolithic components. The precursor components are conveyed in a stream of carrier gas, which can be inert or reactive, and eventually break down on a heated substrate to produce a solid material [44]. The reactant gases employed in the CVD process give a crucial advantage, as they enable the process to take advantage of the various peculiarities of gases. As a result, unlike most other plating/coating techniques, CVD is not a line-of-sight approach and can be exploited to coat restricted access surfaces. The morphology of the product is determined by several parameters, including substrate material and temperature, total gas flow pressure, and reaction gas mixture composition. Gas flow across the reaction chamber can eliminate the volatile by-products created during the process. The CVD technique is used in heat-resistant coating and corrosion-resistant coating. The main benefits of this process include material formation below the melting point, high versatility (any element or compound can be deposited), economical production, uniform distribution over a large surface area, and more selective deposition due to high activation energy for reaction with foresight.

CVD belongs to the atomistic class of vapor-transfer processes, wherein the deposition species are atoms, molecules, or a mixture of these. The reaction begins in the gas phase due to heating, followed by some chemical reactions with the substrate in between the reactive gases to finally undergo a deposition on the substrate. The entire procedure is divided into five steps: (1) Diffusion of reactants through the boundary layer, (2) Adsorption of reactants on the substrate (wafer) surface, (3) Chemical decomposition and surface relocation to attachment points, site incorporation, and other surface reactions, (4) Desorption of adsorbed species from the surface, and (5) Diffusion of by-products back to the mains. The deposition rate υ is commonly represented using Eq. (2.8):

$$\upsilon \; \alpha \; \frac{k_s h_g}{k_s + h_g} \tag{2.8}$$

where k_s is the surface reaction rate and h_g is the mass transfer coefficient. It is evident that when $k_s \ll h_g$ deposition is reaction rate limited, and when $k_s \gg h_g$ deposition is mass transfer limited. Therefore, the deposition rate can be controlled by changing the time of the reaction as well as the volume of the reactive gases and precursor gases inside the chamber. This will in turn be reflected in the thickness of the coating obtained.

There exist different types of CVD techniques based on the operating pressure: (1) atmospheric pressure CVD, (2) ultra-high vacuum CVD (below 10^{-6} Pa ~ 10^{-8} torr), and (3) low-pressure CVD. Furthermore, CVD techniques are categorized based on the physical properties of the vapor such as aerosol-assisted CVD and direct liquid injection CVD [45]. Various other plasma methods are also present: microwave plasma-assisted CVD, plasma-enhanced CVD, remote plasma-enhanced CVD, atomic-layer CVD, combustion CVD, hot filament CVD, hybrid physical–chemical vapour deposition, metalorganic CVD, rapid thermal CVD, and photo-initiated CVD [45].

Recently, a study on the impacts of time and temperature on the structure and characteristics of Co-doped TiO_2 nanotubes, and multiwalled CNTs (Co-TiO_2/CNT) synthesized through the CVD method, showed that longer synthesis time leads to the formation of damaged nanotube arrays resulting in more defective CNTs [46]. Another research group in this context focused on the relationship between the growth condition and the final architecture of nanopyramidal ZnO via the CVD technique to generate materials that are generally unstable and challenging to make. It was claimed that the enhanced surface energy, electrostatic potential, and polarity of pyramidal ZnO results in lower interfacial resistance at the surface, leading to improved photocatalytic activity [48]. [49] studied a high-quality graphene via the CVD process (CVDG) combined with TiO_2 to make CVDG-TiO_2 composite to gain a fundamental understanding of graphene in promoting photocatalysis. The CVDG-TiO_2 catalysts achieved excellent stability yielding a hydrogen evolution rate of 68 molh^{-1}, which is about 26-fold higher than that of blank TiO_2 and 10-fold higher compared to RGO-TiO_2, under the same reaction conditions. The work of Li et al. on tubular g-C_3N_4 grown on the carbon framework using a facile one-step CVD process showed increased surface area of bulk g-C_3N_4 and improved visible light absorption in the composite [47]. Simultaneously, the carbon structure enhanced the efficiency of charge separation and transfer. The twin actions of static adsorption and photodegradation allowed the framework to remove nearly 95% MB in 180 minutes. Also, the SnS_2/SnO_2 heterostructure nanoflakes manufactured using a one-step CVD technique show significant improvement in photodegradation ability of MB. Since the interface of SnO_2/SnS_2 could create an electric field to separate electrons and holes, it resulted in the improved photodegradation property [43].

Typically, copper (Cu), nickel (Ni), iron (Fe), and cobalt (Co) are used as the template or substrate for lateral growth of graphene on a substrate. Work of [50] demonstrated the synthesis of graphene foam using the Ni foam as a 3D scaffold through the CVD method. A transformer-coupled plasma-enhanced CVD method was utilized to achieve a flower-like 3D porous graphene/ZnS photocatalyst. Because of its 3D macroporous framework and high specific surface area, the photocatalyst demonstrated improved interface contact between the flower-like 3D porous graphene and ZnS nanoparticles with outstanding H_2 production activity (~11.6 $mmolg^{-1}h^{-1}$) compared to pure ZnS (1.5 $mmolg^{-1}h^{-1}$) [51]. Similarly, Fe_7S_8/Fe_3O_4@Fe synthesized via the vacuum CVD method showed an intense contact of Fe_7S_8 and Fe_3O_4 components, resulting in a broadened bandgap and enhanced antioxidative capability of the Fe_7S_8. Photoinduced active species created through the photoinduced electron pathway oxidized methyl orange and phenols directly using this photocatalyst [52]. $In_xGa_{1-x}N$/GaN core/shell nanowires were produced spontaneously on the n-type Si (1 1 1) substrate using CVD and a precursor mixture of $Ga(AcAc)_3$ and $In(AcAc)_3$ with an Ni catalyst (CVD). The quantity of H_2 evolved by water splitting with a hole scavenger of CH_3OH was used to assess the photocatalytic activity of the Pt-loaded nanowires [53].

Although researchers have explored this method extensively, it has several disadvantages as well. Major concern is with the release of inflammable gases that can ignite in the presence of oxygen or the release of harmful gases that are hazardous to the environment and the operators. Also, contamination and safety of the instrument is a concern, which may in turn negatively affect the products obtained after the reaction. Further, this material is not always reliable for a mixture of materials as, during composite formation, one material may be reactive with the gases, and the other may remain inert. In such cases, the expected final product cannot be accomplished. Necessary analysis of the reaction conditions should be carried out before using this methodology.

2.2.6 Microemulsion Method

Microemulsions have received much attention because they provide a small heterogeneous environment for nanoparticle production. Microemulsions are thermodynamically stable dispersions of oil, water, surfactant, and often cosurfactant. They can be classified as oil-in-water (O/W), water-in-oil (W/O), or bicontinuous systems. W/O systems are typically 10–100 nm-sized water droplets dispersed in a continuous oil medium and stabilized by surfactant molecules. These droplets act as nanoreactors for nanoparticle production, which increase the interfacial area and interfacial energy. The droplet distribution, as well as the kinetics of interdroplet interaction, influence particle production in these systems. The Gibb's free energy of formation G is given by Eq. (2.9):

$$\Delta G = \gamma \Delta A - T\Delta S \tag{2.9}$$

where γ is the surface tension at the oil–water interface, ΔA is the change in the interfacial area, T is the absolute temperature, and ΔS is the change in entropy.

The role of surfactant and cosurfactant is to lower the interfacial surface tension even to a negative value to maintain an overall negative value for the Gibb's free energy, which shows the spontaneity and thermodynamic stability of the microemulsion. The surfactant is stabilized by microcavities, which produce a cagelike structure that prevents particle nucleation, growth, and agglomeration. Some of the most commonly used surfactants are cationic cetyltrimethylammonium bromide (CTAB), anionic bis(2-ethylhexyl) sulfosuccinate (AOT), sodium dodecyl benzene sulfonate (SDBS), lauryl sodium sulfate (SDS), non-ionic Triton X-100, and sorbitan monooleate Span 80, nonylphenyl ether (NP-5), and polyoxyethylene (2.9) nonylphenyl ether (NP-9).

The microemulsion, among other synthesis methods, is a highly flexible and repeatable approach for regulating the size of nanoparticles and creating nanoparticles with a narrow size distribution. This approach helps to control the size, shape, and crystal structure of metallic nanoparticles in some instances, enabling various seeds with varying dimensions using the same reagents. Controlling reaction duration, temperature, and reaction conditions allow us to build several geometries that can be used in a range of scientific areas [54].

Typically, titania nanoparticles were developed, changing the water concentration in the W/O microemulsion between 5% and 30%. The reverse microemulsion comprises 45–60% cyclohexane as the oil phase, 24–33% CTAB-n-butanol as the surfactant and cosurfactant mixture, and 5–30% water as the water phase. The surfactant to cosurfactant ratio in the microemulsion was kept at 2:1 by weight. Following the workup stages, titanium isopropoxide mixed with isopropanol was added to the transparent microemulsion solution drop by drop while stirring to allow for hydrolysis. The particle size, surface area, and crystalline structure of TiO_2 are all influenced by the reverse microemulsion's composition. A TiO_2 sample with 15% water yielded a high yield [55]. The main highlights of the microemulsion method are the ease of synthesis procedure without much energy input, and high reproducibility. Ghafoor et al. studied the impact of doping based on optical, photocatalytic, and structural characteristics of a series of $La_{1-x}Ce_xNi_{1-y}Fe_yO_3$ nanoparticles synthesized using the microemulsion method. The rhombohedral phase of perovskite with particle size in the range of 60–80 nm was formed exhibiting a significant red shift in the UV–Vis absorption [56]. The bandgap was tuned from 2.77 to 2.64 eV by a. Degradation of Congo red (CR) dye under visible light irradiation was used to test the photocatalytic efficacy of $LaNiO_3$ and $La_{0.80}Ce_{0.20}Ni_{0.80}Fe_{0.20}O_3$ catalysts. The performance of the substituted catalyst was superior to that of the original. Similarly, the effect of dopant content on the properties of $Co_{1-x}Cr_xFeO_3$ prepared via the microemulsion method was carried out. MB degradation rates up to 55.38% were observed in 80 min under solar light illumination [57].

Metallic nanoparticles are frequently generated in W/O microemulsions by integrating two different microemulsions of a metal salt and a reducing agent sequentially, with reactants transferred between micelles via Brownian motion, attractive van der Waals forces, and repulsive osmotic and elastic forces among reverse micelles. A modified microemulsion method was used to prepare nanoscale TiO_2 hollow spheres with exterior diameters of 25–35 nm, a wall thickness of 4–6 nm, and a 15–20 nm diameter inner hollow. Hollow sphere synthesis required optimizing the hydrolysis of $TiCl_4$ as a precursor material at the W/O phase interface. The issue was addressed by incorporating gelatine into the water-filled micelles, which effectively slows down the reaction rate [58]. Titanium tetraisopropoxide (TTIP) and tetraethylorthosilicate (TEOS) were used as precursors for synthesizing nanosized pure TiO_2/SiO_2 particles via a reverse microemulsion with the anionic surfactant sodium bis(2-ethylhexyl)sulfosuccinate (AOT). A reverse microemulsion solution was prepared by dissolving 0.045 mol of surfactant in cyclohexane in an appropriate amount of distilled water. Since the rate of hydrolysis of TEOS is substantially slower than that of TTIP, it was hydrolyzed in the first stage using a premixed reverse microemulsion solution. The two precursors were added to the premixed solution and agitated for four hours to produce TiO_2/SiO_2 particles. The hydrolysis was performed at 30°C, followed by the workup processes [58]. A microemulsion method can also be used for doping. For example, by altering the Fe and Gd concentrations, a variety of $La_{1-x}Gd_xCr_{1-y}Fe_yO_3$ nanoparticles were produced. The solutions were stirred for 7 h and up to 50°C followed by the addition of CTAB and 0.3 M ammonia (solution) and continued stirring for an additional 2 h. After the washing and annealing process, the nanoparticles were subjected to photocatalytic studies [59]. Similarly, a research group tried to compare the photocatalytic efficiency of TiO_2/CNT composite synthesized using both sol-gel and microemulsion methods. It was concluded that the catalyst prepared via microemulsion showed higher photocatalytic activity in MB degradation studies compared with the sol-gel method. The sol-gel process produced a heterogeneous and nonuniform TiO_2 coating with some bare CNT surfaces and random TiO_2 aggregation, whereas the microemulsion method produced a regulated reaction rate with a homogeneous and regular TiO_2 layer on CNTs in the nanometer range with very high surface area [60]. The only disadvantages of this method are the usage of a larger amount of surfactants, temperature stability influenced by microemulsion and pH, and limited solubility capacity for substances with high melting points [61]. This catalyst is used to degrade MB dye under UV light, particularly $NiFeWO_4$, which showed good photocatalytic action. Under UV irradiation, the current research could be used to cleanse wastewater on a large scale in order to eliminate dangerous dyes [62].

2.2.7 One-pot Synthesis

One-pot synthesis is an essential method of enhancing the efficiency of a chemical reaction by subjecting it to several chemical processes in a single reactor. The one-pot approach is recommended because it saves time and resources while increasing chemical yield by eliminating a lengthy extraction procedure and intermediate chemical compound purification. For one-pot synthesis, specific reaction criteria are effective. These reactions include those in which the intermediate compound is unstable, malodorous, or hazardous. Further, there exists an equilibration of starting material, the intermediate compound reaction generates convertible side products into the desired intermediate compound or product, and the reactivity remains constant in both the starting material and the intermediate compound [63].

The one-pot approach was utilized to synthesize Ag-doped TiO_2 immobilized on a cellulose-derived carbon bead ($Ag/TiO_2@C$) photocatalyst to remove ceftriaxone sodium residue from wastewater. Under simulated sunlight, Ag (0.5%)/TiO_2@C showed superior photocatalytic activity in the breakdown of ceftriaxone sodium [64]. The formation of a reduced graphene oxide–zinc oxide sphere composite via simple microwave irradiation provided fast and consistent heating to integrate the formation of ZnO nanostructures and the reduction of graphene oxide (GO) in one pot without a template, complex metalligand, or surfactant. The nanocomposite displayed a tenfold increase in photodegradation activity toward MB dye in water when compared to ZnO or RGO alone under visible light irradiation [65]. BiOX (X = Cl, Br, I) nanoplate microspheres were produced using a one-pot solvothermal approach that involved mixing $Bi(NO_3)_3H_2$ and ethylene glycol comprising KCl, NaBr, and KI with a Bi/X molar ratio of 1. BiOX photocatalytic performances on methyl orange degradation under UV-visible light illumination (>420 nm) were investigated and found to exhibit superior photocatalytic activity [66].

Another benefit of the one-pot synthesis method is the cost-effectiveness of using chemicals and solvents. As the reaction workup at each step has to utilize solvents that will eventually go to waste, multiple-step synthesis in a single flask would only require a final workup at the last stage. A minimal level of effort is required: rather than working up at each stage of the synthesis, which could take up to three steps, it can be done only in a single step at the end. Recently, work was done on a unique low-temperature one-pot synthesis of nitrogen-doped $TiO_2/ZnFe_2O_4$ catalyst in a liquid urea mixture as the nitrogen source. Compared to $ZnFe_2O_4$ and $TiO_2/ZnFe_2O_4$, this catalyst exhibited maximum photocatalytic performance due to the advent of a heterostructure at the interface and the addition of the nitrogen atom. Meanwhile, under UV illumination, the heterostructure successfully separated photoinduced electrons and holes, improving the duration of the photogenerated electrons [67]. This research provided a simple method for fabricating hybrids and a functional and easily recyclable photocatalyst for water purification. Using a one-pot molten salt approach, CdS/C_3N_4 heterojunctions have been created in which thiourea was used as a carbon and nitrogen source and $Cd(NO_3)_2$ as a metal source. The resulting materials contain wurtzite CdS after a 2 h reaction at an initial temperature of 200°C. CdS/C_3N_4 nanojunctions outperformed virgin C_3N_4 in terms of visible light responsiveness and adsorption capabilities. Thus, the composite outperforms standard CdS and C_3N_4 in photocatalytic HER efficiency [68].

The one-pot synthesis method is also very effective in producing ternary systems with ease and high yield. A simple one-pot hydrothermal approach has been established to synthesize a novel $CdS/BiOBr/Bi_2O_2CO_3$ ternary heterostructure photocatalyst. The heterostructure material demonstrated desirable phases, highly regulated morphology, and increased visible light absorption. The morphological characterization analysis revealed that improved microscopic contact between the constituent phases resulted in the heterojunction between the three phases [69]. A simple one-pot anion-exchange procedure was used to make an $AgBr/Ag_3PO_4$ photocatalyst. The material displayed a remarkable photocatalytic capacity for the photodegradation of RhB when exposed to visible light. Under solar light illumination, the optimum 48% composite illustrated the highest photocatalytic efficiency within 25 min [70]. In addition to experimental work, numerous theoretical researchers are also working on the successful preparation of photocatalysts via simple one-pot synthesis and optimization of various energy-related photocatalytic applications. One such research group succeeded in zinc oxide nanoparticle synthesis by proper statistical analysis of the collected results using a central composite design. A mathematical formalism for optimization of input parameters was built to forecast the results at any specified time. The impact of crystal growth on the photocatalytic ability of the produced catalyst was investigated in depth in this groundbreaking work [71].

The flake-shaped $NaBiO_3$/Poly (N-methylaniline) (PNMA) photocatalyst was generated using a unique one-pot technique [72]. The effect of PNMA on the photocatalytic performance and surface properties of $NaBiO_3$ was studied. The FT-IR data revealed a strong bond between $NaBiO_3$ and PNMA via the Bi-O-C bond. Although the adorned PNMA did not affect the structure of $NaBiO_3$, the composite resulted in a more hydrophobic surface. In 40 min, the degradation efficiency of RhB was observed to be 99.3%.

Even though one-pot synthesis is an excellent method for efficiently synthesizing nanoparticles, it also has certain limitations. The same solvent may not be used in each synthesis step, and surplus compounds from previous steps may not be compatible with chemicals from later steps in the reaction mixture. As a result, one of the most significant drawbacks of developing a one-pot synthesis is figuring out how to optimize and ensure the compatibility of chemicals, solvents, starting materials, and results. If the previous stage requires an aqueous solution and the next step requires a solution in, say, dichloromethane, and water and dichloromethane are not miscible for practical purposes, then designing a one-pot synthesis including those two steps will be a challenge.

2.2.8 Dip Coating

The dip coating technique produces a range of thin films; however, metal oxides and porous ceramic materials are the most commonly used. As the name suggests, dip coating indicates immersing a substrate into a precursor solution for a thin liquid film on the substrate. Among the available deposition processes, this method is the oldest and most extensively utilized for industrial and laboratory applications, owing to its simple processing, low cost, and high coating quality. Several steps are involved in the procedure, starting with polishing the porous substrate with a layer of smaller particles. Industrial origins may be traced back to the foundational efforts at Schott in the 1940s, and the first patent based on this process was awarded to Jenar Glaswork Schott and Gen in 1939. Several factors influence how to obtain a defect-free film. Some of the necessary conditions are homogeneous support, clean substrate, sufficiently deaerated sol/suspension, optimum thickness of the film to avoid shrinkage and tensile stress during the drying process, and removal of foreign substances in the air or the coating fluids. Particular caution must be taken in certain steps of the coating process, while others are less crucial.

In general, the film-making process entails several technical processes, which are classified into three sections: (1) immersion and resident time, (2) deposition and drainage, and (3) evaporation. Upon substrate immersion at a constant speed in the coating fluid, a coherent liquid layer is adsorbed when the substrate is removed from the precursor solution, which then dries and undergoes chemical processes to solidify. Specific dwell time is given to leave sufficient substrate interaction time with the coating fluid for complete wetting. In the subsequent stages, the substrate is pulled upward at a constant speed to obtain a thin coating fluid/precursor solution layer. Later the excess fluid will drain out from the surface of the substrate. The final stage is the evaporation of liquid from the fluid, resulting in a thin film deposition. A sintering process/post-treatment is usually required to cause the final coating material to burn out residual organics to induce crystallization. The withdrawal and drying steps are crucial for determining the characteristics of the deposited film. During these stages, the interplay between stabilizing, draining, and drying determines the final film thickness. The limitations of dip coating are such that the process is sluggish and can block the screen, resulting in a significant influence on the final product.

Several other factors can thus affect the dip coating process. Some of them include the concentration of precursor solution, the speed at which the substrate is vertically pulled upward from the precursor solution, the angle of withdrawal, the viscosity of the solution, surface tension, and the competitive forces involved in determining the thickness of the film. A research group examined the geometrical and optical properties of ZnO thin film after successfully synthesizing using the cost-effective sol-gel process. The XRD spectra confirmed the hexagonal structure of the films. As the concentration of precursor solution is dependent on the film thickness, an upsurge in the grain size of crystallites was observed with the rise in Zn concentration. ZnO thin films were obtained on a dried glass substrate. Every time, the substrate was removed from the solution and placed in the furnace, where it was allowed to dry before being preheated for 5 min at 300°C. The process was performed six times, from dip-coating to drying. Finally, the ZnO thin film was annealed in the air for an hour at 550°C [73]. Similarly, another research was conducted to examine how drag force and gravity affected the film thickness of polylactic acid (PLA) deposition on the steel surface. At a low withdrawal velocity regime, it was observed that raising withdrawal velocity caused a spike in PLA film thickness. However, it was found that at the relatively high withdrawal velocity regime, the PLA layer thickness begins to drop as the withdrawal velocity increases. PLA film formation is thought to be dominated by the interplay between drag force and gravity during the dip coating process. The findings revealed that while dealing with the dip coating procedure, the gravity impact must be considered. During the process, the consequence is the competition between drag force (solution film going uphill) and gravity (solution film going downhill) controls the variation in resultant solid PLA film thickness as withdrawal velocity increases [74].

Thin films on a substrate are more practical than powders in practical applications since they do not require recovery and can be exploited indefinitely. Thin films can have a range of structures, and there has not been much research on the material impact on photocatalytic characteristics. Recently Al Farsi et al. used a cost-effective sol-gel dip coating procedure to generate pure ZnO and Al-doped ZnO thin film grown on a glass substrate [75]. The films annealed at 550°C exhibited better crystalline properties. A thickness of 100 nm and uniform deposition were observed. Furthermore, the surface morphology of the films was determined by dense granular nanostructures that exclusively covered the glass substrate. Phenol degradation under sunlight was used to examine the photocatalytic activity. With more effective Al doping of 1–2%, the degradation rate and photocatalytic kinetics were improved, showing phenol degradation of up to 80% in just 5 h of solar light illumination [75]. An easy-to-use dip coating process for the fabrication of extended visible light photoactive surfaces of TiO_2, nitrogen-doped TiO_2, porous TiO_2, and-nitrogen doped porous TiO_2 over glass slides has been shown by Raja et al. [76]. The increasing porosity through the use of a PEG matrix improved the photon scattering ability, and was reflected in a positive effect on photocatalytic activity. Under visible light, porous-N–TiO_2 showed a 1.5-fold increase in OH radical production and a 1.54-fold increase in photodegradation. The porous-N–TiO_2 system, with its surface tailored by PEG and with reduced bandgap, proved to be potentially valuable in capturing sunlight. Such approaches can be used in large-scale dip coating techniques at a lower cost.

Recently, studies have started to address the need for recyclable photocatalysts as alternatives for treating contaminated water, and materials that could be used for autonomous photocatalytic systems intended for industrial applications. In this context, [77] developed TiO_2 nanofiber membranes for water treatment using dip coating techniques. An aqueous humic acid (HA) solution was treated with a TiO_2 nanofiber membrane. A roughly 90% removal rate was achieved. Under UV irradiation, HA removal was improved, most likely due to improved filtration performance and photocatalytic degradation of HA, and the usage of TiO_2 nanofibers resulted in an improved decrease of membrane fouling in the presence of UV light illumination.

The dip coating method has an extensive range of applications in the industrial field. Spray, powder, and other commonly used application methods are not appropriate for some sophisticated elements that require coating. Nowadays, dip coating is the preferred method for finishing materials with nooks, crannies, and other complicated regions, as long as these areas can be drained. Manufacturers have used this method to paint tough metals, ceramics, plastics, PVC, and other geometrically complex parts and products for nearly a century.

2.3 Properties of Ideal Photocatalytic Material

Properties that are ideally expected in photocatalytic material are listed below. Any endeavor leading to material preparation should focus on improving some or most of these properties.

i) *Wide photon absorption range:* Photons available from sunlight / light sources need to be maximally absorbed for generation of electron–hole pairs.
ii) *Efficient separation of charge carriers:* The material should possess better charge transfer properties so that the electron–hole pairs generated reach the surface site before recombination.
iii) *Avoiding recombination of electron–hole pairs:* Recombination of generated electron-hole pairs formed at conduction and valence bands need to be delayed.
iv) *Effective redox potential – R /O /X type potential:* The material needs a favorable band potential for conversion of electron–hole pairs into free radicals and formation of radicals catered to the application.
v) *Large surface area / reactive site – adsorption – electronegativity:* Photocatalytic material is expected to possess a higher surface area. Ideally with higher surface area, reactive sites for electron–hole pair generation would be higher. Adsorption of reactant over the photocatalytic material is expected to be higher. Adsorption of reactant depends upon the electronegativity / surface charge.
vi) *Tunable bandgap and morphology:* Photocatalytic material is expected to allow bandgap and morphological tunability to attain the above said properties.
vii) *Electron transfer properties:* The material should have uniformity in crystal structure to favor electron transfer property and interfacial charge transfer property.

viii) *Physical stability:* Physical stability of the entire system is expected over the application of several cycles – repeatability and reusability constitute important attributes of a photocatalyst, having a bearing on the economy of the catalytic conversion process.
ix) *Chemical stability* – An ideal photocatalyst is expected to be chemically stable in harsh chemical environments for reliable industrial applications.
x) *Photo-stability:* Ideal photocatalytic material is expected to have stable structure upon illumination over a long period of time, avoiding photocorrosion.

2.4 Engineering Photocatalytic Properties

To pursue ideal photocatalytic characteristics of a semiconductor, engineering of the intrinsic properties and reaction environment constitutes an important approach in its development. The factors that influence the characteristics of a photocatalyst and techniques to engineer the bandgap of catalyst material are summarized below.

Photon absorption range: Photon absorption range is determined by bandgap properties. To reduce the bandgap, structural modification by doping and oxygen vacancy creation constitute key factors to modify and improve the photon absorption. Upconversion is an upcoming strategy to extend the photon absorption range, by converting near-deep-visible range photons and near-IR photons (low energy photons) to high-energy photons. Upconverted photons could be reabsorbed by a bandgap matched material. Thus, a broad range of photons could be utilized [78, 79].

High recombination rate: The recombination rate of electron–hole pairs is a rate limiting step in every photocatalytic material. The higher the recombination rate, the lower will be the rate of electron–hole pairs reaching its surface reactive sites. One of the techniques to overcome this limitation is to introduce electron sinks in the form of metal impregnation/oxide composites to undergo the Z-scheme mechanism. Another key technique to improve the electron transfer pathway is by improving the crystal structure, introducing nanostructured material, improving the mean free path of electron–hole pairs to reach the surface, and have long-range graphene-like electron conductors [80, 81, 82, 83].

Free radical generation: Free radical generation is a key factor in improving overall efficiency. Band potential matching and interfacial effects are the two important factors to limit the conversion efficiency. Improving the electronegativity on the surface depending upon the reactant environment could improve the conversion and utilization efficiency [81, 83, 84, 85].

Irradiance: Though solar irradiance has high intensity, it gives a fixed range of the photo spectrum. External illumination systems, however, could be designed to match the required bandgap of the material, leading to an effective utilization of photons for efficient free radical conversion. When external light sources are used, increasing intensity would increase the efficiency. Beyond an extent, however, since the number of reactive sites remain constant, the efficiency shall be independent of intensity.

Catalyst loading: To achieve overall efficiency, catalyst loading has to be optimized according to reactant environment and irradiation intensity. Increasing photocatalytic concentration increases photocatalytic activity up to an optimum range depending upon the reactant environment. Beyond the optimum range, photon flux to reach the surface reduces due to increasing opaqueness of the system, reflection losses, and surface area loss due to agglomeration of particles [82, 86].

Surface area: Surface area of the photocatalytic material plays the key role in improving its efficiency of presenting reactive sites toward facilitating increased adsorption of reactants for photocatalytic activity. Nanostructured materials could improve the surface related properties like specific surface area, reactive site generation, adsorption, and surface charge, together with other photocatalytic properties.

Oxygen adsorption: Oxygen adsorption over the active surface would improve the conversion of conduction band electrons into superoxide radicals [87, 88].

Organic element nature: In hydrogen generation applications, organic matter, namely dye, effluent, and other supporting solvents, could affect the activity of a photocatalyst. Based on the application, the photocatalyst surface and reactant environment have to be optimized to maximize the adsorption over the surface and improve the efficiency of hydrogen conversion.

pH: pH plays a key role in modifying the reactant environment by altering the electrical double layer on the surface of the catalyst. It consequently affects the adsorption characteristics, affects electron–hole pairs reaching its surface, and influences the preference of free radical generation and its pathway [82, 87, 89].

Temperature: Lower temperature improves the adsorption but lags in desorption of degraded products. Higher temperature favors the degradation efficiency, but affects adsorption characteristics. Ideally 40–45°C could improve the trade-off between adsorption and degradation [90].

2.5 Energy Bandgap

When there is a single electron orbit for a single atom, there is only one energy for the valence shell. When two atoms are brought close to each other, however, there is an intermixing of electrons and their energy levels. This leads to formation of a number of permissible energy levels, which merge together to form bands. The electrons being used in binding the atoms together exist in the *valence band* (VB) at energy $\boldsymbol{E}_{\mathrm{v}}$, while the remaining electrons not being used for bonding are free to move around forming a *conduction band* (CB) at energy $\boldsymbol{E}_{\mathrm{c}}$. The energy difference between the top of the VB and the bottom of the CB of a solid material defines its *energy band gap* $\boldsymbol{E}_{\mathrm{g}}$. Essentially, $\boldsymbol{E}_{\mathrm{g}}$ represents the minimum energy that is required to excite an electron up to a state in the CB where it can participate in conduction. Figure 2.5(a) depicts the typical energy diagram of a semiconductor material.

2.5.1 Manifestation of Direct/Indirect Band Gaps

The band gap represents the energy difference (minimum) between the maximum energy level of the VB and the minimum energy level of the CB; however, the energy levels where the electrons remain stable in the VB and the CB are not generally at the same value of the electron momentum. In such materials, their crystal momentum, and its corresponding

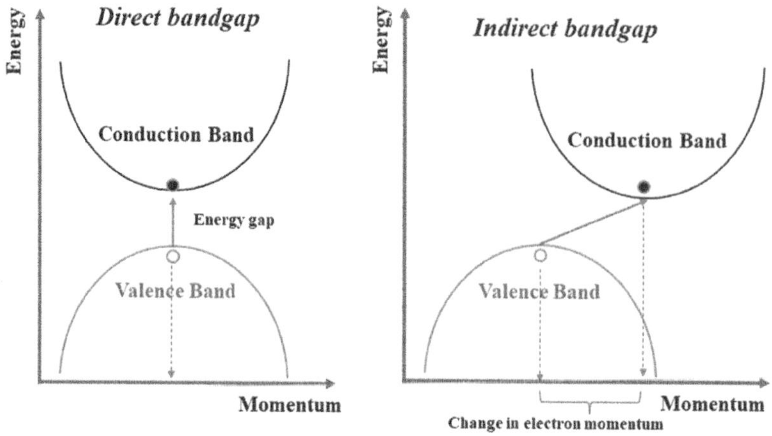

Figure 2.5 Schematic depiction of valance and conduction bands in a semiconductor and formation of *direct* and *indirect* bandgaps.

k-vector in the Brillouin zone, defines the minimum energy state of the CB. The maximum energy level of the VB is defined by another k-vector. Interestingly, two scenarios arise here such that when these two k-vectors are the same, the resulting band gap is referred to as a *direct* band gap, and when they rae different, it is referred to as an *indirect* band gap. The manifestation of *direct* and *indirect* bandgaps is depicted in Figure 2.5(b). In other words, the bandgap is *direct* if an electron can recombine with a hole of the same momentum, emitting a photon. This photon has a relatively large wavelength, and therefore a negligible wavenumber, when compared to the k-vectors of electrons. Therefore, the conservation of total crystal momentum is easily fulfilled [91]. When the maximum and minimum energy levels occur at different points in k-space, the electron has a lower probability for direct transition from CB to VB, so it skips radiative transition. The electron must transfer momentum to the *phonons* of the crystal lattice; thus, a large change in crystal momentum is associated with a transition between the CB maximum and the VB minimum [92].

In brief, a *direct* bandgap semiconductor with E_g shall be excited with a photon of energy similar to E_g without any additional energy requirement/energy loss. In an *indirect* band gap semiconductor, however, upon photon excitation, the excited electron interacts with phonons (lattice vibrations) due to a change in momentum before forming an electron–hole pair.

It is noted that the indirect process proceeds at a much slower rate, as it requires three entities to intersect in order to proceed: an electron, a photon, and a phonon; whereas it is a direct radiative transition for the other case. In effect, the recombination process is much more efficient for a direct band gap semiconductor than for an indirect band gap semiconductor, where the process must be mediated by a phonon. Hence, direct bandgap semiconductors are preferred in applications similar to illumination and energy conversion devices, whereas the indirect bandgap semiconductors are good in photocatalytic reactions.

2.5.2 Bandgap in Composite Materials

When the bandgap of the semiconductor material is very large compared to the energy of an incident photon, the semiconductor VB electrons remain unexcited. When another semiconductor, however, having a smaller energy gap is coupled, the photocatalytic efficiency of the composite increases under UV or visible (including solar) light. This can be understood from the fact that the photogenerated electrons, in order to prevent charge recombination, shall move from more negative to less negative Fermi energy in the CB while holes flow from more positive to less positive Fermi energy in the VB. The ultimate performance of composite photocatalyst also depends on various other factors such as its microstructure, size, morphology, and adsorption behavior on different surfaces of the photocatalyst.

2.5.3 Particle Size Effect on Band Gap

In atoms or molecules, electrons usually occupy distinct energy levels. When atoms get together to form crystal phases, such energy levels merge together and overlap each other. During crystallization, while the particle size remains smaller than about 10 nm, the overlapping energy levels are distinct. Beyond this size range, the atoms' energy levels merge and form a band structure. We thus see in bulk solids these energy levels merge together to form bands. Interestingly, when particle size is of the order of 10 nm or lower, emerging band positions lead to unique photon absorption and emission properties. The variation of bandgap with particle size is depicted in Figure 2.6. The band gap energy increases with decreasing particle size. It must be noted, however, that as material approaches the nanoscale, its surface area increases giving rise to availability of many reactive sites for catalysis. In the optimization of a material's property, then, these opposing characteristics need to be taken into account.

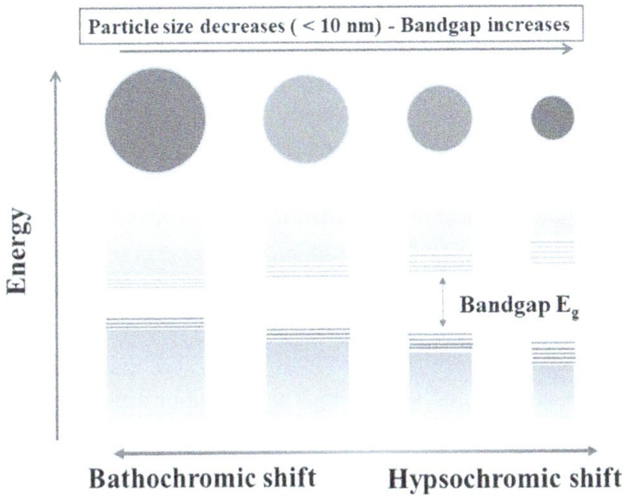

Figure 2.6 Depiction of increase in energy of bandgap of nanoparticles with size.

2.5.4 Band Gap of Quantum Dots and Low-dimensional Systems

Low-dimensional materials such as zero-dimensional particles (quantum dots), one-dimensional (1D) particles (nanoribbons and nanorods), and 2D materials (nanosheets) are considered to be new cost-effective and efficient photocatalysts. Studies on optimizing the photocatalyst properties of these low-dimensional semiconductors show that the morphology and dimensionality of the materials effectively modulates their optical paths and charge carrier diffusion pathways, thereby leading to enhanced photocatalytic properties [93].

In a colloidal solution of nanoparticles, the optical path of light increases as it gets scattered from all sides of the particle. At the same time, isolated nanoparticle nature leads to carrier diffusion length to be limited at the nanoscale. In contrast, in 1D structures such as nanotubes, nanorods, nanowires, etc., the diffusion length extends to the microscale or even longer. Another advantage of 1D nanomaterials is their ability to trap light by internal scattering. Light incident on the nanomorphological features such as nanowires and nanotubes gets reflected between these structures, leading to a dramatic increase in its path length. Obviously, the enhancement of path length is dependent on the geometry of 1D nanostructures. Further, light–matter interactions at subwavelength structures of these nanomaterials gently confine light exhibiting significant resonances [94, 95], referred to as leaky mode resonances, leading to enhanced light absorption [96, 97]. These features were exploited in the development of many high-performance photonic devices, such as photodetectors [98], electro-optic modulators [99], solar cells [96, 98], and photocatalysts [100]. In brief, by controlling the dimension of these structures it could be possible to control the light absorption enhancement at a desired wavelength.

2.5.5 Deriving E_g from the Kubelka–Munk Function

Optical properties of a material are dependent on its energy bandgap. Characterization of the energy bandgap therefore remains as one of the essential requirements of the study of materials. Experimentally, there are many approaches that quantify the value of this energy gap. Electrochemical methods identify the electrical bandgap, which gives information on reduction and oxidation potentials of the material. XPS and ultraviolet photoelectron spectroscopy (UPS) yield the VB potential, from which the electrical band gap can be calculated. More commonly, the bandgap derived from absorption characteristics of the material is known as the *optical bandgap*. Following the Kulbeka–Monk function, the absorption spectra is derived from diffuse reflectance given by the equation

$$\left(F\left(R_{\infty}\right).h\nu\right)=K/S=\left(1-R_{\infty}\right)^{2}/2R \tag{2.10}$$

Where $R_{\infty} = R_{\text{sample}} / R_{\text{standard}}$ is the reflectance of an infinitely thick specimen, h is the Planck constant, ν is the photon's frequency, and K and S are the absorption and scattering coefficients, respectively. The energy-dependent absorption coefficient α proposed by by Mott, Davis, and Tauc is expressed by Eq. (2.11):

$$\left(\alpha.h\nu\right)^{1/n}=B\left(h\nu-E_{g}\right) \tag{2.11}$$

where B is a constant. The factor n in the above equation assumes the value of ½ or 2 depending on the nature of the electron to be in a *direct* or *indirect* band, respectively. Putting $F(R_\infty)$ instead of α yields the form

$$\left(F\left(R_\infty\right).h\nu\right)^{1/n} = B\left(h\nu - E_g\right) \tag{2.12}$$

To experimentally determine the band gap, the linear part of the data is fitted to a straight line using the above equation. Figure 2.7 shows a plot of $F(R_\infty).h\nu$ vs. $h\nu$ using experimental data on absorption as a function of wavelength. From the equation, one can infer that the intersection point of the straight line on the x-axis is the band gap energy.

2.5.6 Deriving E_g from the Electrochemical Approach

Cyclic voltammetry is an important tool utilized for determination of the absolute band edge positions of materials. Electrochemical experiments are performed to derive understanding of the redox characteristics of the semiconductor. In the CV, the potential is cycled linearly with time between two values. When the energy of electrons on the electrode (Fermi level) matches within k_BT the energy level of the analyte species in solution or adsorbed on the electrode surface, electron transfer takes place resulting in a current response. Since the electron transfer processes are mediated by the CB and VB edges, the positions of the respective energy levels directly correlate with its energy band gap. This

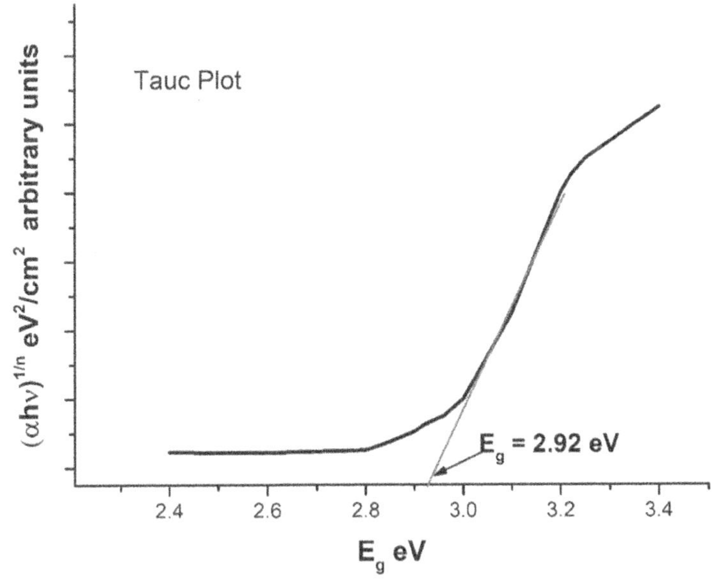

Figure 2.7 Schematic Tauc plot of semiconductor oxide to deduce bandgap (E_g) of the material. The linear part of the plot is extrapolated to the x-axis.

difference between the reduction and oxidation peaks is referred to as the electrochemical band gap. This method is generally used for quantum dot semiconductors and polymer composites [101].

2.6 Engineering the Desired Band Gap

The electronic characteristics of a semiconductor are controlled by its electronic band structure, including light absorption, charge separation, migration, and recombination [102]. The type of electronic band structure of a semiconductor dictates the nature of its optical transition. The optical transition in direct bandgap semiconductors does not involve a change in crystal momentum; it will absorb all incident photons with energy equivalent to the bandgap energy. This leads to short migration of the charge carriers. In contrast, indirect semiconductors require a change in the crystal momentum during optical transition. Since the incident photon possesses little momentum, indirect transition requires the involvement of phonons (lattice vibration). The need for lattice vibration necessitates larger material thickness in indirect bandgap semiconductors to absorb the incident light completely. Consequently, the charge carriers of an indirect semiconductor with a short minority charge diffusion length may be recombined before they tapped out [103]. The photon energy is given by $E = hv = \frac{hc}{\lambda}$ where c is the light velocity in vacuum and λ is the wavelength of light. In addition, the electronic band structure affects the charge carrier mobility, which is dependent on the widths of the CB and VB. The charge carrier mobility is inversely proportional to the effective mass of the carriers (either electrons or holes), which is determined by the curvature of individual bands. Broad bands are largely curved, leading to small effective mass and high charge carrier mobility.

The theoretical maximum solar-to-hydrogen efficiency of a photocatalyst is determined by its electronic band structure. In photoelectrochemical splitting of water, the energy band edge positions of semiconductors must overlap with the energy levels of water reduction and oxidation. Shifting up the CB edge with respect to the water reduction level increases the thermodynamic driving force for water reduction. The most effective approach to increase the conversion efficiency is to reduce the band gap of the material to extend the light absorption spectral range into the visible light region or even the near-infrared light region. It is pertinent to note that about 80% of total solar radiation is made available in visible–near-infrared regions. Having said this, from the point of view of thermodynamics, the minimum energy required to overcome the standard Gibb's free energy change (1.23 eV) for water splitting plus the thermodynamic losses (0.3–0.5 eV) must be made available. Kinetics demands an overpotential of 0.4–0.6 eV to enable a fast reaction. An ideal band gap, therefore, is 1.9–2.3 eV for solar water splitting.

Photocatalytic materials can be broadly classified based on the respective band structure. Transition metal oxides represent the majority of well-studied photocatalysts. Amongst those is titanium oxide (TiO_2), which has three crystal structures: rutile, brookite, and anatase, where anatase and rutile types are often used as photocatalysts. TiO_2 has a wide bandgap (3.2 eV); the photoactive region falls in the ultraviolet region, and the narrow absorption of nearly 5% of the solar spectrum acts as a major drawback for photocatalytic application [104]. Research on modification of TiO_2 for narrowing the bandgap to improve photoreactivity is being pursued in many ways, as depicted in Figure 2.8.

Doping causes the generation of vacancies, interstitial or substitutional defects, which will be able to alter the physical properties like color, magnetic properties, electric properties including conductivity and reactivity, etc. [105]. Doping TiO_2 with nonmetals such as N, C and S results in absorption of longer wavelength light. Raja et al. reported the 1.5-fold increase in visible light photocatalysis due to the red shift in absorption edge achieved by structural as well as bandgap modification in TiO_2. Interestingly, it was also established that the concentration of hydroxyl species increases the same fold for N-doped TiO_2 material under visible light illumination [76]. Further, in TiO_2 the photocatalytic efficiency has been shown to increase by deposition of noble metals, namely Au, Pt, Pd, Rh, Ni, Cu, and Ag. Alamelu et al. have reported a TiO_2 photocatalyst loaded with 5 wt% of Pt exhibited a 4.5-fold increase in photocatalytic pursuance of Rhodamine B (RhB) and 45-fold increase in 4-nitrophenol reduction, in comparison to pristine TiO_2. The increased photocatalytic efficiency is attributed to the strong absorption ability toward RhB and 2-nitrophenol, increased charge carrier separation, high surface area, good light-harvesting ability, and increased electron transport. An increase in weight percentage beyond 5 wt%, however, seems to reduce the catalytic efficiency [106].

The study of nanocomposite formation for increased photocatalytic performance by altering the bandgap has considerably accelerated. Au nanoparticles decorated with sulfonated graphene-TiO_2 nanocomposite catalysts exhibit 3.2-fold higher photocatalytic performance compared to pristine TiO_2 systems. The improved photocatalytic performance is attributed to the synergistic effect of increased interfacial electron transfer between composite, decreased recombination rate of photoexcited electrons and holes, and the plasmonic effect of Au nanoparticles [107]. The cost-effective synthesis of biphasic TiO_2 nanoparticles and its graphene-TiO_2 nanocomposite by a hydrothermal process has been

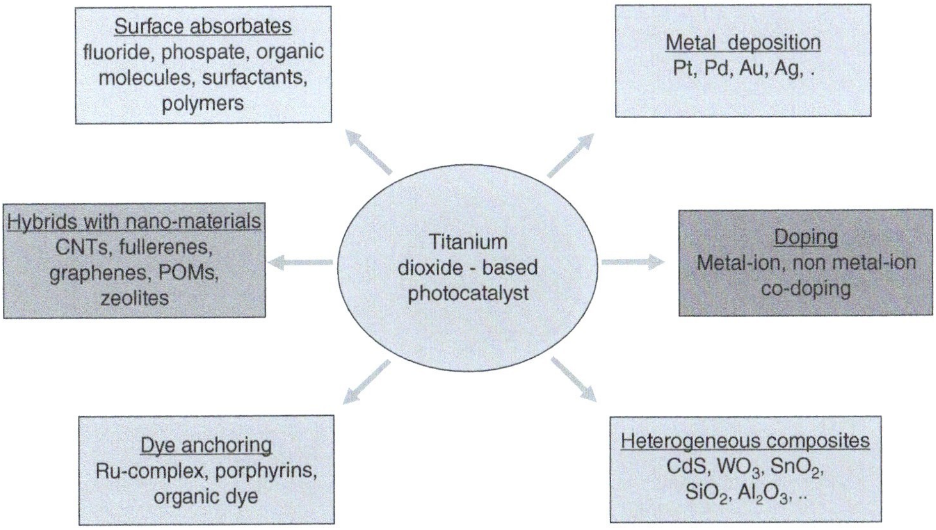

Figure 2.8 Common research approach to achieve bandgap modification in titanium dioxide-based material.

reported, with the high surface area of graphene shown to suppress the electron–hole pair recombination rate in the nanocomposite. This resulted in redshift of the absorption edge and contraction of the band gap from 2.98 eV to 2.85 eV [108]. The studies on TiO_2/Bi_2WO_6 nanocomposite by Shiamala et al. demonstrate harnessing solar energy for performing photocatalytic deconstruction of a biopolymer, namely starch. Smaller molecular fragments thus derived constitute potential organic precursors in the anaerobic methanation and ethanol fermentation [41].

Structural modifications of a catalyst, such as core-shell design, to enhance its catalyst property and functional ability are well studied. Narzary et al. synthesized structurally modified TiO_2 to magnetically retrievable and visible light active $Fe_3O_4@SiO_2@g\text{-}C_3N_4/TiO_2$ core-shell microsphere nanocomposite. The photocatalytic performance examined using RhB and methyl orange at 10 ppm show that a 10 wt% $g\text{-}C_3N_4/TiO_2$ composite catalyst can degrade over 91% of both cationic and anionic dyes. The rate constant of reaction was determined to be 3.2-fold higher for RhB, and 27.7-fold higher for methyl orange in comparison to TiO_2 photocatalyst. This renders $Fe_3O_4@SiO_2@g\text{-}C_3N_4/TiO_2$ core-shell microsphere nanocomposite an efficient photocatalyst for cationic dyes and as a high-rate remover of anionic dyes [39].

Bismuth oxide photocatalytic materials, namely Bi_2O_3, having crystal forms $\alpha\text{-}Bi_2O_3$ and $\beta\text{-}Bi_2O_3$ with respective bandgap energies of 2.85 eV and 2.58 eV, are found to be visible-light active photocatalysts. Even though it is active in the visible light range of the spectrum, this photocatalyst shows low efficiency due to faster recombination of photogenerated carriers. The various strategies to enhance the photocatalytic activity of bismuth oxide are shown in Figure 2.9. Doping is an effective method to reduce the band gap of bismuth based oxide compounds. Xu [109] et al. reported the band gap of Bi_3TiNbO_9 can be reduced from 3.1 eV to 2.6 eV when it is doped with a 10% molar ratio of Ni^{2+}. Another example of

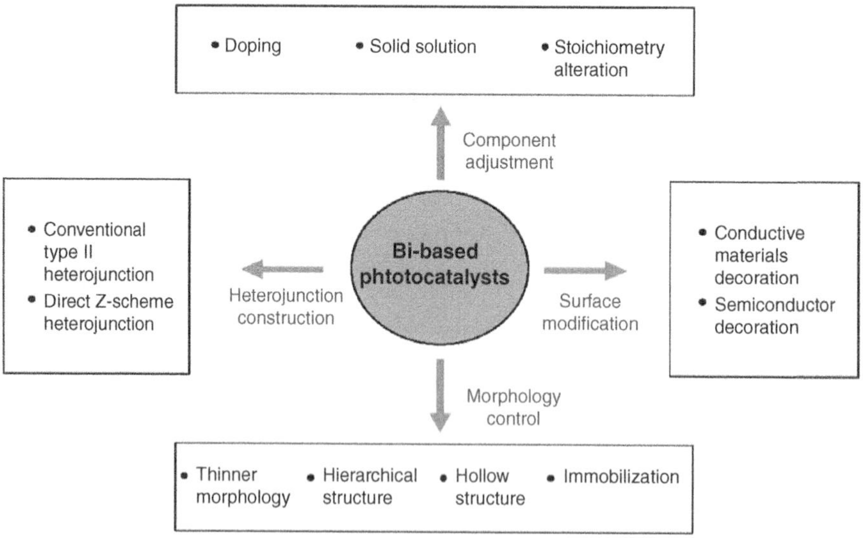

Figure 2.9 Enhancement strategies for nanoscale Bi-based photocatalysts.

doping Bi_2WO_3 with small amount of Br with surfactant cetyltrimethylammonium bromide (CTAB) resulted in narrowing of the bandgap of the initial compound [110]. Even though the doping can further narrow the bandgap, the low utilization efficiency of generated charge carriers is a hurdle in the way of achieving efficient bismuth based photocatalytic systems.

ZnO, WO_3, and Fe_2O_3 are other transition metal oxide semiconductors which are widely used in photocatalysis, having interesting physical properties. The properties such as quantum size effect, efficient UV absorption characteristics, and piezoelectric nature make ZnO a desirable candidate for photocatalysis. ZnO based photocatalysts exhibit strong interaction between oxygen vacancies and absorbed oxygen, which can be very well utilized for photo-oxidation reactions. The bandgap alignment of ZnO with ZnS, ZnSe, and ZnTe is shown in Figure 2.10. Due to the wide band gap (3.2 eV) of ZnO, the excitation wavelength falls in the ultraviolet region (387 nm) of the solar spectrum. Modification of band structure of ZnO for enhanced photocatalytic activity toward the visible range becomes imperative. Figure 2.10 depicts how composite photocatalysts can be utilized in narrowing down the energy bandgap.

Anandan et al. reported on the band engineered solid solution formation along with the structural modification of cocatalyst in a $Cd_xZn_{1-x}O$ system by a hybrid approach where the surface modified Cu^+ ions were under visible light irradiation, and electrons in Cu^{2+}/Cu^+ redox couples cause the efficient reduction of absorbed oxygen molecules [111]. Photocatalytic studies by Beura et al. on tin ion (Sn^+)-doped ZnO showed that up to 99% of methyl orange was degraded under UV light with an optimum doping of 1% for most efficient photodegradation. The resultant Sn^{4+} doped ZnO had lowered bandgap, increased photoluminescent lifetime, as well as increased charge transfer [112]. Bandgap narrowing was realized in ZnO by inducing oxygen vacancies. Ansari et al. reported the modification of ZnO via a biogenic approach, i.e. by means of an electrochemically active biofilm (EAB). This was found to induce oxygen vacancies in pure ZnO (p-ZnO), resulting in the formation of modified ZnO (m-ZnO) with narrow bandgap without using any dopants [113].

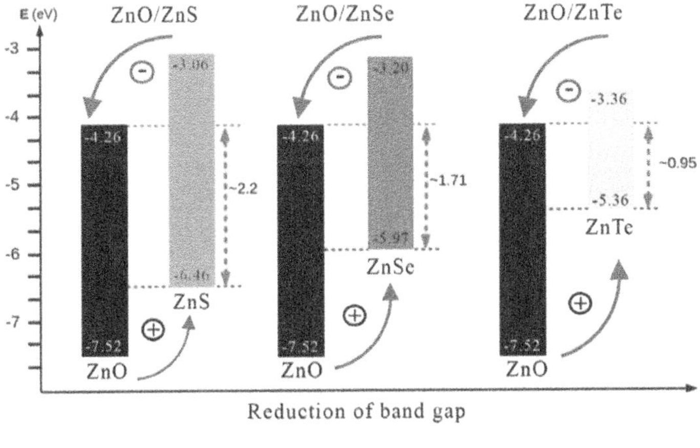

Figure 2.10 Bandgap alignment of Zn-based systems.

A bandgap in the range of 2.4–2.8 eV and material stability allows WO_3 to be used as both main catalyst and cocatalyst in photocatalytic reactions. Further, WO_3 has a large surface area and excellent photo absorbing capability. The low reduction and oxidation potential of the CB of WO_3 and VB of $BiWO_3$ allows the application of these catalysts in selective functional group transformation in organic reactions. To achieve an efficient photocatalytic system, extending the optical response to longer wavelength by bandgap modification, foreign ion doping, noble metal deposition, and heterostructure with other semiconductors have been attempted. The changes in the surface and bulk properties as well as the charge carrier transfer dynamics associated with each of these modifications lead toward high efficiency and enhanced activity. The doping with Cu^{2+} and introduction of more exposed facets increase the degradation rate of these oxidizing agents [114].

Iron (Fe) is widely studied as a catalyst for redox reactions. Among other iron oxides, Fe_2O_3, an n-type semiconductor with band gap of 2.2 eV with visible-light absorption capacity is extensively investigated.

The bandgap modification of Fe_2O_3 is achieved by increasing its surface area and minimizing carrier charge recombination. In such a scenario higher charge mobility is achieved by addition of support material, design of heterojunction, and Z-scheme photocatalysis.

Yu et al. reported hierarchical porous γ-Fe_2O_3@SiO_2@TiO_2 composite microspheres with a sandwich-like structure of SiO_2 encapsulation of γ-Fe_2O_3 core and within a TiO_2 shell. This modification restricts the wide bandgap electronic barrier of the SiO_2 middle layer, and the SiO_2 surface serves as a good absorbent for dye degradation [115]. In the heterostructure modification, hydrothermal synthesis of Fe_2O_3/Bi_2S_3 nanorods were studied by Helal et al. The formation of heterostructure resulted in an efficient separation of electron–hole pairs with reaction rate about three times higher than pure Bi_2S_3 and exhibited good stability and recyclability [116].

In addition to oxide semiconductors, nonoxide photocatalysts, namely copper sulfide (CuS), zinc sulfide (ZnS), molybdenum sulfide (MoS_2), tungsten sulphide (WS), and organic catalyst structures like carbon nitride (C_3N_4) and graphitic carbon nitride (g-C_3N_4), have also shown remarkable physical and chemical properties that qualify them as good candidates for photocatalyst applications. Table 2.1 enumerates some of the most frequently used photocatalysts and their energy bandgap values.

The emerging classification of 2D materials shows unique performance in solar spectrum-driven photocatalysis. Even though the 2D materials have features like graphene, necessary changes are required to achieve highly efficient systems by structural modification, doping, heterostructure formation, etc.

Solid semiconductors possess a band of energy levels forming a bandgap between the VB and CB. Modification of a bandgap with valid energy levels facilitates electrons occupying such additional / modified energy levels. During emission, such nonforbidden energy levels shift the characteristic peak emission toward higher wavelengths, called a bathochromic shift, and to lower wavelengths, called a hypsochromic shift. Consequently, the number of permissible energy levels are expanded, producing a bathochromic shift. Incorporation of additional energy levels accommodates electrons by absorbing or emitting shorter wavelength photons. Similarly, a hypsochromic shift shall be exhibited when the bandgap of the semiconductor is increased. Conventionally, breakdown of an energy band into energy levels leads to an increase in bandgap. Such energy levels shall vary drastically

Table 2.1 Commonly used photocatalysts and their bandgap energy.

Photocatalyst	Bandgap (eV)	Photocatalyst	Bandgap (eV)
Si	1.1	SiC	3.0
WSe_2	1.2	TiO_2 rutile	3.02
α-Fe_2O_3	2.2	Fe_2O_3	3.02
Cds	2.4	TiO_2 anatase	3.2
$NaBiO_3$	2.62	ZnO	3.2
V_2O_5	2.7	$SrTiO_3$	3.4
B_2WO_6	2.78	SnO_2	3.5
WO_3	2.8	ZnS	3.7

with particle size (Figure 2.6). In hypsochromic shifts, electrons are accommodated in a widened bandgap, so that only shorter wavelength photons (high-frequency / high-energy photons) are absorbed or emitted [117].

2.7 Photocatalytic Mechanisms, Schemes and Systems

An ideal photocatalyst material is one that will be capable of performing both oxidation and reduction reactions under appropriate conditions. Semiconductors are selected as photocatalytic materials for having a minimum bandgap of 1.23 eV. In the case of metals, the oxidizing power of the VB and the reducing power of the CB are almost the same. Insulators are not considered because of the high value of the bandgap requiring very high energy photons to create appropriate excitons to take place. During photocatalytic water splitting as depicted in Figure 2.11, the semiconductor absorbs photons from the solar spectrum having energies greater than the bandgap energy ($E > E_g$). The electrons in the semiconductor are excited from the occupied VB to the unoccupied CB. Subsequently, formed electrons and holes independently diffuse into the semiconductor surface and participate in hydrogen evolution reactions (HERs) and oxygen evolution reactions (OERs) on the photocatalytic surface.

$$2H + +2e^- \rightarrow H_2 \tag{2.13}$$

$$2H_2O + 4h^+ \rightarrow 4H^+ + O_2 \tag{2.14}$$

For the reaction to occur, the minimum potential of the CB maximum should be more negative than the H^+/H_2 energy level, whereas the VB maximum (VBM) should be more positive than the O_2/H_2O energy level [118]. The performance and efficiency of a photocatalytic material depends on the conversion efficiency and catalyst lifetime. The selection of semiconductor for photocatalysis depends on the specific application. For example, in the case of photocatalytic water splitting the top of the VB and bottom of CB have to be shifted in the opposite direction for the reduction and oxidation reaction to be facile.

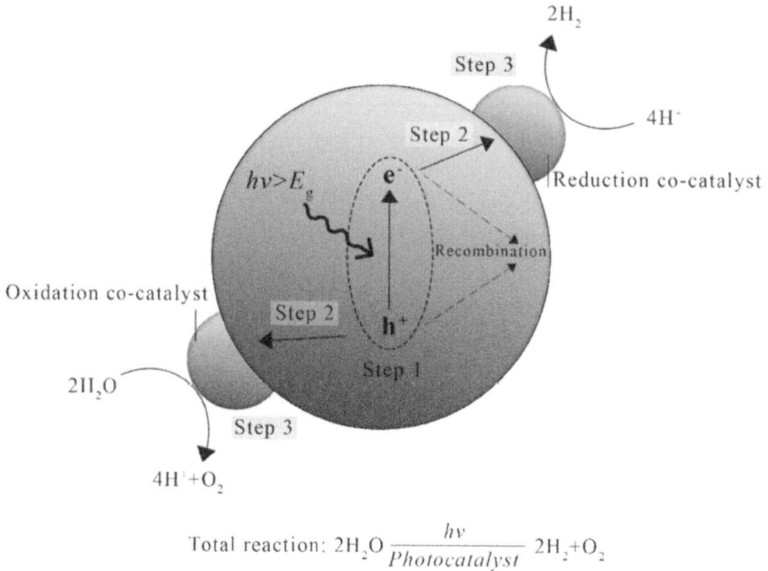

Figure 2.11 Depiction of oxidation–reduction mechanism leading to water splitting by photocatalysis.

Since the discovery of the Honda–Fujishima effect [119], TiO_2 and its various transits were studied for photocatalytic application. Later transition metals with d^0 electronic configurations (such as Ti^{4+}, Zr^{4+}, Nb^{5+}, Ta^{5+}, and W^{6+}), and typical metal cations with d^{10} electronic configurations (such as Ga^{3+}, In 3^+, Ge^{4+}, Sn^{4+}, and Sb^{5+}) were [120] subjected to photocatalytic applications. Several other semiconducting materials including CdS, Fe_2O_3, and WO_3 were also used. Constant improvements in the efficiency and reusability of the systems has been happening with the introduction of new semiconductor materials and combinations of semiconductors and other support structures.

The band alignments of crystalline phases and semiconductor junctions are done to enhance the efficiency of photocatalysis. Wang et al. reported electronic bandgap tuning of α-Ga_2O_3 via addition of suitable dopants, increasing the activity of α-Ga_2O_3 in the deep ultraviolet and ultraviolet regions of the solar spectrum [121].

The anatase crystal structure of titanium is (Ti) is preferably used in photocatalysis due to its abundant trapped states near the VB. These states are present in the rutile phase but completely absent in brookite. Li et al. reported that prolonged irradiation may reconstruct the anatase and brookite surfaces to give a quasirutile phase that allows water splitting reactions [122]. Increasing the crystallinity of the material will diminish the recombination sites formed due to structural defects such as vacancies and dislocations.

Cocatalysts should have a redox reaction site that will catalyse the reaction with much lower activation energy. An ideal cocatalyst shall provide smooth charge migration from photoexcited semiconductors. Pt and TiO_2 constitute good examples of a cocatalyst.

For many years, a number of strategies were explored to achieve increased efficiencies in photocatalyst materials. Doping, cocatalysts, metal loading, and structural and bandgap

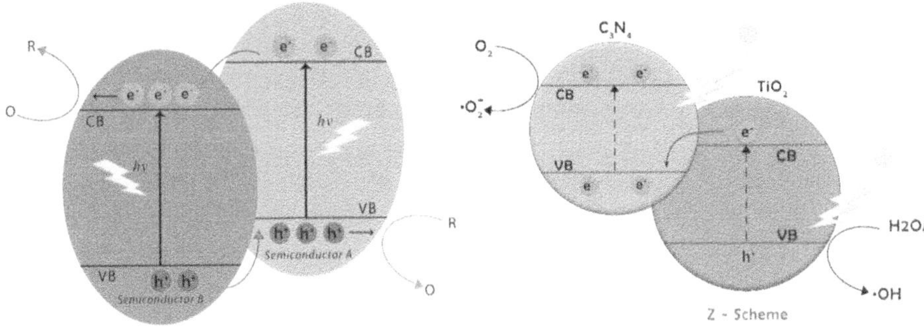

Figure 2.12 Schematic of band alignment composite photocatalyst: a) Type II semiconductor photocatalyst; b) Z-scheme photocatalyst.

modification forming heterojunctions are the processes of choice. The bandgap modification by forming type II heterojunction semiconductors has been introduced where the CB and VB of semiconductor *A* are slightly higher than CB and VB of semiconductor *B*. The charge carrier mechanism between two semiconductors, namely, semiconductor *A* and semiconductor *B* in a heterojunction, takes place by the migration of electrons from the CB of semiconductor *A* to the CB of semiconductor *B* whereas the holes migrate from the VB of semiconductor *B* to the VB of semiconductor *A* on photo-radiation exposure. This way one can spatially separate the charge carriers.

Z-scheme photocatalysis proposed by Yu et al. reported high yield in the formaldehyde (HCHO) degradation performance of the TiO_2/g-C_3N_4 composite. The name Z-scheme was proposed because the charge carrier transportation in this scheme appeared to be in the shape of the letter "Z". Even though type II heterojunction photocatalysis and Z-scheme photocatalysis seem to be identical in schematic representation (Figure 2.12), when it comes to the charge carrier transportation, they have a tremendous difference between them. In Z-scheme photocatalysis, upon photo-irradiation the electrons with lower reduction ability from semiconductor *B* recombine with photogenerated holes of low oxidation ability of semiconductor *A*. Thus, the photogenerated electrons of semiconductor *A* with high reduction ability and holes of semiconductor *B* with high oxidation ability can be maintained, which is due to electrostatic attraction between the charge carriers. This modification will help the material effectively overcome the drawback of type II heterojunction photocatalysis (Figure 2.12).

2.8 Summary and Perspectives

Various approaches to the synthesis of nanoparticles were discussed in the previous sections, including sol-gel, solid-state, hydrothermal, microemulsion, one-pot, and dip coating. Each strategy has its importance and end benefits. The microemulsion method assumes significance in that the surfactant-to-water ratio can control the particle size. Ambient temperature sol preparation, gel processing, high purity of precursors, product homogeneity, low sintering temperature, ease of manufacturing multicomponent materials, and reasonable control over the morphology of powder particles are the main advantages of the sol-gel method. Low reaction temperatures in hydrothermal and solvothermal processes yield high chemical uniformity and the possibility of obtaining novel metastable structures. Unlike the hydrothermal method, the solvothermal method allows for the removal of foreign anions, because organic solutions have a low relative permittivity. Solid-state reactions are simple to perform and cost-effective; they also leave limited side product formation and are compatible with large-scale production. Dip coating has the advantages of being inexpensive and making it easy to modify layer thickness; at the same time, it has the disadvantages of being slow.

Control over energy bandgaps, which manifest in the optical and electrical properties of the material, is an important requirement that dictates the choice of synthesis method for a given photocatalyst system. Specific synthesis routes are taken depending on the need to control the number of layers, form heterostructures, strain engineer the lattice, introduce chemical doping, and regulate alloying, intercalation, substrate engineering, etc. The constituent materials can have dramatically varied properties due to band structure engineering.

Development in visible-light active photocatalysts led to renewed interest in photocatalytic materials usage for energy harvest and environmental remediation, made evident by the enormous number of articles published annually. Material development techniques for different photocatalysts focus on maximizing efficiency by targeting one or more phases in the photocatalytic reaction. Generic synthesis strategies include composite formation with other functional materials or between different semiconductors (heterojunction formation) and their morphological variations impacting their surface-active sites, as well as charge separation, utilizing sensitizers and cocatalysts to enhance UV–Vis absorption, and doping.

Various preparation procedures discussed so far, added together with the idea of compositional alteration, lead to the generation of novel photocatalysts for applications. Additionally, much effort has been expended into developing photocatalytic reactors to improve the yield of the desired product.

It is evident that the electronic structure of the semiconductor controls the photocatalytic reactions. Sound engineering of these structures without deterioration of the stability is a challenging issue to be addressed. Designing efficient photocatalyst systems is usually dependent on trial and error rather than logical scientific thinking until such explicit knowledge becomes available. A recent trend in this direction, however, is to analyze the theoretical models of the novel composition and structure, fine-tune its desired properties prior to synthesis, and finally, validate experimentally through a conventional synthesis route.

From a sustainability aspect, nanomaterial formulations that reduce and recycle their waste and utilize room temperature aqueous solutions to generate their products are the best. These design specifications also increase industrial safety while lowering production costs. It has long been recognized that the environmental impact of nanomaterials must be studied from a variety of perspectives (efficiency, transport, waste, and toxicity). Overall, we çan anticipate the concept of sustainability becoming a design criterion for nanomaterial manufacturing.

References

1 Nakanishi, K. (2015). Properties and applications of sol–gel materials: functionalized porous amorphous solids (Monoliths). In David Levy and Marcos Zayat (Edts) *The Sol-Gel Handbook*, p745–766. Germany: Wiley-CH Verlag GmbH & Co. KGaA.

2 Landau, M.V. (2008). Sol–gel process. In: *Handbook of Heterogeneous Catalysis*, pp. 119–160. Wiley Online Library.

3 Ding, X.Z., Qi, Z.Z., and He, Y.Z. (1995). Effect of hydrolysis water on the preparation of nano-crystalline titania powders via a sol-gel process. *Journal of Materials Science Letters* 14 (1): 21–22.

4 Brinker, C.J. (1988). Hydrolysis and condensation of silicates: effects on structure. *Journal of Non-crystalline Solids* 100 (1–3): 31–50.

5 Yoon, K.H., Noh, J.S., Kwon, C.H., and Muhammed, M. (2006). Photocatalytic behavior of TiO2 thin films prepared by sol–gel process. *Materials Chemistry and Physics* 95 (1): 79–83.

6 Hernández, A., Maya, L., Sánchez-Mora, E., and Sánchez, E.M. (2007). Sol-gel synthesis, characterization and photocatalytic activity of mixed oxide ZnO-Fe 2 O 3. *Journal of Sol-Gel Science and Technology* 42 (1): 71–78.

7 Danish, M., Ambreen, S., Chauhan, A., and Pandey, A. (2015). Optimization and comparative evaluation of optical and photocatalytic properties of TiO2 thin films prepared via sol–gel method. *Journal of Saudi Chemical Society* 19 (5): 557–562.

8 Owens, G.J. et al. (2016). Sol–gel based materials for biomedical applications. *Progress in Materials Science* 77: 1–79.

9 Moussaoui, R. et al. (2017). Sol–gel synthesis of highly TiO2 aerogel photocatalyst via high temperature supercritical drying. *Journal of Saudi Chemical Society* 21 (6): 751–760.

10 Chen, L. et al. (2009). Preparation and photocatalytic properties of strontium titanate powders via sol–gel process. *Journal of Crystal Growth* 311 (3): 746–748.

11 Portia, S., Parthibavarman, M., and Ramamoorthy, K. (2021). Unpredicted visible light induced advanced photocatalytic performance of Eu doped CaTiO3 nanoparticles prepared by facile sol–gel technique. *Journal of Cluster Science* 1–10.

12 Nunocha, P. et al. (2021). A new route to synthesizing La-doped SrTiO3 nanoparticles using the sol-gel auto combustion method and their characterization and photocatalytic application. *Materials Science in Semiconductor Processing* 134: 106001.

13 Seetharaman, S. (2005). *Fundamentals of Metallurgy*. (Woodhead Publishing Series in Metals and Surface Engineering). United Kingdom: Woodhead Publishers.

14 Cho, S.J., Uddin, M.J., and Alaboina, P. (2017). Review of nanotechnology for cathode materials in batteries. In: *Emerging Nanotechnologies in Rechargeable Energy Storage Systems* (ed. L.M. Rodriguez-Martinez and N. Omar), 83–129. s.l.: Elsevier.

15 Shtarev, D.S. et al. (2021). Revisiting the BaBiO3 semiconductor photocatalyst: synthesis, characterization, electronic structure, and photocatalytic activity. *Photochemical and Photobiological Sciences* 20 (9): 1147–1160.

16 Vyas, M.K. and Chandra, A. (2019). Synergistic effect of conducting and insulating fillers in polymer nanocomposite films for attenuation of X-band. *Journal of Materials Science* 54 (2): 1304–1325.

17 Ali, H. et al. (2021). Solid-state synthesis of direct Z-scheme Cu2O/WO3 nanocomposites with enhanced visible-light photocatalytic performance. *Catalysts* 11 (2): 293.

18 Hu, J., Cao, Y., Wang, K., and Jia, D. (2017). Green solid-state synthesis and photocatalytic hydrogen production activity of anatase TiO 2 nanoplates with super heat-stability. *RSC Advances* 7 (20): 11827–11833.

19 Xie, J. et al. (2019). Simple solid-state synthesis of BiOCl/Bi2O2CO3 heterojunction and its excellent photocatalytic degradation of RhB. *Journal of Alloys and Compounds* 784: 377–385.

20 Jia, X. et al. (2011). Solid state synthesis of tin-doped ZnO at room temperature: characterization and its enhanced gas sensing and photocatalytic properties. *Journal of Hazardous Materials* 193: 194–199.

21 Reda, G.M., Fan, H., and Tian, H. (2017). Room-temperature solid state synthesis of Co3O4/ZnO p–n heterostructure and its photocatalytic activity. *Advanced Powder Technology* 28 (3): 953–963.

22 Faraji, G., Kim, H.S., and Kashi, H.T. (2018). *Severe Plastic Deformation: Methods, Processing and Properties*. Amsterdam: Elsevier.

23 Wei, X. et al. (2021). Facile ball-milling synthesis of CeO2/g-C3N4 Z-scheme heterojunction for synergistic adsorption and photodegradation of methylene blue: characteristics, kinetics, models, and mechanisms. *Chemical Engineering Journal* 420: 127719.

24 Zhu, X. et al. (2021). Environmental-friendly synthesis of heterojunction photocatalysts g-C3N4/BiPO4 with enhanced photocatalytic performance. *Applied Surface Science* 544: 148872.

25 Yin, W. et al. (2021). In-situ synthesis of MoS2/BiOBr material via mechanical ball milling for boosted photocatalytic degradation pollutants performance. *ChemistrySelect* 6 (5): 928–936.

26 Mandal, R.K. and Pradhan, S.K. (2021). Superior photocatalytic performance of mechanosynthesized Bi2O3–Bi2WO6 nanocomposite in wastewater treatment. *Solid State Sciences* 115: 106587.

27 Hewitt, S.A. and Kibble, K.A. (2009). Effects of ball milling time on the synthesis and consolidation of nanostructured WC–Co composites. *International Journal of Refractory Metals & Hard Materials* 27 (6): 937–948.

28 Cao, W. (2007). Synthesis of nanomaterials by high energy ball milling. https://www.understandingnano.com/nanomaterial-synthesis-ball-milling.html (accessed 28 July 2022).

29 Cao, Z. et al. (2021). High-energy ball milling assisted one-step preparation of g-C3N4/TiO2@Ti3C2 composites for effective visible light degradation of pollutants. *Journal of Alloys and Compounds*, 899: 161771.

30 Muramatsu, Y. et al. (2005). Gas contamination due to milling atmospheres of mechanical alloying and its effect on impact strength. *Materials Transactions* 46 (3): 681–686.

31 Adschiri, T., Hakuta, Y., and Arai, K. (2000). Hydrothermal synthesis of metal oxide fine particles at supercritical conditions. *Industrial & Engineering Chemistry Research* 39 (12): 4901–4907.

32 Liu, N., Chen, X., Zhang, J., and Schwank, J.W. (2014). A review on TiO2-based nanotubes synthesized via hydrothermal method: formation mechanism, structure modification, and photocatalytic applications. *Catalysis Today* 225: 34–51.

33 Li, Z. et al. (2005). Hydrothermal synthesis, characterization, and photocatalytic performance of silica-modified titanium dioxide nanoparticles. *Journal of Colloid and Interface Science* 288 (1): 149–154.

34 Zhang, Y.C., Du, Z.N., Li, K.W., and Zhang, M. (2011). Size-controlled hydrothermal synthesis of SnS2 nanoparticles with high performance in visible light-driven photocatalytic degradation of aqueous methyl orange. *Separation and Purification Technology* 81 (1): 101–107.

35 He, J. et al. (2021). Ionic liquid-hydrothermal synthesis of Z-scheme BiOBr/Bi2WO6 heterojunction with enhanced photocatalytic activity. *Journal of Alloys and Compounds* 865: 158760.

36 Wang, F. et al. (2017). S-TiO2 with enhanced visible-light photocatalytic activity derived from TiS2 in deionized water. *Materials Research Bulletin* 87: 20–26.

37 Huang, Z. et al. (2017). Facile synthesis of S-doped reduced TiO2-x with enhanced visible-light photocatalytic performance. *Chinese Journal of Catalysis* 38 (5): 821–830.

38 Narzary, S., Alamelu, K., Raja, V., and Ali, B.J. (2020). Visible light active, magnetically retrievable Fe3O4@ SiO2@ g-C3N4/TiO2 nanocomposite as efficient photocatalyst for removal of dye pollutants. *Journal of Environmental Chemical Engineering* 8 (5): 104373.

39 Jamjoum, H.A.A. et al. (2021). Synthesis, characterization, and photocatalytic activities of graphene oxide/metal oxides nanocomposites: A review. *Frontiers in Chemistry* 9: 752276.

40 Shiamala, L., Alamelu, K., Raja, V., and Ali, J. (2018). Synthesis, characterization and application of TiO2–Bi2WO6 nanocomposite photocatalyst for pretreatment of starch biomass and generation of biofuel precursors. *Journal of Environmental Chemical Engineering* 62 (2): 3306–3321.

41 Sajid, M.M. et al. (2021). Photocatalytic performance of ferric vanadate (FeVO4) nanoparticles synthesized by hydrothermal method. *Materials Science in Semiconductor Processing* 129: 105785.

42 Jin, C., Feng, W., and Zhao, X. (2021). In-situ hydrothermal synthesis of SnS2/SnO2/rGO nanocomposites with enhanced photogenerated electron transfer for photoreduction of CO2 to CH4. *IOP Conference Series: Earth and Environmental Science* 766 (1): 012051.

43 Wang, C.Y. et al. (2021). Visible light photocatalytic properties of one-step SnO2-templated grown SnO2/SnS2 heterostructure and SnS2 nanoflakes. *Nanaotechnology* 32 (30): 305706.

44 Hernández-Ramírez, A. and Medina-Ramírez, I. (2016). *Photocatalytic Semiconductors*. Springer International Pu.

45 Bhattacharyya, P. and Basu, S. (2011). CVD grown materials for high temperature electronic devices: a review. *Transactions of the Indian Ceramic Society* 70 (1): 1–9.

46 Guaglianoni, W.C. et al. (2021). Influence of CVD parameters on Co-TiO2/CNT properties: a route to enhance energy harvesting from sunlight. *International Journal of Applied Ceramic Technology* 18 (4): 1297–1306.

47 Li, H. et al. (2021). Tubular g-C_3N_4/carbon framework for high-efficiency photocatalytic degradation of methylene blue. *RSC Advances* 11 (30): 18519–18524.

48 Lim, T. et al. (2021). High-index crystal plane of ZnO nanopyramidal structures: stabilization, growth, and improved photocatalytic performance. *Applied Surface Science* 536: 147326.

49 Huang, Z. et al. (2021). Insight into the real efficacy of graphene for enhancing photocatalytic efficiency: A case study on CVD graphene-TiO2 composites. *ACS Applied Energy Materials*, 4: 8755–87.

50 Sirat, M.S. et al. (2017). Growth conditions of graphene grown in chemical vapour deposition (CVD). *Sains Malays* 46: 1033–1038.

51 Chang, C.J., Wei, Y.H., and Huang, K.P. (2017). Photocatalytic hydrogen production by flower-like graphene supported ZnS composite photocatalysts. *International Journal of Hydrogen Energy* 42 (37): 23578–23586.

52 Niu, H. et al. (2021). Single-crystalline Fe7S8/Fe3O4 coated zero-valent iron synthesized with vacuum chemical vapor deposition technique: enhanced reductive, oxidative and photocatalytic activity for water purification. *Journal of Hazardous Materials* 401: 123442.

53 Luo, Q., Yuan, R., Hu, Y.L., and Wang, D. (2021). Spontaneous CVD growth of InxGa1-xN/GaN core/shell nanowires for photocatalytic hydrogen generation. *Applied Surface Science* 537: 147930.

54 Zielińska-Jurek, A., Reszczyńska, J., Grabowska, E., and Zaleska, A. (2012). Nanoparticles preparation using microemulsion systems. In: *Microemulsions-an Introduction to Properties and Applications* (ed. R. Nijjar), 229–250. London: IntechOpen.

55 Mohapatra, P., Mishra, T., and Parida, K.M. (2006). Effect of microemulsion composition on textural and photocatalytic activity of titania nanomaterial. *Applied Catalysis A: General* 310: 183–189.

56 Ghafoor, A. et al. (2021). Ce and Fe doped LaNiO3 synthesized by micro-emulsion route: effect of doping on visible light absorption for photocatalytic application. *Materials Research Express* 8 (8): 085009.

57 Bibi, I. et al. (2021). Effect of dopant on ferroelectric, dielectric and photocatalytic properties of chromium-doped cobalt perovskite prepared via micro-emulsion route. *Results in Physics* 20: 103726.

58 Zurmühl, C., Popescu, R., Gerthsen, D., and Feldmann, C. (2011). Microemulsion-based synthesis of nanoscale TiO2 hollow spheres. *Solid State Sciences* 13 (8): 1505–1509.

59 Aamir, M. et al. (2021). Micro-emulsion approach for the fabrication of La1− xGdxCr1−yFeyO3: magnetic, dielectric and photocatalytic activity evaluation under visible light irradiation. *Results in Physics* 23: 104023.

60 Li, Y. et al. (2012). Carbon nanotube/titania composites prepared by a micro-emulsion method exhibiting improved photocatalytic activity. *Applied Catalysis A: General* 427: 1–7.

61 Modan, E.M. and Plăiașu, A.G. (2020). Advantages and disadvantages of chemical methods in the elaboration of nanomaterials. *The Annals of "Dunarea de Jos" University of Galati. Fascicle IX, Metallurgy and Materials Science* 43 (1): 53–60.

62 Abdelbasir, S.M. et al. (2021). Superior UV-light photocatalysts of nano-crystalline (Ni or Co) FeWO 4: structure, optical characterization and synthesis by a microemulsion method. *New Journal of Chemistry* 45 (6): 3150–3159.

63 Hayashi, Y. (2016). Pot economy and one-pot synthesis. *Chemical Science* 7 (2): 866–880.

64 Yang, J. and Luo, X. (2021). Ag-doped TiO2 immobilized cellulose-derived carbon beads: one-pot preparation, photocatalytic degradation performance and mechanism of ceftriaxone sodium. *Applied Surface Science* 542: 148724.

65 Tien, H.N. et al. (2013). One-pot synthesis of a reduced graphene oxide–zinc oxide sphere composite and its use as a visible light photocatalyst. *Chemical Engineering Journal* 229: 126–133.

66 Zhang, X., Ai, Z., Jia, F., and Zhang, L. (2008). Generalized one-pot synthesis, characterization, and photocatalytic activity of hierarchical BiOX (X= Cl, Br, I) nanoplate microspheres. *The Journal of Physical Chemistry C* 112 (3): 747–753.

67 Yao, Y. et al. (2015). One-pot approach for synthesis of N-doped TiO2/ZnFe2O4 hybrid as an efficient photocatalyst for degradation of aqueous organic pollutants. *Journal of Hazardous Materials* 291: 28–37.

68 Zhao, W. et al. (2022). One-pot molten salt method for constructing CdS/C3N4 nanojunctions with highly enhanced photocatalytic performance for hydrogen evolution reaction. *Journal of Environmental Sciences* 112: 244–257.

69 Majhi, D. et al. (2020). One pot synthesis of CdS/BiOBr/Bi2O2CO3: a novel ternary double Z-scheme heterostructure photocatalyst for efficient degradation of atrazine. *Applied Catalysis B: Environmental* 260: 118222.

70 Su, J. et al. (2021). A one-pot synthesis of $AgBr/Ag_3PO_4$ composite photocatalysts. *RSC Advances* 11 (17): 9865–9873.

71 Noman, M.T. et al. (2020). One-pot sonochemical synthesis of ZnO nanoparticles for photocatalytic applications, modelling and optimization. *Materials* 13 (1): 14.

72 Wu, W. and Zhou, H. (2021). One-pot preparation of NaBiO3/PNMA composite: surface properties and photocatalytic performance. *Applied Surface Science* 544: 148910.

73 Amutha, C. et al. (2014). Influence of concentration on structural and optical characteristics of nanocrystalline ZnO thin films synthesized by Sol-Gel dip coating method. *Progress in Nnaotechnology and Nanomaterials* 3 (1): 13–18.

74 Fang, H.W. et al. (2008). Dip coating assisted polylactic acid deposition on steel surface: film thickness affected by drag force and gravity. *Materials Letters* 62 (21–22): 3739–3741.

75 Al Farsi, B. et al. (2021). Structural and optical properties of visible active photocatalytic Al doped ZnO nanostructured thin films prepared by dip coating. *Optical Materials* 113: 110868.

76 Raja, V., Shiamala, L., Alamelu, K., and Ali, B.J. (2016). A study on the free radical generation and photocatalytic yield in extended surfaces of visible light active TiO2 compounds. *Solar Energy Materials and Solar Cells* 152: 125–132.

77 Zhang, Q. et al. (2016). Fabrication of TiO2 nanofiber membranes by a simple dip-coating technique for water treatment. *Surface & Coatings Technology* 298: 45–52.

78 Chen, G., Qiu, H., Prasad, P., and Chen, X. (2014). Upconversion nanoparticles: design, nanochemistry, and applications in theranostics. *Chemical Reviews* 114 (10): 5161–5214.

79 De Wild, J. et al. (2011). Upconverter solar cells: materials and applications. *Energy & Environmental Science* 4 (12): 4835–4848.

80 Di Paola, A., García-López, E., Marcì, G., and Palmisano, L. (2012). A survey of photocatalytic materials for environmental remediation. *Journal of Hazardous Materials* 211: 3–29.

81 Henderson, M. (2011). A surface science perspective on TiO2 photocatalysis. *Surface Science Reports* 66 (6–7): 185–297.

82 Reza, K.M., Kurny, A.S.W., and Gulshan, F. (2017). Parameters affecting the photocatalytic degradation of dyes using TiO2: a review. *Applied Water Science* 7 (4): 1569–1578.

83 Wen, J. et al. (2015). Photocatalysis fundamentals and surface modification of TiO2 nanomaterials. *Chinese Journal of Catalysis* 36 (12): 2049–2070.

84 Brookes, I.M., Muryn, C.A., and Thornton, G. (2001). Imaging water dissociation on TiO 2 (110). *Physical Review Letters* 87 (26): 266103.
85 Pritchard, H.O. and Skinner, H.A. (1955). The concept of electronegativity. *Chemical Reviews* 55 (4): 745–786.
86 Neppolian, B., Choi, H.C., Sakthivel, S., Arabindoo, B., and Murugesan, V. (2002). Solar/UV-induced photocatalytic degradation of three commercial textile dyes. *Journal of Hazardous Materials* B89: 303–317.
87 Fujishima, A., Zhang, X., and Tryk, D.A. (2008). TiO2 photocatalysis and related surface phenomena. *Surface Science Reports* 63 (12): 515–582.
88 Hernández-Alonso, M.D., Fresno, F., Suárez, S., and Coronado, J.M. (2009). Development of alternative photocatalysts to TiO2: challenges and opportunities. *Energy & Environmental Science* 2 (12): 1231–1257.
89 Litter, M. (1999). Heterogeneous photocatalysis: transition metal ions in photocatalytic systems. *Applied Catalysis B: Environmental* 89–114 (2–3): 23.
90 Kanheya, M., Yablonsky, G.S., and Ray, A.K. (2005). Macro kinetic studies for photocatalytic degradation of benzoic acid in immobilized systems. *Chemosphere* 60: 1427–1436.
91 Kittel, C. (1996). *Introduction to Solid State Physics*, 7e, 308. New York: John Willey and Sons Inc.
92 Ashcroft, N. and Mermin, N. (1976). *Solid State Physics*. New York: Holt, Rinehart and Winston.
93 Voiry, D., Suk Shin, H., Loh, K.P., and Chhowalla, M. (2018). Low-dimensional catalysts for hydrogen evolution and CO 2 reduction. *Nature Reviews Chemistry* 2 (1): 1–17.
94 Bohren, C.F. and Huffman, D.R.J. (2008). *Absorption and Scattering of Light by Small Particles*. New York: Wiley & Sons.
95 N. Bohr. (1935). Can quantum-mechanical description of physical reality be considered complete? *Physical Review* 48: 696.
96 Atwater, H.A. and Polman, A. (2011). Plasmonics for improved photovoltaic devices. *Nature Materials* 9: 205-213.
97 Novotny, L. and Huls, N. (2011). Antennas for light. *Nature Photonics* 5 (2): 83–90.
98 Cao, L., Park, J.S., Fan, P., Clemens, B., and Brongersma, M.L. (2010). Resonant germanium nanoantenna photodetectors. *Nano Letters* 10 (4): 1229–1233.
99 Reed, G.T., Mashanovich, G., Gardes, F.Y., and Thomson, D.J. (2010). Silicon optical modulators. *Nature Photonics* 4 (8): 518–526.
100 Han, P. et al. (2018). Metal nanoparticle photocatalysts: synthesis, characterization, and application. *Particle & Particle Systems Characterization* 35 (6): 1700489.
101 Ingole, P. (2019). A consolidated account of electrochemical determination of band structure parameters in II–VI semiconductor quantum dots: a tutorial review. *Physical Chemistry Chemical Physics* 21 (9): 4695–4716.
102 Hoffmann, M.R., Martin, S., Choi, W., and Bahnemann, D.W. (1995). Environmental applications of semiconductor photocatalysis. *Chemical Reviews* 95 (1): 69–96.
103 Cardona, M. and Yu, P. (2005). *Fundamentals of Semiconductors*, 619. Berlin Heidelberg: Springer-Verlag.
104 Zang, Y. et al. (2014). Hybridization of brookite TiO 2 with gC 3 N 4: a visible-light-driven photocatalyst for As 3+ oxidation, MO degradation and water splitting for hydrogen evolution. *Journal of Materials Chemistry A* 2 (38): 15774–15780.

105 Ganduglia-Pirovano, M.V., Hofmann, A., and Sauer, J. (2007). Oxygen vacancies in transition metal and rare earth oxides: current state of understanding and remaining challenges. *Surface Science Reports* 62 (6): 219–270.

106 Alamelu, K. and Jaffar Ali, B.M. (2018). Sunlight impelled photocatalytic pursuance of Ag-TiO2-SGO and Pt-TiO2-SGO ternary nanocomposites on rhodamine B degradation. *AIP Conference Proceedings* 1942: 050025.

107 Alamelu, K. and Ali, B.J. (2020). Au nanoparticles decorated sulfonated graphene-TiO2 nanocomposite for sunlight driven photocatalytic degradation of recalcitrant compound. *Solar Energy* 211: 1194–1205.

108 Alamelu, K., Raja, V., Shiamala, L., and Jaffar Ali, B.M. (2018). Biphasic TiO2 nanoparticles decorated graphene nanosheets for visible light driven photocatalytic degradation of organic dyes. *Applied Surface Science* 430: 45–154.

109 Xu, L., Wan, Y., Xie, H., Huang, Y., Qiao, X., Qin, L., and Seo, H.J. (2016). On structure, optical properties and photodegradated ability of Aurivillius-type Bi3TiNbO9 nanoparticles. *Journal of the American Ceramic Society* 99 (12): 3964–3972.

110 Zheng, H., Guo, W., Li, S., Yin, R., Wu, Q., Feng, X., Ren, N., and Chang, J.S. (2017). Surfactant (CTAB) assisted flower-like Bi2WO6 through hydrothermal method: unintentional bromide ion doping and photocatalytic activity. *Catalysis Communications* 88: 68–72.

111 Anandan, S., Ohashi, N., and Miyauchi, M. (2010). ZnO-based visible-light photocatalyst: band-gap engineering and multi-electron reduction by co-catalyst. *Applied Catalysis B: Environmental* 100 (3–4): 502–509.

112 Beura, R., Pachaiappan, R., and Thangadurai, P. (2018). A detailed study on Sn4+ doped ZnO for enhanced photocatalytic degradation. *Applied Surface Science* 433: 887–898.

113 Ansari, S.A. et al. (2013). Oxygen vacancy induced band gap narrowing of ZnO nanostructures by an electrochemically active biofilm. *Nanoscale* 5 (19): 9238–9246.

114 Kumar, S. and Rao, K. (2015). Tungsten-based nanomaterials (WO3 & Bi2WO6): modifications related to charge carrier transfer mechanisms and photocatalytic applications. *Applied Surface Science* 355 939–958.

115 Yu, X., Liu, S., and Yu, J. (2011). Superparamagnetic γ-Fe2O3@ SiO2@ TiO2 composite microspheres with superior photocatalytic properties. *Applied Catalysis B: Environmental* 104 (1–2): 12–20.

116 Helal, A. et al. (2017). Hydrothermal synthesis of novel heterostructured Fe2O3/Bi2S3 nanorods with enhanced photocatalytic activity under visible light. *Applied Catalysis B: Environmental* 213: 18–27.

117 Qiu, X., Li, L., Zheng, J., Liu, J., Sun, X., and Li, G. (2008). Origin of the enhanced photocatalytic activities of semiconductors: a case study of ZnO doped with Mg2+. *The Journal of Physical Chemistry C* 112 (32): 12242–12248.

118 Kudo, A. and Miseki, Y. (2009). Heterogeneous photocatalyst materials for water splitting. *Chemical Society Reviews* 38 (1): 253–278.

119 Fujishima, A.H.K. (1972). Electrochemical photolysis of water at a semiconductor electrode. *Nature* 238 (5358): 37–38.

120 Tong, H. et al. (2012). Nano-photocatalytic materials: possibilities and challenges. *Advanced Materials* 24 (2): 229–251.

121 Wang, X. et al. (2012). Photocatalytic overall water splitting promoted by an α–β phase junction on Ga2O3. *Angewandte Chemie* 124 (52): 13266–13269.

122 Li, R. et al. (2015). Achieving overall water splitting using titanium dioxide-based photocatalysts of different phases. *Energy & Environmental Science* 8 (8): 2377–2382.

3

Photocatalytic Decontamination of Organic Pollutants from Water

Mehdi Al Kausor[1]* *and Dhruba Chakrabortty*[2]

[1]*Department of Chemistry, Science College, Kokrajhar, BTR, Assam, India*
[2]*Department of Chemistry, B.N. College, Dhubri, Assam, India*
**Corresponding author*

3.1 Introduction

In various daily activities of our social life such as drinking, cleansing, washing, industrial manufacturing, and irrigation, pure water is an essential requirement. Unfortunately, due to inappropriate treatment, industrial wastewater constantly released into aquatic systems severely contaminates the quality of those natural systems. The release of organic pollutants into the natural water system as industrial effluent creates a threat to the organisms that live in this environment and also to mankind. Such industrial effluent includes dyes, phenols, pesticides, and drugs such as antibiotics, which are constantly being discharged from various industries such as leather, textile, paper, printing, photography, paint, woven, food processing, agricultural research, cosmetics, pharmaceutical, etc. The textile industries alone use around 66% [1] of approximately 0.1 million synthetic dyes available across the globe [2]. The dyeing process releases approximately 10–15% of the dyes used into the wastewater [3, 4]. Some commonly used dyes include acid orange 7, acid blue 25, acid yellow 36, brilliant blue, congo red, crystal violet, methylene blue, methyl orange, malachite green, Rhodamine B, and many more. These dyes are toxic, carcinogenic [5, 6] to living beings, highly photostable, nonbiodegradable in nature [7, 8], and persist in aquatic environments. The photosynthesis in aquatic plants is affected badly due to dye-contaminated water that hinders sunlight penetration through it [9]. Another class of organic water pollutants that causes severe environmental problems is that of the phenol and phenol derivatives [10]. Such compounds are being released from different industrial sources such as refineries, pesticides, insecticides, pharmaceutical, pulp and paper industries, etc. Phenolic compounds include nitrophenols, chlorophenols, aminophenols, chlorocatechols, and methylphenols. These compounds exist in the environment over a longer period of time and are also toxic and carcinogenic to humans and animals [11–13]. The phenolic compounds can donate free electrons resulting in the formation of phenoxy radicals and intermediates, which may enter a living cell and damage membranes of the endoplasmic reticulum, mitochondria, and nucleus and their components like enzymes and nucleic acids. Pharmaceutical effluents such as antibiotics released from pharmaceutical industries and hospitals are hazardous to aquatic life and humans [14]. Antibiotics in wastewater can increase the development of the antibiotic resistant bacteria known as multidrug

Photocatalysts and Electrocatalysts in Water Remediation: From Fundamentals to Full Scale Applications, First Edition. Edited by Prasenjit Bhunia, Kingshuk Dutta, and S. Vadivel.

resistant (MDR) bacterial strains even at very low concentrations [15, 16]. Similarly, some pesticides widely used in agriculture activities can also pollute ground water by leaching through soil. Some commonly used pesticides include organophosphates, organochlorines, carbamates, dithiocarbamates, paraquats, diquates, captan, methyl bromide, and ethylene dibromide [17]. On decomposition of these organic pollutants, the dissolved oxygen (DO) in the water system is consumed sharply, which may cause depletion of oxygen leading to severe consequences for the aquatic system.

Decontamination of such organic pollutants from sewage before discharging them into the water system is therefore an ongoing area of research. The general methods employed for the decontamination of these toxic chemicals from wastewater are classified into three types viz. physical, chemical, and biological methods. Physical methods include adsorption [18], ion exchange [19], membrane filtration [20], and sedimentation [21]. Chemical methods, such as the use of chemical reagents like Fenton's reagent and ozonation [22, 23], oxidation [24], electrochemical oxidation [25], and photocatalytic degradation [26, 27], have been employed for removal of these pollutants from wastewater. In biological treatments, some living or dead biological agents such as bacteria, fungi, and algae are used to degrade or remove toxic chemicals from industrial wastes [6, 28]. Among all these methods, the decontamination of organic water pollutants by photocatalytic degradation processes has been regarded as an attractive method in recent times.

In photocatalytic degradation reactions, the catalysts used are known as photocatalysts. A photocatalyst can be defined as a substance that can be activated by a photon having a definite energy that initiates the photoreaction. A photoreaction may be either a homogenous or heterogeneous reaction. Heterogeneous photocatalysis using semiconductor materials has been widely used to treat organic pollutants in wastewater. The most commonly employed heterogeneous photocatalysts are generally semiconductor metal oxides, and such materials have seen increasing attention in recent years due to their prospective applications in solar energy conversion and environmental purification. The removal of organic pollutants by heterogeneous photocatalysis has materialized as a prominent method of wastewater decontamination, because this method is low cost, generates nontoxic substances like CO_2 and H_2O, and is easy in execution as it uses free and inexhaustible solar energy. Such photocatalysts can be recycled and regenerated for repeated cycles of reaction, resulting in minimum catalyst loss, and it also has a higher number of active sites, which provide higher catalytic activity. The most attractive feature in this technique is that the reactive oxygen species and active holes produced during the irradiation of the material in the photocatalytic system allow the destruction of a number of organic pollutants.

3.2 Photocatalytic Degradation Mechanisms of Organic Contaminants

The general mechanism of a photocatalytic reaction in the contaminated water involves irradiation of the photocatalyst with visible light (VL) or ultraviolet light (UVL) having energy ($h\nu$) greater than the band gap of the material. This results in the excitation of the electrons present in the valence band (VB) to the conduction band (CB) of the semiconductor photocatalyst leaving behind some holes in the VB. The photogenerated electrons thus

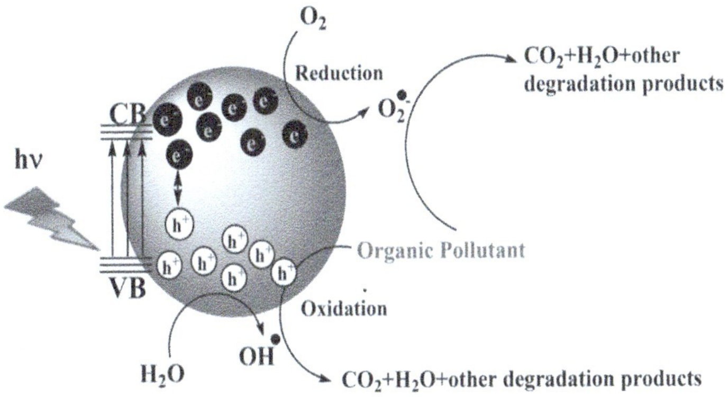

Figure 3.1 Schematic diagram of a photocatalytic reaction.

produced react with the oxygen molecules in the solution to form a superoxide anion (O_2^-), and the holes react with the water molecules adsorbed on the surface of the material to produce the hydroxyl radical (OH•). The photocatalytic activity of the material depends on its ability to maintain a separation of the photogenerated electron–hole pairs by avoiding its recombination, which enables the electrons and holes to migrate to the surface of the catalyst and take part in the reaction with adsorbed pollutants on the surface of the photocatalyst. The mechanism of the photoreaction is shown in Figure 3.1 and the reactions are summarized below.

$$\text{Photocatalyst} + h\nu \rightarrow e^- + h^+ \tag{3.1}$$

$$O_2 + e^- \rightarrow O_2^- O_2 + e^- \rightarrow O_2^- \tag{3.2}$$

$$H_2O + h^+ \rightarrow OH\bullet + H^+ \tag{3.3}$$

$$O_2^- / OH\bullet / h^+ + \text{Organic pollutant} \rightarrow CO_2 + H_2O + \text{other degradation products} \tag{3.4}$$

3.3 Advanced Photocatalytic Materials for Decontamination of Organic Pollutants

Heterogeneous photocatalysis using semiconductor materials has been widely used to treat organic pollutants in water and in air. This process was developed in the 1970s and is commonly known as advanced oxidation process (AOP). The heterogeneous photocatalytic oxidation process has been applied in the past two decades with a view to decompose and mineralize the organic compounds that cause water pollution. The process is carried out using a semiconductor photocatalyst that accelerates the photoreaction [29]. Since Fujishima and Honda [30] reported TiO_2 as a photocatalyst in O_2 generation from water splitting in 1972, inumerous materials like doped TiO_2 [31–34], metal oxides such as SnO_2 [35, 36], Bi_2O_3 [37], ZnO [38], Fe_2O_3 [39], WO_3 [40], multimetal oxides such as $CaBi_2O_4$ [41], $InMO_4$ (M = V, Nb, Ta) [42], $BiVO_4$ [43], Bi_2WO_6 [44, 45], and many more have been applied as photocatalyst in degradation of organic contaminants from wastewater.

Mekprasart and coworkers [46] synthesized Nitrogen-doped TiO_2 nanopowders by homogenization and conventional mechanically stirred techniques in deionised water. The amount of nitrogen loaded in the TiO_2 matrix was varied and the as-obtained intermediate products were annealed at 100° C for 30, 60 and 120 min. The photocatalytic activity of TiO_2 and N-doped TiO_2 (50 g) was dispersed in a RhB (3 mM) dye solution, and the degradation of the dye was investigated under visible light for 1 hr. It was observed that both N-doped TiO_2 and TiO_2 P25 can effectively degrade and decolorize RhB. The constant rate of RhB degradation was found to be 0.051 min^{-1} and 0.105 min^{-1} in presence of TiO_2 P25 and N-doped TiO_2 photocatalyst respectively. These results of the investigation revealed that doping by nitrogen in TiO_2 can significantly improve the photocatalytic degradation as compared to pure TiO_2 due to narrowing of the band gap energy that leads to the enhancement of optical absorption of the photocatalyst under visible illumination.

Kusiak-Nejman and coworkers [47] synthesized ZnO nanoparticles by microwave solvothermal synthesis and investigated the effect of the ZnO nanoparticles' size on the photocatalytic degradation of phenol in water solution under the influence of UV and visible light. The photocatalytic reactions were carried out by taking a catalyst dose of 0.2 g/L with an initial phenol concentration of 10 ppm and illuminated by UV and visible light. The result of the investigation revealed that the photocatalytic performance of ZnO NPs increases with the increase of the particle size and the decrease of the specific surface area. The highest reaction rate for 100% phenol degradation ($k = \sim 0.03\ min^{-1}$) was found for a ZnO sample with the average particle size of 71 nm after 90 min of UV radiation. The authors have reported that the hugher photocatalytic activity was found to be due to the decrease in the charge carrier recombination rate.

Atchudan and coworkers [48] have constructed TiO_2 nanoparticles nicely grafted on GO by a simple solvothermal method. The photocatalytic activity of TiO_2-GO nanocomposite was investigated on the degradation of toxic organic dyes MB and MO under UV-light irradiation. TiO_2-GO achieves a maximum degradation efficiency of 100% and 84% on MB and MO in a neutral solution within 25 and 240 min with rate constant of 0.268 and 0.00675 min^{-1} for MB and MO, respectively. It was found that the photocatalytic activity of TiO_2-GO on MB is 12 times higher than that of bare TiO_2. The results of this investigation revealed that GO plays an important role in the enhancement of photocatalytic performance. The high photocatalytic efficiency and rapid degradation may be attributed to the increased light absorption and the reduced charge recombination with the introduction of GO.

These semiconductor photocatalysts have limitations, however, and show lower degradation efficiency due to their wide band gap, poor quantum efficiency under visible light, possibility of photocorrosion, and faster reunion of photogenerated electron–hole pairs. To address these issues, the synthesis of highly photonic efficient photocatalyst material has become an extensive area of research in heterogeneous photocatalysis. Many research groups are working on fabrication of semiconductor photocatalysts with a variety of other materials to improve the photocatalytic performance to a considerable extent by depositing various metals/nonmetals on its surface, like Ag/Ag_3PO_4 [49], Cr/TiO_2, V/TiO_2 [50], M-Bi_2O_4 (M = Cu,Zn) [51], Ce-doped ZnO [52], B-doped ZnO [53], N-doped TiO_2 [54]; by coupling the semiconductor with other semiconductor materials such as CeO_2/SiO_2 [55], MoS_2/ZnS [56], TiO_2/ZnO [57]; or by coupling with some two-dimensional (2D) materials such as Graphene Oxide (GO) nanosheets, Graphitic carbon nitride (g-C_3N_4) polyaniline (PANI), molybdenum disulphide (MoS_2) viz. Ag_3PO_4/GO [58], $BiVO_4$/g-C_3N_4 [59], ZnO/PANI [60], CdS@MoS_2, [61] to design some type II heterojunctions or Z-scheme system nanocomposites.

3.4 Solar/Visible-light Driven Photocatalytic Decontamination of Organic Pollutants

The most widely studied TiO_2-photocatalyst and many other semiconductor photocatalysts in degradation of organic pollutants have become an emerging solution in environmental applications. However, a major drawback of these materials is that such materials absorb only a small portion of the solar spectrum in the UV region due to the higher band gap energy. In order to solve these issues so that the more abundant source of light energy coming from the sun can be utilized, some modified binary semiconductor systems were developed. Such solar/visible-light driven photocatalysts are found to be safe to handle, more efficient, and the process becomes less expensive in decontamination of organic pollutants in wastewater.

The synthesis, characterization, and application of a series of reduced graphene oxide (rGO) and TiO_2 nanofibers (rGO/TiO_2 NFs) with varying wt% of rGO was reported by Nasr and coworkers [62]. The materials were synthesized by electrospinning methods and characterized by various spectroscopic techniques. The research group reported the photocatalytic degradation performance of the material with respect to the degradation of methyl orange (MO) dye under visible light. The rGO(2 wt%)/TiO_2 NFs exhibit higher photocatalytic activity of about 90% degradation of MO after 120 min of visible light irradiation, whereas only 35% degradation was achieved by commercial TiO_2. It was observed that with increase in the amount of GO up to 7 wt% the photocatalytic activity gradually decreases to 26%. With increase in the amount of GO in the GO (7 wt%) /TiO_2 composite, the light absorption is reduced on the TiO_2 surface. This leads to a decrease in the generation of photoexcited electrons [63]. This excess of GO also increased the opportunity for the recombination of the photogenerated electron–hole pairs by increasing the collision among the photogenerated electrons and holes [64]. The rate constant exhibits a maximum of 0.0186 min^{-1} for rGO (2 wt%) / TiO_2 nanocomposites, which is about 5.5-fold higher than that of commercial TiO_2-P25. The augmented photocatalytic activity is attributed to improved visible light absorption due to incorporation of GO and the quick transfer of photo-electrons to the rGO layers, which reduces the probability of the electron–hole pair recombination in TiO_2. Thus, more photo-electrons were available compared to pure TiO_2 NFs, to take part in the photodegradation process. The results of this study revealed that incorporation of GO into TiO_2 NFs is an efficient and simple way for enhancing the visible light photocatalytic activity. The coupling of various semiconductor photocatalysts with g-C_3N_4 has become an efficient way to improve the photocatalytic performance of the semiconductor photocatalysts. Boonprakob et al. [65] reported the synthesis of g-C_3N_4/TiO_2 composites by direct heating of a mixture of melamine and presynthesized TiO_2

nanoparticles in Ar gas flow with varying contents of g-C_3N_4/TiO_2. The photocatalytic activities of the material were investigated by measuring the decolorization of methylene blue (MB) under visible light. It was observed that self photolysis of MB is negligible and about 35% of the dye was degraded in presence of pure TiO_2 films within 3 h of irradiation. The photocatalytic activity of the g-C_3N_4/TiO_2 composite was observed as 68% in the presence of 50 wt% g-C_3N_4/TiO_2 film. The improved photocatalytic activity of the g-C_3N_4/TiO_2 films may be due to the red shift of the light absorption range and the strong visible light absorption intensity of the g-C_3N_4/TiO_2 composites, and also due to an improved charge separation efficiency of photogenerated electron–hole pairs due to the suitably matching CB and VB levels and the close interfacial connection between TiO_2 and g-C_3N_4 in the composite material [66, 67].

Saravanan and coworkers [60] synthesized a series of PANI/ZnO composites with different mole ratios of PANI to ZnO nanocomposites systems by precipitation followed by sonication methods. The materials were used in photocatalytic degradation of MO and MB dye under visible light. It was observed that in both cases of the dyes, the PANI/ZnO nanocomposite with a PANI to ZnO mole ratio of 1.5 exhibits the highest photocatalytic activity, and almost 100% of the dyes were degraded in 180 min of visible light irradiation. The first-order rate constants were found to be 2.325×10^{-2} min^{-1} for MO and 2.575×10^{-2} min^{-1} for MB dye, which are about 178 and 172 times higher than that of pure ZnO. The nanocomposite system displays superior photodegradation efficiency due to better crystallinity. It was also suggested that with increase in the amount of PANI, the ZnO particles undergo aggregation that eventually reduces the degradation efficiency [68, 69].

Z-scheme CdTe/TiO_2 heterostructure composites were prepared via hydrothermal methods by Gong and coworkers [70]. The photodegradation efficiency of the material was investigated by degradation of tetracycline hydrochloride (TC-H) under visible light illumination. The heterostructure photocatalyst CdTe/TiO_2 grown at 120°C exhibits the best degradation efficiency, which can degrade 78% of TC-H after 30 min of visible light illumination. The improved percentage degradation was attributed to the synergistic effects of better visible light absorption, proficient charge separation, and comparatively high redox potentials of e^-s and h^+s associated with the Z-scheme charge transfer mechanism. Ramos-Delgado and coworkers [71] synthesized 2%WO_3/TiO_2 composite by a sol-gel method, and the photocatalytic activity of the material was investigated under solar irradiation using a broad spectrum pesticide malathion as a model contaminant. The 2%WO_3/TiO_2 composite exhibits better degradation efficiency than pure TiO_2. The degradation results showed that the material can degrade completely (100%) malathion using solar light irradiation for 2 h, while the TOC abatement was 63% after 5 h. The result of this study revealed that 2%WO_3/TiO_2 composite can become a promising material for solar photocatalytic degradation of water contaminated by pesticides. The augmented photocatalytic performance may be attributed to the incorporation of very reactive, well-dispersed WO_3 clusters over anatase TiO_2 surfaces, which results in increased surface area for better adsorption of the pollutant on the surface of the material and reduced charge recombination.

In all these binary semiconductor photocatalytic systems, the basic mechanism of photocatalytic degradation of the organic pollutants can be summarized as shown in Figure 3.2.

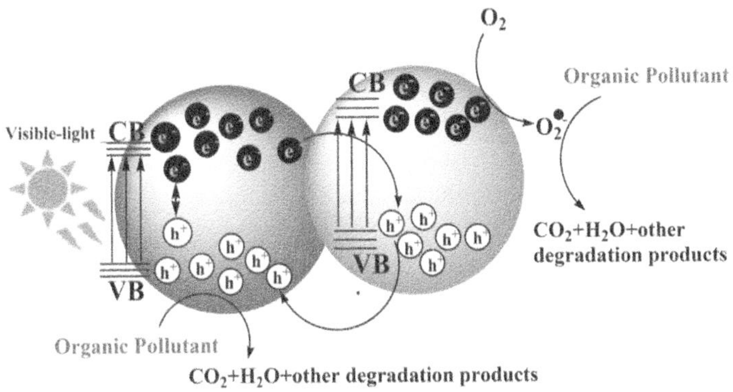

Figure 3.2 Mechanism of photodegradation of organic pollutants by a Z-Scheme binary composite.

It is believed that when the material is illuminated by visible light, the electrons from the VB of the photocatalyst-I (PC-I), which may be TiO_2 or ZnO, get excited to the higher energy CB leaving behind some holes (h^+) in the VB. Since the VB minima of PC-II is higher than that of PC-I, the photo-induced electrons can now migrate from the CB of PC-I to the VB of PC-II while the holes of the VB of PC-II migrate toward the VB of PC-I. This process minimizes photocorrosion, and electron–hole pair recombination thereby enhances photocatalytic efficiency. The holes can now directly degrade organic pollutants to CO_2 and H_2O or indirectly with the help of $\bullet O_2^-$ radicals generated by reduction of O_2 in the medium utilizing the photoexcited electrons of PC-II.

3.5 Emerging Scientific Opportunities of Photocatalysts in Removal of Organic Pollutants

The unique heterostructure nanocomposites fabricated by various research groups have become excellent materials in the case of decontamination of organic pollutants from water due to the quick transfer of charge from the semiconductor (PC-I) to the supporting material, which is also a semiconductor (PC-II), improved electron–hole pair separation efficiencies, narrow band gap, enhanced visible light response, as well as higher stability. The combination of heterojuctions with desired materials and controllable morphology has become a potential area of research for the design and development of some novel photocatalytic materials. In recent years, some highly efficient dual channel ternary composites such as ZnS–Ag_2S–RGO [72], N-TiO_2/Ag_3PO_4/GO [73], g-C_3N_4/WO_3/MoS_2 [74], MoS_2/gC_3N_4/TiO_2 [75], Ag_2MoO_4/Ag_2S/MoS_2 [76], etc. have been reported with respect to their applications in degradation of organic water pollutants. The design and development of some Z-scheme ternary heterostructure nanocomposites utilizing some 2D materials such as graphene oxide (GO) nanosheets, graphitic carbon nitride (g-C_3N_4) polyaniline (PANI), molybdenum disulphide (MoS_2), etc. may become an emerging area of research in environmental applications. GO has a huge specific surface area, excellent electrical conductivity, and high chemical stability [77]. The fabrication of semiconductor materials with GO enhanced the photocatalytic activity, because the photogenerated electrons can easily transfer from semiconductor materials to the GO sheets, which prevents the recombination of electron–hole pairs [78, 79]. In addition to that, GO can also increase the visible light absorption ability of the composite materials. Many research groups utilized a new type of 2D material known as graphitic carbon nitride (g-C_3N_4). The unique electric, optical, structural, and physiochemical properties such as moderate band gap (~2.7 eV), appropriate electronic band structure, absorption of visible light, non-toxicity, low cost, and good stability [80, 81], make g-C_3N_4 a new class in environmental and energy applications [82, 83]. The tuneable band gaps and efficient intercalation of various compounds lead to profound use of this novel material in heterogeneous catalysis and support. The material can be readily utilized to fabricate some hybrid photocatalysts with controllable size and pore structures, size distributions, and morphologies [84]. Other promising visible light responsive photocatalysts such as MoS_2 or conducting polymer polyaniline (PANI) can also be utilized in this regard.

Mousavi and coworkers [85] have prepared magnetically separable g-C_3N_4/Fe_3O_4/Ag_3VO_4 nanocomposites by fabricating Fe_3O_4 and Ag_3VO_4 on the surface of the 2D g-C_3N_4 sheets. The ternary heterostructure nanocomposite, g-C_3N_4/Fe_3O_4/Ag_3VO_4 exhibits higher photocatalytic activity in photodegradation of RhB under visible light than bare g-C_3N_4 and g-C_3N_4/Fe_3O_4 binary composites. The result of the study revealed that the nanocomposite with 60% of Ag_3VO_4 i.e. g-C_3N_4/Fe_3O_4/Ag_3VO_4 (60%) can degrade about 75% RhB dye in 300 min with a rate constant of 118×10^{-4} min^{-1}, which was about 14, 8, and 3-fold higher than those of the bare g-C_3N_4, binary g-C_3N_4/Fe_3O_4, and bare Ag_3VO_4 samples. The higher photocatalytic activity of the nanocomposites was attributed to the efficient separation of the charge carriers and increased visible light absorption capacity of the material. Potle and coworkers [86] have synthesized a reduced graphene oxide based ternary (rGO)-ZnO-TiO_2 nanocomposite by sonochemical methods in which ZnO and TiO_2 nanoparticles were grafted on rGO sheets. The photocatalytic activity of the material was investigated by taking crystal violet (CV) as model dye with an initial concentration of 50 ppm and catalyst loading of 0.10 g/L at temperature of 35°C and pH 6.5. The material can degrade 87.06% of CV after 20 min in presence of UV-irradiation. The results reveal that the adsorption of CV dye on the synthesized rGO-ZnO-TiO_2 ternary nanocomposite was significantly enhanced due to the higher surface area of rGO, due to which the CV dye degradation was enhanced in presence of UV light.

Zhou and coworkers [87] have synthesized a series of AgCl/Ag_3PO_4/g-C_3N_4 composites with different molar ratios of AgCl by a facile anion exchange method. The materials are used in photodegradation of sulfamethoxazole (SMX) antibiotics. It was found that the composite photocatalyst AgCl/Ag_3PO_4/g-C_3N_4 (43%) showed highest photocatalytic activity with an efficiency of 100% degradation of SMX ($C_0 = 1$ mg/L and $w = 0.1$ g/L) after 90 min irradiation under visible light, while 95.0% SMX was degraded within the first 2 h from 10:30 a.m. to 12:30 p.m. under natural sunlight. The enhanced photocatalytic activity may be attributed to the quick transfer of photogenerated h^+s on the valance band of g-C_3N_4 and producing zero valent Cl-atoms (Cl^0), which act as the main reactive oxidation species for SMX photodegradation. Dehghan and coworkers [88] have synthesized a heterogeneous photocatalyst, (rGO)/Fe_3O_4/ZnO nanocatalyst, for photodegradation of a selected fungicide, Metalaxyl (MX), in aqueous solution under visible light irradiation. The material can degrade about 92.11% MX (C_0 = 30 ppm, w = 0.5 mg/mL) within 120 min with a pseudo-first-order rate constant of 2.28×10^{-2} min^{-1}. The material was found to be stable and reusable up to five cycles and exhibited better photocatalytic activity because of the synergistic effect between rGO, Fe_3O_4, and ZnO under visible light irradiation.

In all the dual channel Z-scheme ternary composites discussed above, the mechanism of the photodegradation process may be summarized as shown in Figure 3.3. In such photocatalytic systems, the electrons from the VB of both PC-I and PC-II are simultaneously excited to their respective CB on irradiation by light leaving behind holes in the VB of both. The electrons from the CB of PC-II are transferred to the CB of PC-I, while the holes of the VB of PC-I are transferred to the VB of PC-II, because both the CB potential and VB potential of PC-II are more negative than that of PC-I [89]. The photo-excited electrons combine with absorbed O_2 to form $\bullet O_2^-$ radicals, which can promptly degrade organic pollutants. The resultant holes can now deliberately degrade the pollutants because of their stronger oxidizing ability. The PC-III acts as a supported material which further promotes the separation efficiency of electron–hole pairs. The fabrication of better photocatalysts is an ongoing research in photocatalysis of organic water pollutants. Many novel materials have been developed so far (Table 3.1). The process depends on lowering of the band gap energy of the material to such an extent that electron–hole pair recombination is minimized. In addition to this, there should be an efficient charge transfer, and the material must exhibit photophysical properties with respect to absorption of a wide range of solar radiation. It is also important to know the surface reactions or what is happening on the material surface. The overall target is thus to utilize a facile approach to control the morphology as well as the crystal structure to develop a highly efficient material.

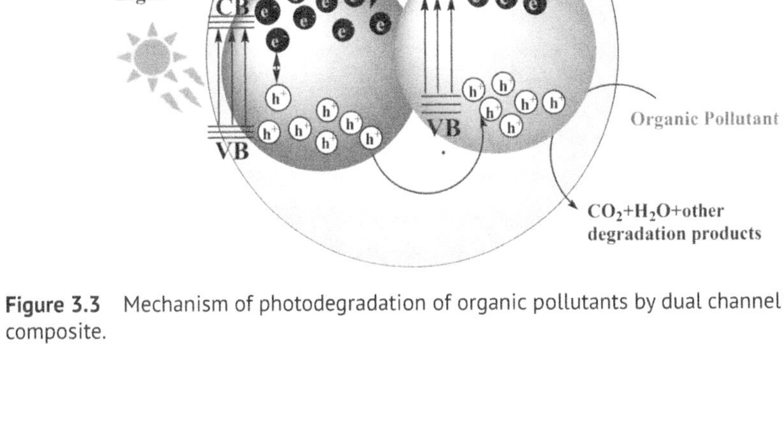

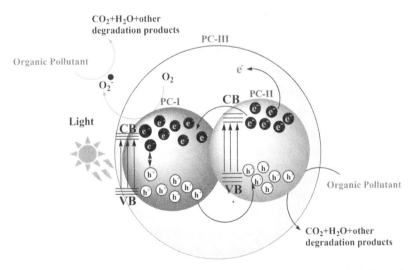

Figure 3.3 Mechanism of photodegradation of organic pollutants by dual channel ternary composite.

Table 3.1 Performance of photocatalytic degradation of limited organic pollutants by some emerging ternary nanocomposite materials.

Heterostructure	Method of synthesis	Pollutant selected	Experimental conditions	Efficiency	Rate constant	Increment w.r.t. bare material	Ref.
Ag_3PO_4-TiO_2-rGO	Photocatalytic reduction and ion exchange	MO	$C_0 = 20$ ppm $w = 1$ g/L $t = 120$ min Light Source = VL	80%	83×10^{-4}	–	[90]
(GO)/Ag_3PO_4/AgBr	In situ anion-exchange process	RhB	$C_0 = 2.5 \times 10^{-5}$ M $w = 1$ g/L $t = 40$ min Light Source = VL	90%	5.4×10^{-2}	3.6-fold w.r.t. Ag_3PO_4	[91]
TiO_2/Ag_3PO_4/GO	Chemical exfoliation and hydrothermal	RhB	$C_0 = 20$ ppm $w = 1$ g/L $t = 20$ min Light Source = VL	–	12.81×10^{-2}	4-fold w.r.t. Ag_3PO_4	[92]
Ag_3PO_4/PANI/GO	Coprecipitation	RhB	$C_0 = 12$ mg/L $w = 2$ g/L $t = 12$ min Light Source = VL	98%	18.7×10^{-2} min^{-1}	2.1-fold (Ag_3PO_4)	[93]
TiO_2/PANI/GO	In situ oxidation polymerization method	RB	$C_0 = 25$ mg/L $w = 800$ mg/L $t = 150$ min Light Source = UVL	97%	30.5×10^{-3}	4.7-fold w.r.t. TiO_2	[94]
TiO_2/$CoMoO_4$/PANI (30%)	Chemical Oxidative polymerization	RhB	$C_0 = 1.0 \times 10^{-5}$ M $w = 0.4$ g/L $t = 150$ min Light Source = VL	100%	287×10^{-4} min^{-1}	21.7-fold w.r.t. TiO_2	[95]

PANI-TiO_2/rGO	In situ polymerization of PANI	RhB	$C_0 = 1 \times 10^{-5}$ M $w = 0.5$ g/L $t = 90$ min Light Source = VL	90.5 %	0.020 min^{-1}	3.1-fold w.r.t. TiO2	[96]
Cu_2O/ZnO-PANI	One-pot solvothermal method and in-situ polymerization	CR	$C_0 = 30$ mg/L $w = 1$ g/L $t = 30$ min Light Source = VL	100%	0.10 min^{-1}	1.23-fold w.r.t. Cu_2O	[97]
ZnS/CdS/PANI	Chemical oxidative polymerization	RhB	$C_0 = 20$ mg/L $w = 0.7$ mg/L $t = 135$ min Light Source = VL	96.5 %	1.7×10^{-2} min^{-1}	3.4-fold w.r.t. ZnS	[98]
PANI-TiO_2-Fe_3O_4 (10%)	One-pot method via chemical oxidative polymerization	RB-5	$C_0 = 10$ mg/L $w = 0.15$ g/L $t = 180$ min Light Source = VL	94.4%	1.589×10^{-2} min^{-1}	20.4-fold w.r.t. TiO2	[99]
MoS_2/Ag_3PO_4/Ag		Phenol	$C_0 = 5$ mg/L $w = 1$ g/L $t = 200$ min Light Source = VL	95%	–	–	[100]
Ag/Ag_3PO_4/TiO_2 NT		2-CP	$C_0 = 20$ mg/L w = 1 g/L $t = 240$ min Light Source=....	60%	4.7×10^{-3}	–	[101]

(Continued)

Table 3.1 (Continued)

Heterostructure	Method of synthesis	Pollutant selected	Experimental conditions	Efficiency	Rate constant	Increment w.r.t. bare material	Ref.
Carbon nitride/ PANI/ ZnO	In-situ polymerization and evaporation induced assembly (EIA) process	4-CP	$C_0 = 120\ \mu L$ w = 0.5 g/L $t = 120$ min Light Source = UVL	50%	$0.0049\ min^{-1}$	5.3-fold w.r.t ZnO	[102]
PANI/TiO_2 -rGH hydrogel	Chemical reduction	BPA	$C_0 =$ w = $t = 40$ min Light Source = UVL	99 %	–	–	[103]
PANI/TiO_2/ZnO	In situ chemical oxidative polymerization	p-Cresol	$C_0 = 100$ ppm w = - $t = 360$ min Light Source = UVL	99.3%	$0.2607\ min^{-1}$	3.6-fold w.r.t. ZnO 2.9-fold w.r.t. TiO_2	[104]
g-C_3N_4/MoS_2-PANI	In-situ polymerization	BPA	$C_0 = 20$ mg/L w = - $t = 60$ min Light Source = VL	92.6%	$0.0395\ min^{-1}$	1.5-fold w.r.t. g-C_3N_4/ MoS_2	[105]
FeO/ZnO@PANI	Chemical reduction and in-situ polymerization	3-AMP	$C_0 = 80$ mg/L w = - $t = 120$ min Light Source = VL	92%	–	1.7-fold w.r.t. TiO_2	[106]

CuO/GN@PANI	Radio frequency (RF) magnetron sputtering of CuO over GN@ Pani film	2-CP	$C_0 = 25$ mg/L $w = 0.2$ g/L $t = 340$ min Light Source = VL	100%	1.080×10^{-2} min^{-1}	1.8-fold w.r.t. GN/ PANI	[107]
$BiVO_4/Ag_3PO_4$/PANI	In-situ oxidative polymerization	CIP	$C_0 = 10$ mg/L $w = 0.1$ g/L $t = 60$ min Light Source = UVL	85.92%	0.00894 L mg^{-1} min^{-1}	10-fold w.r.t. $BiVO_4$ 1.6-fold w.r.t. $BiVO_4/Ag_3PO_4$	[108]
rGO-PANI assisted c-ZnO	In-situ oxidative polymerization	ACP	$C_0 = 6.8 \times 10^{-3}$ M $w = 2$ g/L $t = 100$ min Light Source = NSL	47%	0.0055 min^{-1}	–	[109]
Ag_3PO_4/PANI@g-C_3N_4	Hydrothermal method	MCP	$C_0 = 15$ mg/L $w = 0.08$ g/L $t = 50$ min Light Source = VL	99.6%	–	–	[110]
PANI/g-C_3N_4/CeO_2	In-situ chemical oxidative polymerization method	DZN	$C_0 = 10$ mg/L $w = 0.1$ g/L $t = 180$ min Light Source = UVL	96.78%	0.102 min^{-1}	–	[111]
CuO/TiO_2/PANI	In-situ chemical polymerization method	CPS	$C_0 = 5$ mg/L $w = 0.45$ g/L $t = 90$ min Light Source = VL	95%	0.0318 min^{-1}	–	[112]

(Continued)

Table 3.1 (Continued)

Heterostructure	Method of synthesis	Pollutant selected	Experimental conditions	Efficiency	Rate constant	Increment w.r.t. bare material	Ref.
AgI/Ag_3PO_4/g-C3N4	In-situ ion exchange	Nitenpyram (NTP)	$C_0 = 5$ ppm $w = 0.5$ g/L $t = 4$ min Light Source = VL	95%	0.76 min^{-1}	16.2-fold w.r.t. g-C_3N_4 2.4-fold w.r.t. Ag_3PO_4,	[113]
Fe_3O_4-SnO_2-gC_3N_4	Hydrothermal	Carbofuran	$C_0 = 1.5$ mg/L $w = 3.5$ mg/L $t = 150$ min Light Source = VL	89%	0.015 min^{-1}	2.5-fold w.r.t. SnO_2	[114]

3.6 Limitations of Photocatalytic Decontamination and Key Mitigation Strategies

The research done by various groups reveals that different combinations of semiconductor photocatalysts can be fabricated to synthesize some binary or even ternary heterojunction photocatalysts, and the materials thus synthesized possess improved photocatalytic activity toward the degradation of organic pollutants. The enhanced photocatalytic activity is mainly due to the improved light absorption range, increased specific surface area, higher absorption capacity, narrower band gap energy, faster charge transfer, minimal electron–hole pair recombination, and negligible photocorrosion. Significant progress has been achieved in this regard; however, there are some limitations that must be resolved in the near future so that a desired material can be obtained with excellent photoactivity. These include:

i) In most of the works mentioned, the optimization of the reaction system parameters were not investigated. It is well known that the reaction system parameters such as photocatalyst dose, initial concentration of the pollutant, pH of the test solution, temperature, light intensity, presence of oxidizing agents or electron acceptors, and the presence of ionic components play an important role in photocatalytic degradation. Thus, the optimization of these factors should be investigated properly.
ii) In many studies reported so far, the kinetics of the photodegradation has not been studied. The study of the reaction kinetics by various kinetic models must be applied, determining the rate constants by changing the system parameters under different conditions.
iii) It is well known that for better photocatalytic activity, the particles of the pollutant must be adsorbed on the surface of the photocatalyst material. Therefore, the level of adsorption of the pollutant on the catalyst surface must be investigated by batch adsorption studies. The kinetics of the adsorption should also be investigated by changing various system parameters applying various kinetic models to find the maximum adsorption capacity of the material.
iv) Most of the works focussed on designing the material without knowing the reactions that take place on the surface of the material. The reactions on the surface are critical and important to understand the degradation process and a possible mechanism should be elucidated.
v) The photocatalytic degradation of materials are tested with artificially prepared effluent containing limited organic pollutants only; however, a good number of cationic and anionic dyes, phenolic compounds, antibiotics, and pesticides may be present in real effluents. Degradation of those pollutants, therefore, should also be investigated properly. Moreover, in real practice the effluents may contain a mixture of dyes, phenolic compounds, antibiotics, pesticides, etc. Study on the photocatalytic degradation of simultaneous coexisting pollutants should therefore also be studied.
vi) The photocatalytic degradation of pollutants must be studied properly either by using direct methods using chromatographic techniques or by using indirect methods like examining the maximum reduction of the carbon content in the effluent, which can be done by estimating the Chemical Oxygen Demand (COD) before and after the photodegradation.

It is expected that if these issues and challenges are addressed properly, photocatalytic degradation of organic pollutants will be an excellent technique in large scale industrial applications. The researchers may adopt the following strategies to improve the degradation performance of a material:

i) The performance of any binary or ternary composites depends on the combination of the materials during the fabrication process. Proper choice of the other material(s), therefore, is an important factor while designing the nanocomposite materials. There are a number of materials available, and using those, a desired binary or ternary composites can be designed and tested for the photocatalytic degradation of water polluting organic compounds.
ii) Many research groups have concentrated on material design without knowing the changes that happen on the material surface. It may be noted that the surface reactions are crucial and important to understand the reaction properly. Thus, an insight into the mechanism of the degradation reaction, and understanding the reaction and reaction by-products are the prominent points to be remembered in designing novel photocatalyst materials.
iii) The research target is to synthesize a novel material, and due importance should be given to the development of a low-cost, environmentally friendly, stable, and recyclable binary or ternary material for large scale industrial applications with efficient photocatalytic performance.

3.7 Summary and Future Directions

The treatment of organic pollutants in micropolluted water has always been paid attention to by many researchers. The removal of these pollutants from contaminated water is essential for a green environment. There are many materials that have been designed, and their performances in degradation of organic pollutants have been tested by various research groups. The photocatalytic performance of many commonly used photocatalysts like nanostructured metal oxides, sulphides, and phosphates were greatly improved by fabricating with some 2D materials like GO, RGO, g-C_3N_4, MoS_2, PANI, etc. to develop some binary as well as ternary Z-scheme heterojunction photocatalysts. The augmented photocatalytic performance is accredited to the small-sized well-dispersed particles on the two-dimensional sheets, higher adsorption of the pollutants on the active sites, and lower electron–hole pair recombination. Many improved materials are thus designed and developed day by day for application toward degradation of toxic organic pollutants from wastewater.

The motivation behind the rapid development of an efficient material is to combine the outstanding properties of various supporting materials with other constituent nanomaterials to fabricate a novel material for the photocatalytic degradation of organic water pollutants. More efforts may be made, therefore, on the development of well-organized, environmentally friendly, low-cost and highly efficient fabrication methods for attaining improved and extraordinary photocatalytic performance. The cost factor is very much important in determining the selection of the suitable photocatalyst for large scale industrial applications. The lack of efficient and low-cost harvesting techniques and inadequate scale-up strategies pose an obstacle to the industrial applications of the photocatalytic materials. Future research work, therefore, should aim at addressing these issues.

References

1 Gita, S., Hussan, A., and Choudhury, T.G. (2017). Impact of textile dyes waste on aquatic environments and its treatment. *Environment and Ecology* 35 (3C): 2349–2353.

2 Raval, N.P., Shah, P.U., and Shah, N.K. (2017). Malachite green "a cationic dye" and its removal from aqueous solution by adsorption. *Applied Water Science* 7 (7): 3407–3445.

3 Chowdhury, S. and Saha, P. (2011). Adsorption kinetic modeling of safranin onto rice husk biomatrix using pseudo-first-and pseudo-second-order kinetic models: comparison of linear and non-linear methods. *CLEAN–Soil, Air, Water* 39 (3): 274–282.

4 Adegoke, K.A. and Bello, O.S. (2015). Dye sequestration using agricultural wastes as adsorbents. *Water Resources and Industry* 12: 8–24.

5 Asfaram, A., Ghaedi, M., Ghezelbash, G.R., and Pepe, F. (2017). Application of experimental design and derivative spectrophotometry methods in optimization and analysis of biosorption of binary mixtures of basic dyes from aqueous solutions. *Ecotoxicology and Environmental Safety* 139: 219–227.

6 Bekçi, Z., Seki, Y., and Cavas, L. (2009). Removal of malachite green by using an invasive marine alga Caulerpa racemosa var. cylindracea. *Journal of Hazardous Materials* 161 (2–3): 1454–1460.

7 Lü, X.F., Ma, H.R., Zhang, Q., and Du, K. (2013). Degradation of methyl orange by UV, O_3 and UV/O 3 systems: analysis of the degradation effects and mineralization mechanism. *Research on Chemical Intermediates* 39 (9): 4189–4203.

8 Yao, Y., Bing, H., Feifei, X., and Xiaofeng, C. (2011). Equilibrium and kinetic studies of methyl orange adsorption on multiwalled carbon nanotubes. *Chemical Engineering Journal* 170 (1): 82–89.

9 Raghuvanshi, S.P., Singh, R., Kaushik, C.P., and Raghav, A. (2004). Kinetics study of methylene blue dye bioadsorption on baggase. *Applied Ecology and Environmental Research* 2 (2): 35–43.

10 Chiou, C.H., Wu, C.Y., and Juang, R.S. (2008). Influence of operating parameters on photocatalytic degradation of phenol in UV/TiO_2 process. *Chemical Engineering Journal* 139 (2): 322–329.

11 Schweigert, N., Hunziker, R.W., Escher, B.I., and Eggen, R.I. (2001). Acute toxicity of (chloro-) catechols and (chloro-) catechol-copper combinations in Escherichia coli corresponds to their membrane toxicity in vitro. *Environmental Toxicology and Chemistry: An International Journal* 20 (2): 239–247.

12 Bruce, R.M., Santodonato, J., and Neal, M.W. (1987). Summary review of the health effects associated with phenol. *Toxicology and Industrial Health* 3 (4): 535–568.

13 Matos, J., Laine, J., and Herrmann, J.M. (1998). Synergy effect in the photocatalytic degradation of phenol on a suspended mixture of titania and activated carbon. *Applied Catalysis B: Environmental* 18 (3–4): 281–291.

14 Durán-Álvarez, J.C., Avella, E., Ramírez-Zamora, R.M., and Zanella, R. (2016). Photocatalytic degradation of ciprofloxacin using mono-(Au, Ag and Cu) and bi-(Au–Ag and Au–Cu) metallic nanoparticles supported on TiO2 under UV-C and simulated sunlight. *Catalysis Today* 266: 175–187.
15 Walter, M.V. and Vennes, J.W. (1985). Occurrence of multiple-antibiotic-resistant enteric bacteria in domestic sewage and oxidation lagoons. *Applied and Environmental Microbiology* 50 (4): 930–933.
16 Alexy, R., Kümpel, T., and Kümmerer, K. (2004). Assessment of degradation of 18 antibiotics in the closed bottle test. *Chemosphere* 57 (6): 505–512.
17 Abdollahi, M., Ranjbar, A., Shadnia, S., Nikfar, S., and Rezaiee, A. (2004). Pesticides and oxidative stress: a review. *Medical Science Monitor* 10 (6): RA141–RA147.
18 Chiou, M.S., Ho, P.Y., and Li, H.Y. (2004). Adsorption of anionic dyes in acid solutions using chemically cross-linked chitosan beads. *Dyes and Pigments* 60 (1): 69–84.
19 Kiseleva, M.G., Lubov'V, R., and Nesterenko, P.N. (2001). Ion-exchange properties of hypercrosslinked polystyrene impregnated with methyl orange. *Journal of Chromatography A* 920 (1–2): 79–85.
20 Xu, Y., Lebrun, R.E., Gallo, P.J., and Blond, P. (1999). Treatment of textile dye plant effluent by nanofiltration membrane. *Separation Science and Technology* 34 (13): 2501–2519.
21 Perelo, L.W. (2010). In situ and bioremediation of organic pollutants in aquatic sediments. *Journal of Hazardous Materials* 177 (1–3): 81–89.
22 Robinson, T., McMullan, G., Marchant, R., and Nigam, P. (2001). Remediation of dyes in textile effluent: a critical review on current treatment technologies with a proposed alternative. *Bioresource Technology* 77 (3): 247–255.
23 Lin, S.H. and Lin, C.M. (1993). Treatment of textile waste effluents by ozonation and chemical coagulation. *Water Research* 27 (12): 1743–1748.
24 Ahn, D.H., Chang, W.S., and Yoon, T.I. (1999). Dyestuff wastewater treatment using chemical oxidation, physical adsorption and fixed bed biofilm process. *Process Biochemistry* 34 (5): 429–439.
25 Yong, K.O., Wang, Z.L., Yu, W.A., Jia, Y.U., and Chen, Z.D. (2011). Degradation of methyl orange in artificial wastewater through electrochemical oxidation using exfoliated graphite electrode. *New Carbon Materials* 26 (6): 459–464.
26 Gottam, R., Srinivasan, P., La, D.D., and Bhosale, S.V. (2017). Improving the photocatalytic activity of polyaniline and a porphyrin via oxidation to obtain a salt and a charge-transfer complex. *New Journal of Chemistry* 41 (23): 14595–14601.
27 Joshi, M. and Purwar, R. (2004). Developments in new processes for colour removal from effluent. *Review of Progress in Coloration and Related Topics* 34 (1): 58–71.
28 Lakshmi, K., Varadharajan, V., and Kadirvelu, K.G. (2020). Photocatalytic decontamination of organic pollutants using advanced materials. *Modern Age Wastewater Problems*: 195–212.
29 Gaya, U.I. and Abdullah, A.H. (2008). Heterogeneous photocatalytic degradation of organic contaminants over titanium dioxide: a review of fundamentals, progress and problems. *Journal of Photochemistry and Photobiology C: Photochemistry Reviews* 9 (1): 1–2.
30 Fujishima, A. and Honda, K. (1972). Electrochemical photolysis of water at a semiconductor electrode. *Nature* 238 (5358): 37–38.

31 Wu, J.C. and Chen, C.H. (2004). A visible-light response vanadium-doped titania nanocatalyst by sol–gel method. *Journal of Photochemistry and Photobiology A: Chemistry* 163 (3): 509–515.

32 Li, Y., Hwang, D.S., Lee, N.H., and Kim, S.J. (2005). Synthesis and characterization of carbon-doped titania as an artificial solar light sensitive photocatalyst. *Chemical Physics Letters* 404 (1–3): 25–29.

33 Kim, S., Hwang, S.J., and Choi, W. (2005). Visible light active platinum-ion-doped TiO_2 photocatalyst. *The Journal of Physical Chemistry B* 109 (51): 24260–24267.

34 Ao, Y., Xu, J., Fu, D., and Yuan, C. (2009). A simple method to prepare N-doped titania hollow spheres with high photocatalytic activity under visible light. *Journal of Hazardous Materials* 167 (1–3): 413–417.

35 Zhang, D.F., Sun, L.D., Yin, J.L., and Yan, C.H. (2003). Low-temperature fabrication of highly crystalline SnO_2 nanorods. *Advanced Materials* 15 (12): 1022–1025.

36 Wu, S., Cao, H., Yin, S., Liu, X., and Zhang, X. (2009). Amino acid-assisted hydrothermal synthesis and photocatalysis of SnO_2 nanocrystals. *The Journal of Physical Chemistry C* 113 (41): 17893–17898.

37 Zhang, L., Wang, W., Yang, J., Chen, Z., Zhang, W., Zhou, L., and Liu, S. (2006). Sonochemical synthesis of nanocrystallite Bi_2O_3 as a visible-light-driven photocatalyst. *Applied Catalysis A: General* 308: 105–110.

38 Sun, J.H., Dong, S.Y., Wang, Y.K., and Sun, S.P. (2009). Preparation and photocatalytic property of a novel dumbbell-shaped ZnO microcrystal photocatalyst. *Journal of Hazardous Materials* 172 (2–3): 1520–1526.

39 Zhang, G.Y., Feng, Y., Xu, Y.Y., Gao, D.Z., and Sun, Y.Q. (2012). Controlled synthesis of mesoporous α-Fe_2O_3 nanorods and visible light photocatalytic property. *Materials Research Bulletin* 47 (3): 625–630.

40 Wang, F., Di Valentin, C., and Pacchioni, G. (2012). Rational band gap engineering of WO_3 photocatalyst for visible light water splitting. *ChemCatChem* 4 (4): 476–478.

41 Tang, J., Zou, Z., and Ye, J. (2004). Efficient photocatalytic decomposition of organic contaminants over $CaBi_2O_4$ under visible-light irradiation. *Angewandte Chemie* 116 (34): 4563–4566.

42 Oshikiri, M., Boero, M., Ye, J., Zou, Z., and Kido, G. (2002). Electronic structures of promising photocatalysts $InMO_4$ (M= V, Nb, Ta) and $BiVO_4$ for water decomposition in the visible wavelength region. *The Journal of Chemical Physics* 117 (15): 7313–7318.

43 Zhou, L., Wang, W., Liu, S., Zhang, L., Xu, H., and Zhu, W. (2006). A sonochemical route to visible-light-driven high-activity $BiVO_4$ photocatalyst. *Journal of Molecular Catalysis A: Chemical* 252 (1–2): 120–124.

44 Zhang, C. and Zhu, Y. (2005). Synthesis of square Bi_2WO_6 nanoplates as high-activity visible-light-driven photocatalysts. *Chemistry of Materials* 17 (13): 3537–3545.

45 Wu, J., Duan, F., Zheng, Y., and Xie, Y. (2007). Synthesis of Bi_2WO_6 nanoplate-built hierarchical nest-like structures with visible-light-induced photocatalytic activity. *The Journal of Physical Chemistry C* 111 (34): 12866–12871.

46 Mekprasart, W. and Pecharapa, W. (2011). Synthesis and characterization of nitrogen-doped TiO_2 and its photocatalytic activity enhancement under visible light. *Energy Procedia* 9: 509–514.

47 Kusiak-Nejman, E., Wojnarowicz, J., Morawski, A.W., Narkiewicz, U., Sobczak, K., Gierlotka, S., and Lojkowski, W. (2021). Size-dependent effects of ZnO nanoparticles on the photocatalytic degradation of phenol in a water solution. *Applied Surface Science* 541: 148416.

48 Atchudan, R., Edison, T.N., Perumal, S., Karthikeyan, D., and Lee, Y.R. (2017). Effective photocatalytic degradation of anthropogenic dyes using graphene oxide grafting titanium dioxide nanoparticles under UV-light irradiation. *Journal of Photochemistry and Photobiology A: Chemistry* 333: 92–104.

49 Teng, W., Li, X., Zhao, Q., Zhao, J., and Zhang, D. (2012). In situ capture of active species and oxidation mechanism of RhB and MB dyes over sunlight-driven Ag/Ag_3PO_4 plasmonic nanocatalyst. *Applied Catalysis B: Environmental* 125: 538–545.

50 Khakpash, N., Simchi, A., and Jafari, T. (2012). Adsorption and solar light activity of transition-metal doped TiO_2 nanoparticles as semiconductor photocatalyst. *Journal of Materials Science: Materials in Electronics* 23 (3): 659–667.

51 Zhang, J., Jiang, Y., Gao, W., and Hao, H. (2015). Synthesis and visible photocatalytic activity of new photocatalyst MBi_2O_4 (M= Cu, Zn). *Journal of Materials Science: Materials in Electronics* 26 (3): 1866–1873.

52 Rezaei, M. and Habibi-Yangjeh, A. (2013). Simple and large scale refluxing method for preparation of Ce-doped ZnO nanostructures as highly efficient photocatalyst. *Applied Surface Science* 265: 591–596.

53 Pascariu, P., Airinei, A., Olaru, N., Olaru, L., and Nica, V. (2016). Photocatalytic degradation of Rhodamine B dye using ZnO–SnO_2 electrospun ceramic nanofibers. *Ceramics International* 42 (6): 6775–6781.

54 Chakrabortty, D. and Gupta, S.S. (2013). Photo-catalytic decolourisation of toxic dye with N-doped Titania: a case study with Acid Blue 25. *Journal of Environmental Sciences* 25 (5): 1034–1043.

55 Phanichphant, S., Nakaruk, A., and Channei, D. (2016). Photocatalytic activity of the binary composite CeO_2/SiO_2 for degradation of dye. *Applied Surface Science* 387: 214–220.

56 Joy, M., Mohamed, A.P., Warrier, K.G., and Hareesh, U.S. (2017). Visible-light-driven photocatalytic properties of binary MoS_2/ZnS heterostructured nanojunctions synthesized via one-step hydrothermal route. *New Journal of Chemistry* 41 (9): 3432–3442.

57 Khaki, M.R., Shafeeyan, M.S., Raman, A.A., and Daud, W.M. (2018). Evaluating the efficiency of nano-sized Cu doped TiO_2/ZnO photocatalyst under visible light irradiation. *Journal of Molecular Liquids* 258: 354–365.

58 Wang, P.Q., Chen, T., Yu, B., Tao, P., and Bai, Y. (2016). Tollen's-assisted preparation of Ag_3PO_4/GO photocatalyst with enhanced photocatalytic activity and stability. *Journal of the Taiwan Institute of Chemical Engineers* 62: 267–274.

59 Guo, F., Shi, W., Lin, X., and Che, G. (2014). Hydrothermal synthesis of graphitic carbon nitride–$BiVO_4$ composites with enhanced visible light photocatalytic activities and the mechanism study. *Journal of Physics and Chemistry of Solids* 75 (11): 1217–1222.

60 Saravanan, R., Sacari, E., Gracia, F., Khan, M.M., Mosquera, E., and Gupta, V.K. (2016). Conducting PANI stimulated ZnO system for visible light photocatalytic degradation of coloured dyes. *Journal of Molecular Liquids* 221: 1029–1033.

61 Liu, X., Li, J., and Yao, W. (2020). CdS@ MoS_2 hetero-structured nanocomposites are highly effective photo-catalysts for organic dye degradation. *ACS omega* 5 (42): 27463–27469.

62 Nasr, M., Balme, S., Eid, C., Habchi, R., Miele, P., and Bechelany, M. (2017). Enhanced visible-light photocatalytic performance of electrospun rGO/TiO_2 composite nanofibers. *The Journal of Physical Chemistry C* 121 (1): 261–269.

63 Zhou, K., Zhu, Y., Yang, X., Jiang, X., and Li, C. (2011). Preparation of graphene–TiO 2 composites with enhanced photocatalytic activity. *New Journal of Chemistry* 35 (2): 353–359.

64 Zhang, X.Y., Li, H.P., Cui, X.L., and Lin, Y. (2010). Graphene/TiO 2 nanocomposites: synthesis, characterization and application in hydrogen evolution from water photocatalytic splitting. *Journal of Materials Chemistry* 20 (14): 2801–2806.

65 Boonprakob, N., Wetchakun, N., Phanichphant, S., Waxler, D., Sherrell, P., Nattestad, A., Chen, J., and Inceesungvorn, B. (2014). Enhanced visible-light photocatalytic activity of g-C3N4/TiO2 films. *Journal of Colloid and Interface Science* 417: 402–409.

66 Wang, Y., Shi, R., Lin, J., and Zhu, Y. (2011). Enhancement of photocurrent and photocatalytic activity of ZnO hybridized with graphite-like C 3 N 4. *Energy & Environmental Science* 4 (8): 2922–2929.

67 Yan, S.C., Lv, S.B., Li, Z.S., and Zou, Z.G. (2010). Organic–inorganic composite photocatalyst of gC_3 N_4 and TaON with improved visible light photocatalytic activities. *Dalton Transactions* 39 (6): 1488–1491.

68 Saravanan, R., Gupta, V.K., Narayanan, V., and Stephen, A. (2013). Comparative study on photocatalytic activity of ZnO prepared by different methods. *Journal of Molecular Liquids* 181: 133–141.

69 Jassby, D., Farner Budarz, J., and Wiesner, M. (2012). Impact of aggregate size and structure on the photocatalytic properties of TiO2 and ZnO nanoparticles. *Environmental Science & Technology* 46 (13): 6934–6941.

70 Gong, Y., Wu, Y., Xu, Y., Li, L., Li, C., Liu, X., and Niu, L. (2018). All-solid-state Z-scheme CdTe/TiO_2 heterostructure photocatalysts with enhanced visible-light photocatalytic degradation of antibiotic wastewater. *Chemical Engineering Journal* 350: 257–267.

71 Ramos-Delgado, N.A., Hinojosa-Reyes, L., Guzman-Mar, I.L., Gracia-Pinilla, M.A., and Hernández-Ramírez, A. (2013). Synthesis by sol–gel of WO_3/TiO_2 for solar photocatalytic degradation of malathion pesticide. *Catalysis Today* 209: 35–40.

72 Reddy, D.A., Ma, R., Choi, M.Y., and Kim, T.K. (2015). Reduced graphene oxide wrapped ZnS–Ag_2S ternary composites synthesized via hydrothermal method: applications in photocatalyst degradation of organic pollutants. *Applied Surface Science* 324: 725–735.

73 Al Kausor, M. and Chakrabortty, D. (2020). Facile fabrication of N-TiO_2/Ag_3PO_4@ GO nanocomposite toward photodegradation of organic dye under visible light. *Inorganic Chemistry Communications* 116: 107907.

74 Beyhaqi, A., Zeng, Q., Chang, S., Wang, M., Azimi, S.M., and Hu, C. (2020). Construction of g-C_3N_4/WO_3/MoS_2 ternary nanocomposite with enhanced charge separation and collection for efficient wastewater treatment under visible light. *Chemosphere* 247: 125784.

75 Jaleel, U.J., Devi, K.S., Madhushree, R., and Pinheiro, D. (2021). Statistical and experimental studies of MoS_2/gC_3N_4/TiO_2: a ternary Z-scheme hybrid composite. *Journal of Materials Science* 56 (11): 6922–6944.

76 Li, L., Yin, D., Deng, L., Xiao, S., Ouyan, Y., Khaing, K.K., Guo, X., Wang, J., and Luo, Z. (2021). Fabrication of a novel ternary heterojunction composite Ag_2MoO_4/Ag_2S/MoS_2 with significantly enhanced photocatalytic performance. *New Journal of Chemistry* 45 (1): 223–234.

77 Zhuang, S., Xu, X., Feng, B., Hu, J., Pang, Y., Zhou, G., Tong, L., and Zhou, Y. (2014). Photogenerated carriers transfer in dye–graphene–SnO_2 composites for highly efficient visible-light photocatalysis. *ACS Applied Materials & Interfaces* 6 (1): 613–621.

78 Zhang, N., Yang, M.Q., Liu, S., Sun, Y., and Xu, Y.J. (2015). Waltzing with the versatile platform of graphene to synthesize composite photocatalysts. *Chemical Reviews* 115 (18): 10307–10377.

79 Liu, X., Pan, L., Lv, T., and Sun, Z. (2013). Investigation of photocatalytic activities over ZnO–TiO_2–reduced graphene oxide composites synthesized via microwave-assisted reaction. *Journal of Colloid and Interface Science* 394: 441–444.

80 Ye, S., Wang, R., Wu, M.Z., and Yuan, Y.P. (2015). A review on g-C_3N_4 for photocatalytic water splitting and CO2 reduction. *Applied Surface Science* 358: 15–27.

81 Mamba, G. and Mishra, A.K. (2016). Graphitic carbon nitride (g-C_3N_4) nanocomposites: a new and exciting generation of visible light driven photocatalysts for environmental pollution remediation. *Applied Catalysis B: Environmental* 198: 347–377.

82 Xu, J., Wang, G., Fan, J., Liu, B., Cao, S., and Yu, J. (2015). g-C_3N_4 modified TiO_2 nanosheets with enhanced photoelectric conversion efficiency in dye-sensitized solar cells. *Journal of Power Sources* 274: 77–84.

83 Xu, J., Brenner, T.J., Chabanne, L., Neher, D., Antonietti, M., and Shalom, M. (2014). Liquid-based growth of polymeric carbon nitride layers and their use in a mesostructured polymer solar cell with V oc exceeding 1 V. *Journal of the American Chemical Society* 136 (39): 13486–13489.

84 Wen, J., Xie, J., Chen, X., and Li, X. (2017). A review on g-C_3N_4-based photocatalysts. *Applied Surface Science* 391: 72–123.

85 Mousavi, M. and Habibi-Yangjeh, A. (2015). Ternary g-C_3N_4/Fe_3O_4/Ag_3VO_4 nanocomposites: novel magnetically separable visible-light-driven photocatalysts for efficiently degradation of dye pollutants. *Materials Chemistry and Physics* 163: 421–430.

86 Potle, V.D., Shirsath, S.R., Bhanvase, B.A., and Saharan, V.K. (2020). Sonochemical preparation of ternary rGO-ZnO-TiO_2 nanocomposite photocatalyst for efficient degradation of crystal violet dye. *Optik* 208: 164555.

87 Zhou, L., Zhang, W., Chen, L., Deng, H., and Wan, J. (2017). A novel ternary visible-light-driven photocatalyst AgCl/Ag_3PO_4/g-C_3N_4: synthesis, characterization, photocatalytic activity for antibiotic degradation and mechanism analysis. *Catalysis Communications* 100: 191–195.

88 Dehghan, S., Jafari, A.J., FarzadKia, M., Esrafili, A., and Kalantary, R.R. (2019). Visible-light-driven photocatalytic degradation of Metalaxyl by reduced graphene oxide/Fe_3O_4/ZnO ternary nanohybrid: influential factors, mechanism and toxicity bioassay. *Journal of Photochemistry and Photobiology A: Chemistry* 375: 280–292.

89 Lin, H., Ye, H., Xu, B., Cao, J., and Chen, S. (2013). Ag_3PO_4 quantum dot sensitized $BiPO_4$: a novel p–n junction Ag3PO4/BiPO4 with enhanced visible-light photocatalytic activity. *Catalysis Communications* 37: 55–59.

90 Sheu, F.J., Cho, C.P., Liao, Y.T., and Yu, C.T. (2018). Ag_3PO_4-TiO_2-graphene oxide ternary composites with efficient photodegradation, hydrogen evolution, and antibacterial properties. *Catalysts* 8 (2): 57.

91 Wang, H., Zou, L., Shan, Y., and Ternary, W.X. (2018). GO/Ag_3PO_4/AgBr composite as an efficient visible-light-driven photocatalyst. *Materials Research Bulletin* 97: 189–194.

92 Lu, B., Ma, N., Wang, Y., Qiu, Y., Hu, H., Zhao, J., Liang, D., Xu, S., Li, X., Zhu, Z., and Cui, C. (2015). Visible-light-driven $TiO_2/Ag_3PO_4/GO$ heterostructure photocatalyst with dual-channel for photo-generated charges separation. *Journal of Alloys and Compounds* 630: 163–171.

93 Zhang, J., Bi, H., He, G., Zhou, Y., and Chen, H. (2014). Fabrication of Ag_3PO_4– PANI–GO composites with high visible light photocatalytic performance and stability. *Journal of Environmental Chemical Engineering* 2 (2): 952–957.

94 Kumar, A. and Pandey, G. (2018). Comparative photocatalytic degradation of rose bengal dye under visible light by TiO_2, TiO_2/PAni and TiO_2/PANI/GO nanocomposites. *International Journal for Research in Applied Science and Engineering Technology* 6: 339–344.

95 Feizpoor, S., Habibi-Yangjeh, A., Yubuta, K., and Vadivel, S. (2019). Fabrication of $TiO_2/CoMoO_4$/PANI nanocomposites with enhanced photocatalytic performances for removal of organic and inorganic pollutants under visible light. *Materials Chemistry and Physics* 224: 10–21.

96 Ma, J., Dai, J., Duan, Y., Zhang, J., Qiang, L., and Xue, J. (2020). Fabrication of PANI-TiO_2/rGO hybrid composites for enhanced photocatalysis of pollutant removal and hydrogen production. *Renewable Energy* 156: 1008–1018.

97 Mohammed, A.M., Mohtar, S.S., Aziz, F., Aziz, M., and Ul-Hamid, A. (2021). Cu_2O/ZnO-PANI ternary nanocomposite as an efficient photocatalyst for the photodegradation of Congo Red dye. *Journal of Environmental Chemical Engineering* 9 (2): 105065.

98 Ali, H. and Mansor, E.S. (2020). Co-sensitization of mesoporous ZnS with CdS and polyaniline for efficient photocatalytic degradation of anionic and cationic dyes. *Colloid and Interface Science Communications* 39: 100330.

99 Jumat, N.A., Khor, S.H., Basirun, W.J., Juan, J.C., and Phang, S.W. (2021). Highly visible light active ternary polyaniline-TiO_2-Fe_3O_4 nanotube/nanorod for photodegradation of reactive black 5 dyes. *Journal of Inorganic and Organometallic Polymers and Materials* 31 (5): 2168–2181.

100 Zhu, C., Zhang, L., Jiang, B., Zheng, J., Hu, P., Li, S., Wu, M., and Wu, W. (2016). Fabrication of Z-scheme Ag_3PO_4/MoS_2 composites with enhanced photocatalytic activity and stability for organic pollutant degradation. *Applied Surface Science* 377: 99–108.

101 Teng, W., Li, X., Zhao, Q., and Chen, G. (2013). Fabrication of $Ag/Ag_3PO_4/TiO_2$ heterostructure photoelectrodes for efficient decomposition of 2-chlorophenol under visible light irradiation. *Journal of Materials Chemistry A* 1 (32): 9060–9068.

102 Pandiselvi, K., Fang, H., Huang, X., Wang, J., Xu, X., and Li, T. (2016). Constructing a novel carbon nitride/polyaniline/ZnO ternary heterostructure with enhanced photocatalytic performance using exfoliated carbon nitride nanosheets as supports. *Journal of Hazardous Materials* 314: 67–77.

103 Chen, F., An, W., Li, Y., Liang, Y., and Fabricating, C.W. (2018). 3D porous PANI/TiO_2–graphene hydrogel for the enhanced UV-light photocatalytic degradation of BPA. *Applied Surface Science* 427: 123–132.

104 Brooms, T.J., Otieno, B., Onyango, M.S., and Ochieng, A. (2018). Photocatalytic degradation of P-Cresol using TiO_2/ZnO hybrid surface capped with polyaniline. *Journal of Environmental Science and Health, Part A* 53 (2): 99–107.

105 Ahamad, T., Naushad, M., Alzaharani, Y., and Alshehri, S.M. (2020). Photocatalytic degradation of bisphenol-A with g-C_3N_4/MoS_2-PANI nanocomposite: kinetics, main active species, intermediates and pathways. *Journal of Molecular Liquids* 311: 113339.

106 Sharma, S., Sharma, G., Kumar, A., Naushad, M., Mola, G.T., Kumar, A., Al-Misned, F.A., El-Serehy, H.A., and Stadler, F.J. (2020). Visibly active FeO/ZnO@PANI magnetic nano-photocatalyst for the degradation of 3-aminophenol. *Topics in Catalysis* 63 (11): 1302–1313.

107 Ansari, M.O., Kumar, R., Alshahrie, A., Abdel-wahab, M.S., Sajith, V.K., Ansari, M.S., Jilani, A., Barakat, M.A., and Darwesh, R. (2019). CuO sputtered flexible polyaniline@graphene thin films: a recyclable photocatalyst with enhanced electrical properties. *Composites Part B: Engineering* 175: 107092.

108 Chen, S., Huang, D., Zeng, G., Xue, W., Lei, L., Xu, P., Deng, R., Li, J., and Cheng, M. (2020). In-situ synthesis of facet-dependent $BiVO_4$/Ag_3PO_4/PANI photocatalyst with enhanced visible-light-induced photocatalytic degradation performance: synergism of interfacial coupling and hole-transfer. *Chemical Engineering Journal* 382: 122840.

109 Shankar, E.G., Aishwarya, M., Khan, A., Kumar, A.K., and Yu, J.S. (2021). Efficient solar light photocatalytic degradation of commercial pharmaceutical drug and dye using rGO-PANI assisted c-ZnO heterojunction nanocomposites. *Ceramics International* 47 (17): 23770–23780.

110 Balasubramanian, J., Ponnaiah, S.K., Periakaruppan, P., and Kamaraj, D. (2020). Accelerated photodeterioration of class I toxic monocrotophos in the presence of one-pot constructed Ag_3PO_4/polyaniline@g-C_3N_4 nanocomposite: efficacy in light harvesting. *Environmental Science and Pollution Research* 27 (2): 2328–2339.

111 Hussen, A. (2021). PANI/g-C_3N_4/CeO_2 nanocomposite for photodegradation of diazinon in aqueous solution. *Middle East Journal of Applied Science & Technology* 4 (1): 01–22.

112 Nekooie, R., Shamspur, T., and Mostafavi, A. (2021). Novel CuO/TiO_2/PANI nanocomposite: preparation and photocatalytic investigation for chlorpyrifos degradation in water under visible light irradiation. *Journal of Photochemistry and Photobiology A: Chemistry* 407: 113038.

113 Tang, M., Ao, Y., Wang, C., and Wang, P. (2020). Facile synthesis of dual Z-scheme g-C_3N_4/Ag_3PO_4/AgI composite photocatalysts with enhanced performance for the degradation of a typical neonicotinoid pesticide. *Applied Catalysis B: Environmental* 268: 118395.

114 Mohanta, D. and Ahmaruzzaman, M. (2021). Facile fabrication of novel Fe_3O_4-SnO_2-gC_3N_4 ternary nanocomposites and their photocatalytic properties toward the degradation of carbofuran. *Chemosphere* 285: 131395.

4

Photocatalytic Removal of Heavy Metal Ions from Water

Sin Ling Chiam[1], Mohsen Ahmadipour[2], and Swee-Yong Pung[1]

[1] *School of Materials and Mineral Resources Engineering, Universiti Sains Malaysia, Nibong Tebal, Malaysia*
[2] *Institute of Microengineering and Nanoelectronics, Universiti Kebangsaan Malaysia, Bangi, Selangor, Malaysia*

4.1 Introduction

Water pollution has significantly affected the environment and the health of living beings [1, 2]. Globally, the main challenge encountered by the water treatment industries is the comprehensive elimination of contaminants such as organic dyes and heavy metals (HMs) from water. This is attributed to the highly complex effluent matrices with contaminants of diverse properties and concentrations. The effluent matrix is dictated by the industrial activities and is site-specific; however, the same cluster of contaminants could be displaying similar effects wherever they coexist [3]. The concentration of metal ions could affect the environment and living organisms [4].

There is no standard definition for HMs, although the term typically refers to dense metal that is usually toxic at low concentrations. Common HMs of concerns are the metals and metalloids of chromium (Cr), cadmium (Cd), mercury (Hg), zinc (Zn), arsenic (As), platinum (Pt), manganese (Mn), nickel (Ni), and lead (Pb). The domestic processes contributing to HM release are automobile exhaust, smelting, burning crude oil, incineration of waste, and utilization of sewage sludge. However, industrial processes such as metal plating, refining ore, mining, tanning, painting, and paper, fertilizer, and pesticide production are the major contributors to the release of HMs into the environment. Generally, HMs are present in soils, sediments, rocks, and anthropogenic materials [4]. The most familiar HMs from ore minerals are galena (PbS), sphalerite (ZnS), chalcopyrite ($CuFeS_2$), and chromite ($FeCr_2O_4$) [5].

In a report by United Nations Education, Scientific, Cultural Organization (UNESCO), human activities have produced more than 80.0% of wastewater worldwide. This wastewater is eventually released to the environment without proper treatment [6]. The growing utilization of a wide range of metals and the illegal discharge of untreated waste into water streams such as rivers, lakes and oceans have caused the spike of HM concentration in

Photocatalysts and Electrocatalysts in Water Remediation: From Fundamentals to Full Scale Applications, First Edition. Edited by Prasenjit Bhunia, Kingshuk Dutta, and S. Vadivel.

water sources [7]. Moreover, as forecast by World Health Organization (WHO), about 50.0% of the world's total population will live in water-stressed areas by the year 2025. Figure 4.1 shows the schematic representation of water pollution originating from industrial effluent.

The HMs at trace levels serve as microelements that are needed by plants and animals, particularly in the cells and tissue components. At higher concentrations, however, the HMs show destructive effects on the health of living beings [8, 9]. HMs are toxic to living organisms and show persistence because of their nonbiodegradability. They also impose bioaccumulation problems in living tissues [10]. Carcinogenesis/cancer, mutagenesis/mutation, neurological disorder, and multiorgan failure are among those severe health damages resulting from HM exposure. Furthermore, an indirect exposure to HM-contaminated water sources could disturb the ecosystem by biological accumulation, thereby resulting in the biomagnification of HMs in food chains [11]. Additionally, with the presence of HMs in water and sediment, it was reported that the concentrations of Zn, Hg, Pb, Cd, Cr, and As could accumulate up to 18.8, 0.1, 0.3, 0.1, 0.1, and 0.1 mg/kg, respectively [12]. Hence, consumers are at high risk of HM exposure by consuming the aquatic-based foods – fish, prawns, crabs, oysters, and other seafood. Consequently, consumers may suffer from diseases/symptoms such as cancer, kidney failure, anaemia, nephritic syndrome, hepatitis, miscarriages, dermatitis/skin irritation, and vomiting. Several disasters arisen from the contamination of HMs in aquatic streams have also been reported. Table 4.1 presents the industrial sources and health effects due to the exposure to different HMs.

Hence, effective wastewater treatment, particularly the removal of HMs from industrial wastewater prior to discharge into the environment, is of utmost importance. To date, various wastewater treatment methods are employed to eliminate and/or recover the toxic metals or metalloids from water, namely ion exchange [15], membrane filtration [16], adsorption [17], chemical precipitation [18], coagulation–flocculation [19], and electrochemical treatment [20]. The comparison of the advantages and disadvantages of these treatments is presented in Table 4.2. These treatments share common limitations, however, which include being tedious and costly, and having the potential to trigger secondary pollution.

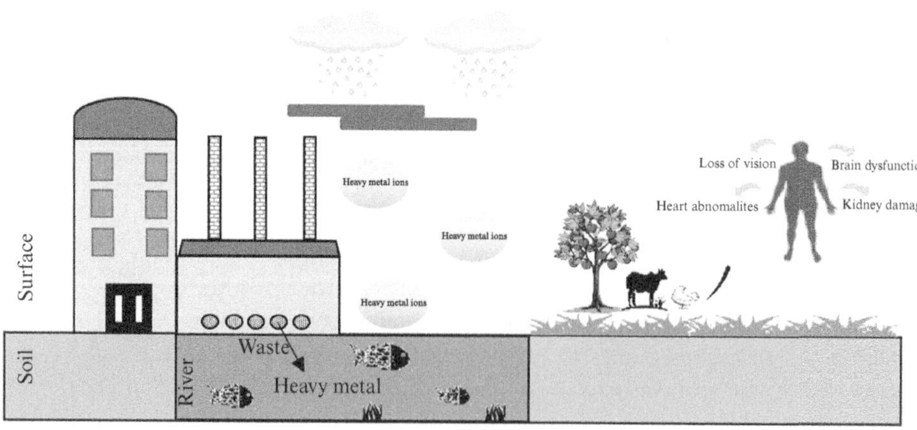

Figure 4.1 How water pollution originates from industrial waste.

Table 4.1 Sources and health effects due to exposure to different HMs. Adapted from [13, 14].

Heavy metal	Sources	Health effects
Chromium (Cr)	- Paint industry - Electroplating - Lather tanning - Data storage - Ferro-alloys manufacturing	- Incidence of respiratory cancer - Skin tumors and cancer - Central nervous system and lung - Hepatic, renal - Stomach cancer and sometimes death
Cadmium (Cd)	- Polymerization - Plastic production - Pigment manufacture - Anti corrosive metal coatings - Plastic stabilizers - Cd-ion batteries	- Bone damage and cancer - Rise in blood pressure - Risks of miscarriage - Damage to DNA and possibly death
Mercury (Hg)	- Pesticides - Petrochemical industry - Chlorine-alkali industry - X-ray tubes - Scientific instruments - Hg vapor lamps - Dental fillings - Pharmaceuticals and fungicides - Catalysts	- Neurobehavioral disorder - Intellectual disability - Attention deficit hyperactivity disorder - Skin rashes - Vomiting - High blood pressure - Tremors - Accumulation in muscle, liver and gills, and even death
Zinc (Zn)	- Mining - Manufacturing processes - Anticorrosion coatings - Rubber industry - PVA stabilizers	- Fever - Restlessness - Diarrhea - Nausea - Asthma
Arsenic (As)	- Wood preservatives - Mining - Animal feed additives, algaecides, herbicides - Pigments - Ceramics, special glasses - Metallurgy - Light filters - Electrical generation - Fireworks	- Acute poisoning - Skin cancer - Liver - Tumors
Platinum (Pt)	- Catalytic converter of modern vehicles - Mining	- Respiratory hypersensitive reaction - Allergic reaction due to halogenated platinum salt

(Continued)

Table 4.1 (Continued)

Heavy metal	Sources	Health effects
Manganese (Mn)	- Mining, industrial discharges and landfill leaching	- Headache - Anorexia - Apathy - Leg cramps - Speech disturbance
Nickel (Ni)	- Electroplating - Paint formulation - Mineral processing - Alloys in steel industry - Computer items - Ni–Cd batteries - Dental and surgical - Arc-welding, rod, and pigments	- Lung cancer - Nasal, sinus - Chronic bronchitis
Lead (Pb)	- Alloys - Ceramics - Plastics - Glassware - Batteries - Cable covers - Sheets, pipes and tubing - Solder	- Effect on brain and nervous system - Abdominal pain - Birth defects - Damage to liver, kidney, and brain - Coma

Table 4.2 Treatment methods for HM removal. Adapted from [21, 22].

Treatment method	Advantage	Disadvantage
Ion exchange	- High removal efficiency - Fast kinetics - High treatment capacity	- Regeneration requirement of resins - Expensive for large amount of wastewater with low concentration of metal ions
Adsorption	- Design flexibility - High efficiency - Simple operation - Space saving - Regeneration possible by proper desorption process	- Expensive active carbon - Performance is absorbent dependent - Works better at low concentration of HMs in wastewater
Membrane filtration	- High efficiency - Simple operation - Space-saving	- Expensive - Complicated process - Low permeate flux - High power consumption

(Continued)

Table 4.2 (Continued)

Treatment method	Advantage	Disadvantage
Chemical precipitation	- Easy and less costly - Sulphide precipitation is better than hydroxide precipitation	- Sulphide precipitation may produce toxic H_2S fumes - Large volume of relatively low density sludge - Mix metal effluent is difficult to remove
Coagulation–flocculation	- Better sludge dewatering	- Complete removal is impossible - Use of various type of chemicals - Sludge generation
Electrochemical	- Rapid and well controlled method - Less sludge generation - 70–100% yield	- High cost investment - Costly electricity supply

Lately, an emerging treatment technique known as heterogeneous photocatalysis has been acknowledged for its removal and recovery of HMs from wastewater. This method utilizes light energy to stimulate semiconductor material to produce electron–hole pairs for the detoxification of contaminants. There are numerous photocatalysts well known for their capability for HM removal from polluted water – for instance, TiO_2 [23], ZnO [24], SnO_2 [25], SnS_2 [25], ZrO_2 [26], CdS [27], and WO_3 [28]. Among them, TiO_2 is the most widely studied photocatalyst [14]. Heterogeneous photocatalysis is a well-known green technology as it can be performed under natural sunlight at ambient temperature and pressure. Besides, photocatalysts used in this method are affordable, less hazardous, and reusable, and have good chemical stability [29, 30], contributing to the popularity of heterogeneous photocatalysts for HM ion removal. The most attractive part of photocatalysis in HM removal is that the HMs are deposited onto the surface of semiconductor material during the photocatalytic process and subsequently can be extracted from the slurry by mechanical and/or chemical means. The photocatalyst may be reused for the subsequent HM removal cycles. Furthermore, precious metals such as silver could be recovered from the wastewater. HM removal by photocatalytic technology is thus an environmentally friendly approach to complement the conventional wastewater techniques.

Great research effort has been channelled into the removal of either HMs or organic contaminants. Nevertheless, these contaminants coexist in industrial wastewater and the environment/natural waters. In the past 11 years, a total of 6951 studies on the simultaneous removal of HMs and organic contaminants have been published by Elsevier. The numbers have accelerated since 2018, as illustrated in Figure 4.2, suggesting growing attention on this area globally.

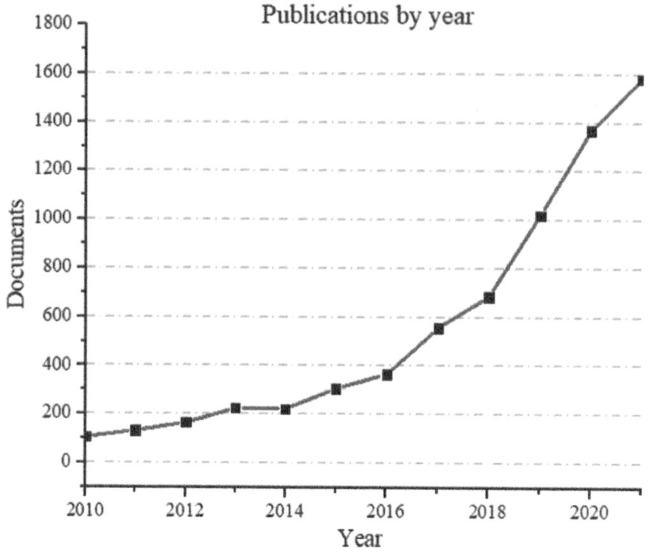

Figure 4.2 Number of publications based on Elsevier database (2010–2021) on the simultaneous elimination of HMs and organic contaminants from the environment.

4.2 Mechanistic Insights on Photocatalytic Removal of Heavy Metal Ions

Photocatalysis is a phenomenon that connects both photochemistry and the catalysis process [1]. In the photocatalysis process, light energy is converted into chemical energy. This is the most crucial feature for this technology. By using appropriate optical excitation sources, the semiconductor photocatalysts are activated, and the charge carriers' transfer within the photocatalyst itself are induced. Contaminants such as HM ions can then be removed through oxidation or reduction. Generally, the entire photocatalytic process involves three steps: charge generation, separation, and consumption [2–4].

The illustration of the general photocatalytic process is shown in Figure 4.3. Photocatalysis reaction starts when a light source with energy larger than the bandgap energy of the semiconductor photocatalyst is irradiated on its surface. Subsequently, optical excitation occurs where electrons from the filled valence band (VB) are excited to the empty conduction band (CB), forming electron–hole pairs. The electron–hole pairs are then migrated to the surface of the photocatalyst for a series of redox reactions. In this process, the recombination of the electron–hole pairs must be minimized so that the photoinduced reactions between the generated holes and the reductant, or the active electrons and the oxidant, can be maximized. The amount and type of reactive radicals created in a photocatalytic reaction are photocatalyst system-dependent. The surface defects in the photocatalyst, morphological and compositional properties of the photocatalysts, types of the target metal ions, and the optical excitation sources could affect the amounts and the types of reactive radicals generated [31].

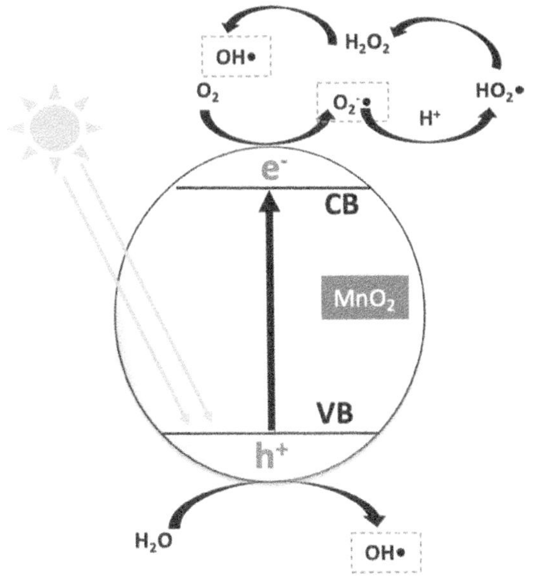

Figure 4.3 Schematic diagram showing a general photocatalytic process [31].

The mechanism of the photocatalytic removal of HM ions involves (a) diffusion of target HM ions from the liquid wastewater toward the surface of the photocatalyst, (b) adsorption of HM ions on the surface of the photocatalyst, and (c) occurrence of a series of photoredox reactions (reduction/oxidation) on the surface of the photocatalyst [8]. Firstly, the HM ions adsorption process is greatly influenced by the surface charge of the photocatalyst. In general, semiconductor metal oxide photocatalysts are negatively charged due to the presence of negatively charged -OH groups; hence, they are able to attract the positively charged metal ions. The pH of wastewater, however, could alter the surface charge properties of a photocatalyst and hence the effectiveness of HM ion adsorption on its surface. In a strong acid medium, the positively charged surface of a photocatalyst tends to repel the positively charged metal ions. On the contrary, the surface charge of a photocatalyst is negative in alkaline solution; this favors the attraction of positively charged metal ions. For instance, it is noted that a highly acidic condition such as pH 2 could produce positively charged β-MnO_2 particles, deteriorating the removal efficiency of Ag^+ ions from solution. Although the absorption process is favored in alkaline solutions, the recombination of electron–hole pairs is likely, which is not desired. Striking a balance between the electrostatic attraction and the photoredox reaction is thus the key factor in the removal of HM ions by photocatalysis.

Secondly, the photoredox reaction by photocatalyst either through reduction or oxidation is dependent on the reduction potential of target metallic couples and the energy levels of the semiconductor. The energy levels of semiconductor photocatalysts can be categorized into wide bandgap e.g. TiO_2, ZnO, SnO_2, ZrO_2, and ZnS, or narrow bandgap e.g. Fe_2O_3, WO_3, CdS, and CdSe. The relative energy levels of major semiconductor photocatalysts are shown in Figure 4.4.

The electrons either migrate to the surface of the photocatalyst or to the reacting solution to reduce the cationic metal ions to a lower oxidation state, i.e. from +1 oxidation state to 0 oxidation state (M^+ to M^0) (Figure 4.5); whereas, the oxidation of metal ions selectively occurs when the oxidation potential of the metal is less positive than the VB level of the semiconductor photocatalysts [9]. In this case, the photogenerated holes or hydroxyl radicals ($^{\bullet}OH$) can oxidize metal ions to a higher oxidation state. In short, for a photoredox reaction to happen, the desired redox reaction potential must lie within the band edge of the photocatalyst. Besides, the electrons and holes can be generated continuously as long

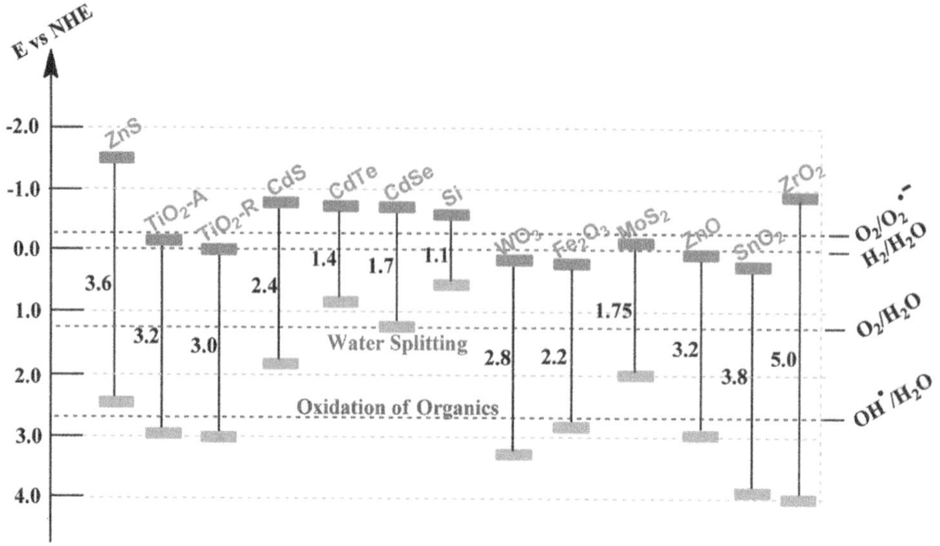

Figure 4.4 Bandgap energy, VB, and CB for a range of semiconductors on a potential scale (E) versus the normal hydrogen electrode (NHE) [32].

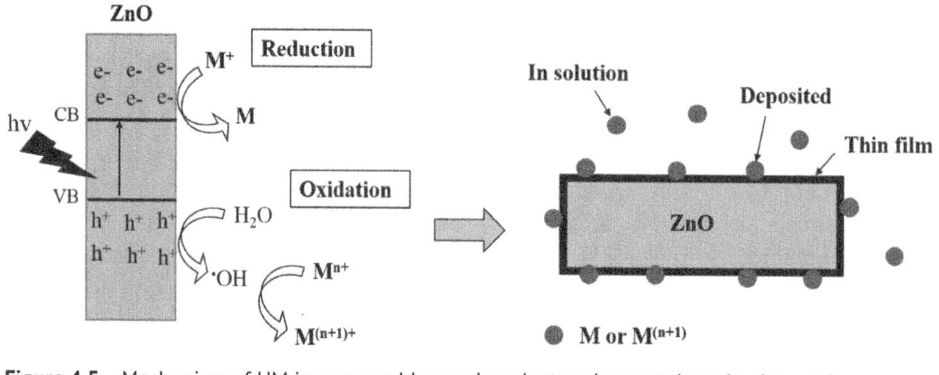

Figure 4.5 Mechanism of HM ion removal by semiconductor photocatalyst via photoredox reactions. Adapted from [33].

as there is an optical excitation source available for the reduction or oxidation of metal ions. Then, the targeted metal ions that are being reduced or oxidized through photocatalytic reactions could either form a thin film on the surface of the photocatalyst, deposit as particles on the surface of the photocatalyst, or precipitate as particles in the solution. Later, they can be recovered from the slurry by mechanical and/or chemical means, such as biphasic catalysis and ion exchange [6, 10], for reuse. Also, precious metals of economical values such as silver can be retrieved from wastewater [7, 10].

In cases where the energy of the light source is insufficient, photoredox reactions will not occur by light irradiation. This is mainly due to the incompatibility of the standard reduction potential of the target metal ions with the photocatalyst. Under such circumstances, adsorption processes will be dominant. The negatively charged semiconductor metal oxide

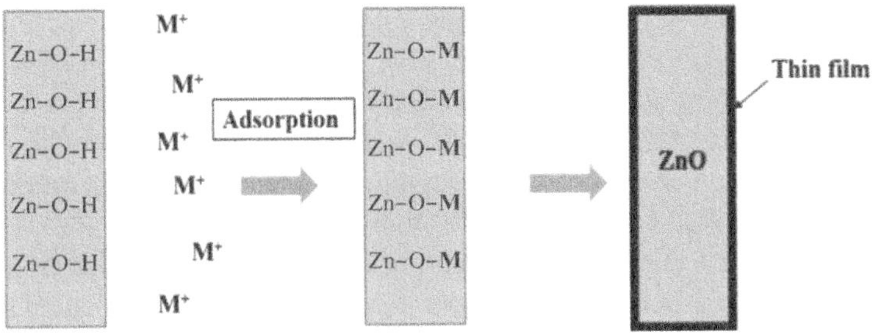

Figure 4.6 Mechanism of HM ion removal by semiconductor photocatalyst via adsorption process. Adapted from [34].

photocatalysts contributed by the -OH groups become the actively adsorptive sites. The cationic HMs in the aqueous solution tend to react with -OH groups and become deposited on the surface of the photocatalyst as illustrated in Figure 4.6. For instance, Thi et al. have reported poor removal of Cd^{2+} or Ni^{2+} ions by a ZnO photocatalyst under UV and visible light irradiation [34]. The UV irradiation did not result in effective removal of Cd^{2+} or Ni^{2+} as compared to other metal ions being studied. The negative standard reduction potential of Ni^{2+}/Ni and Cd^{2+}/Cd were −0.43 V and −0.25 V respectively [35, 36]. As a consequence, the photochemical reduction of Cd^{2+} or Ni^{2+} could not take place. The slight removal of Cd^{2+} or Ni^{2+} ions was due to the adsorption process. When the adsorption process was limited by the number of negative adsorptive sites on the surface of ZnO NRs, the adsorption efficiency was usually poor and might reach a saturation level after a period of time.

In short, the removal of metal ions by a photocatalyst through adsorption, photoredox, or a combination of both depends on the light source and the standard reduction potential between the target metal ions and the photocatalyst.

4.3 Solar/Visible-light Driven Photocatalysts for the Removal of Heavy Metal Ions

The use of various types of photocatalysts for the removal of organic and inorganic contaminants in aqueous solutions is extensively reported. Several semiconductors including TiO_2 [23], ZnO [24], SnO_2 [25], SnS_2 [25], ZrO_2 [26], CdS [27], and WO_3 [28] have been studied as photocatalysts for the removal of toxics in water and wastewater through photocatalytic processes [37–40]. Owing to the attractive characteristics, i.e. being less toxic, economical, available in abundance, environmentally friendly, chemical and thermally stable, as well as having satisfactory optical and electrical properties, titania (TiO_2) and zinc oxide (ZnO) are the two most widely explored photocatalysts. Under UV light exposure, both TiO_2 and ZnO exhibit good photocatalytic activities and are found to be comparably effective as other semiconductor materials [41].

Although the TiO_2 and ZnO photocatalysts display good stability and excellent properties, they have limited activity under visible light exposure due to their wide bandgap (~3.3–3.4 eV). This is one of the major drawbacks when employing these photocatalysts, since UV light only makes up a small portion of the solar spectrum i.e. about 4.0%. The development of an active photocatalyst driven by visible light has been greatly desired by researchers. In order to activate photocatalysts under visible light, it is essential to modify their structure or composition. For this purpose, a few approaches have been developed. These approaches include doping [42], sensitizing with organic dyes [43], and coupling with another narrow bandgap semiconductor [44]. By doping, either the bandgap energy of photocatalysts could be reduced [44] or an energy level induced in the bandgap [45–47]. Examples of coupled photocatalysts that are responsive to visible light are WO_3-doped TiO_2 [48], Fe_2O_3-doped ZnO [44], SnO_2-doped SnS_2 [25], Bi_2O_3-doped ZrO_2 [49], ZnO-doped CdS [50], and rGO-doped WO_3 [28].

Dye sensitization is one of the approaches to enable visible light absorption on the surface of wide bandgap semiconductors [51, 52]. The electrons of adsorbed dye molecules are excited to the CB of a host semiconductor upon visible light illumination. This injection is favorable only when the lowest unoccupied molecular orbital (LUMO) of the dyes has a more negative potential as compared to the CB potential of TiO_2. Various dyes such as 8-hydroxyquinone (HOQ), methylene blue (MB), acid red 44, reactive red dye 198 (RR 198), eosin-Y, mer-bromine, 2′,7′-dichlorofluorescein, Rhodamine B and Rhodamine 6G have been used to sensitize TiO_2 particles to visible light [53–57]. Dye sensitization is a powerful tool to make TiO_2 activated under visible light.

For HM removal, coupling of TiO_2 and ZnO with narrow bandgap semiconductors that are able to absorb visible light is one of the promising approaches; possible choices include Bi_2S_3 [30], CdS [13, 14], CdSe [21], and V_2O_5 [22]. This is a type of surface modification where the principle is similar to dye sensitization except that the sensitizer is a narrow bandgap semiconductor. The coupled narrow bandgap semiconductor is responsible for absorbing the visible light and injecting electrons into the CB of visible-light inactive TiO_2 and ZnO. The injected electrons then migrate to the surface of the TiO_2 photocatalyst for a series of photoredox reaction as shown in Figure 4.7.

The visible-light absorption activity of coupled semiconductors is greatly affected by two main factors: (a) the position of the band edges of the two coupled semiconductors and (b) the visible-light absorption ability of the narrow bandgap semiconductor (sensitizer). Firstly, the efficiency of electron injection between the coupled semiconductors depends on the difference between the respective CB and VB potentials. For an efficient coupled system, the CB of the semiconductor as sensitizer should be higher than that of its

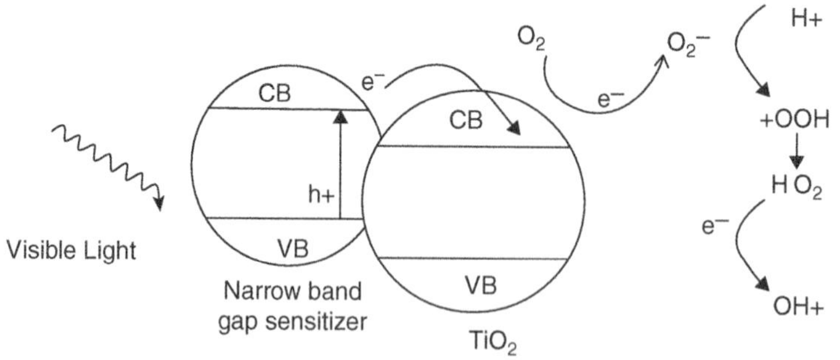

Figure 4.7 Visible light activation of TiO_2 by coupling with a narrow bandgap semiconductor [58] / Elsevier.

counterpart semiconductor, and its VB needs to be lower than that of its counterpart semiconductor [21, 29, 30]. The larger the difference in the band potentials of the coupled semiconductors, the higher is the interfacial charge transfer and vice versa. Secondly, the higher the wavelength at which the semiconductor (sensitizer) is able to absorb visible light, the higher the visible light activity of the coupled system. For instance, the CB potentials of TiO_2, Bi_2S_3, CdS, and CdSe versus NHE at pH 7 are −0.50 V, −0.76 V, −0.95 V, and −1.0 V, respectively [21, 30]. The CB minima of CdS is much higher than that of Bi_2S_3, but the visible light absorption activity of a CdS/TiO_2 coupled system is reported to be much lower than that of Bi_2S_3/TiO_2 [30]. This is due to the fact that Bi_2S_3 can absorb light up to wavelengths of 800 nm, but CdS only absorbs light up to 600 nm.

Despite the merits, the photocorrosion issue of a coupled semiconductor system under UV irradiation needs to be considered. Both semiconductors are activated under UV light, increasing the number of electrons (from self-excitation and injection from sensitizer) in the CB of the host semiconductor. Meanwhile, due to the position of band edges, photogenerated holes of the host semiconductor can be transferred to the sensitizer, increasing the holes concentration in the sensitizer VB [30]. This results in electron–hole recombination due to an excess of charge carriers and significantly reduces the photocatalytic activity. Besides, holes that are left behind in the VB of the sensitizer tend to photocorrode. Photocorrosion of CdS to Cd has been reported for a CdS/TiO_2 system when irradiated with visible light [13, 14].

Thirdly, modification of the bandgap of TiO_2 and ZnO by doping can enhance the visible light absorption activity. In general, doping methods can be classified into two groups: nonmetal doping and transition metal doping. Nonmetal doping uses nitrogen [59–62], halogen [63, 64], sulphur [65, 66], boron [67, 68], and carbon [69, 70] as dopant, whereas transition metal doping typically uses Cu, Co, Mn, and Ni [71–74]. Doping effectiveness is primarily determined by the chemical state and location of the dopants. Between them, nonmetal doping has gotten more research attention. Through doping, the electronic structure of a host semiconductor is modified to extend its absorption into the visible light region. In most cases, however, although more light can be absorbed, doping weakens the photocatalytic activity of the host semiconductor compared to that under UV light irradiation. This could be due to several reasons. Firstly, the introduced dopants can act as recombination sites that reduce the number of active electrons and/or holes for photocatalysis. Secondly, the discrete interstitial states or novel CB and/or VB edges can reduce the reduction or oxidation potential of the modified semiconductor compared to the undoped material [75].

Codoping of nonmetals has produced appealing results for attempts to increase the visible light response of wide bandgap semiconductor photocatalysts. Meanwhile, semiconductor coupling techniques are greatly dependent on the amount of sensitizer semiconductor. The drastic decrease in sensitizer due to photocorrosion will decrease the visible light activity of a host semiconductor. Overall, the progress made till now is promising and has further broaden codoping's applicability in photocatalysis.

Generally, photocatalysis demonstrates numerous benefits, which include zero secondary waste production, high degradation rate, time efficiency, mineralization of toxic compounds, ambient operating conditions, recyclability of the photocatalyst, utilization of renewable sources (solar light), lower cost, and "green" approach. The removal efficiency of semiconductor photocatalysts is influenced considerably by a number of factors such as the initial concentration of metal ions, pH of solution, dosage of catalyst, and contact time. The performance of various photocatalysts is summarized in Table 4.3. The following sections will discuss the factors that affect the HM removal efficiency of semiconductor photocatalysts.

Effect of initial concentration: Initial concentration of metal ions is one of the crucial factors that need to be taken into consideration. Several research works have reported decreased metal ion removal with the increase of the initial concentration of metal ions [36, 76]. For example, in a study on screen printed TiO_2 film for Cr removal, the removal efficiency of Cr ion was reduced by 13.0% (from 99.0% to 86.0%) when the initial concentration of Cr ion was increased from 10.0 to 50.0 mg/l [36]. By using a fixed amount of catalyst, the reactive sites for the adsorption of metal ions and the concentration of reactive species (hydroxyl radicals, •OH and super-oxide, $\bullet O_2$) created during the photocatalytic process were preserved. Hence, the removal efficiency was predicted to reduce with an increase in the initial concentrations of metal ions. This could possibly be due to the full coverage of the photocatalyst surface by the metal ions, and subsequently the light was absorbed by the metal ions instead of by the photocatalyst during the irradiation exposure. This phenomenon limits the number of photons that reach the surface of the photocatalyst, thereby reducing the number of charge carriers such as electrons and holes and •OH generation, since the reactive sites were filled by metal ions and inaccessible for photocatalysis to take place [77, 78].

Effect of pH: The pH of the solution is another vital parameter in determining the removal efficiency of HM ions. The photocatalytic reduction of Cr at different pH is shown in Table 4.3, and the variation of pH shows a visible effect on the photocatalytic reduction of Cr. A higher removal efficiency was recorded in the acidic solutions in contrast to the alkaline solutions. The photocatalytic reduction efficiency remarkably decreased with increasing solution pH. For instance, Wu and Dhimana observed 99.5% and 89.0% of reduction efficiency respectively at pH 2 after 60 min of visible light illumination, whereas 77.0% and 15.0% of Cr (VI) was reduced at pH 10 within the same time period [44, 76]. In another study, Emadian et al. reported that the Cr solution adsorbed more efficiently by $CoFe_2O_4$-doped ZrO_2 in acidic conditions (pH 2) with 68.3% removal efficiency, whereas in alkaline conditions (pH 8) [76], the removal efficiency was 39.8%. This indicates acidic conditions are more favorable for Cr photoreduction. The pH is a fundamental factor in chromium reduction as it governs the Cr (VI) species present: $HCrO_4^-$, CrO_4^{2-}, and $Cr_2O_7^{2-}$. The improvement in photocatalytic reduction in acidic environments is due to the high concentration of H^+ ions, which promotes the photoreaction as shown in Eqs. (4.1)–(4.2). On the contrary, low concentration of H^+ ions in basic media restrains the above reactions. Besides, $Cr(OH)_3$ formed in alkaline solutions may also precipitate on the surface of the photocatalyst and decrease the reduction efficiency by covering the active sites and reducing the light penetration to the surface of the photocatalyst as shown in Eq. 4.3.

$$14H^+ + Cr_2O_7^{2-} + 6e^- \rightarrow 2Cr^{3+} + 7H_2O \tag{4.1}$$

$$7H^+ + HCrO_4^- + 3e^- \rightarrow Cr^{3+} + 4H_2O \tag{4.2}$$

$$CrO_4^{2-} + 4H_2O \rightarrow Cr(OH)_3 + 5OH^- \tag{4.3}$$

Table 4.3 Performance of various photocatalysts for HM removal in the present of sunlight and visible light.

Material	Shape	Method	Size(nm)	Metal removal	Initial concentration(mg/l)	pH	Light	Dosage (gl^{-1})	Removal efficiency (%)	Time(min)	Ref.
CNC/TiO_2	Necklace-like	Hydrolysis	-	Cr	10	3	VL	1.0	85.0	80	[23]
								2.0	92.0		
								3.0	100.0		
								3.0	33.0	10	
									55.0	20	
									81.0	40	
									95.0	60	
WO_3/TiO_2	-	Sol-gel	-	Cr	-	4	VL	0.1	90.0	360	[48]
Ag_2O/TiO_2	-	Chemical precipitation	-	Ag	10	4	VL	0.3	60.0	10	[79]
									55.0	30	
									71.0	60	
									76.0	90	
									83.0	120	
Ga_2O_3/TiO_2	Rod-like	Hydrothermal	11–22	Cr	-	-	Sunlight	2.0	10.0	-	[81]
				Pb					17.0		
				Zn					25.0		
N/TiO_2	Nanofiber	Electrospinning	-	As	10	7	VL	50 mg	100.0	90	[78]
G/TiO_2	-	Hydrothermal	-	Zn	21	4	Sunlight	2.0	51.0	60	[80]
ZnO	Nanoparticle	Sol-gel	8–18	Hg	100	4	VL	0.4	10.0	50	[24]

(Continued)

Table 4.3 (Continued)

Material	Shape	Method	Size(nm)	Metal removal	Initial concentration(mg/l)	pH	Light	Dosage (gl^{-1})	Removal efficiency (%)	Time(min)	Ref.
Pd/ZnO	-	Sol-gel	-	Hg	100	4	VL	0.4	40.0	50	[24]
CNT/Pd/ZnO	-	Sol-gel	-	Hg	100	4	VL	0.4	100.0	50	[24]
CNT/Pd/ZnO	-	Sol-gel	-	Hg	100	4	VL	0.2	45.0	30	[24]
								0.4	59.0	30	
								0.8	75.0	30	
								1.2	100.0	30	
Fe_2O_3/ZnO	Nanoparticles	Solution combustion	-	Cr	20	2	VL	-	89.0	60	[44]
						4			72.0		
						6			27.0		
						10			15.0		
SnO_2	Nanoparticles	Hydrothermal	-	Cr	50	-	VL	0.3	3.0	80	[25]
F/SnO_2	Nanoparticles	Hydrolysis	75–100	Cr	-	-	VL	20 mg	50.0	150	[82]
SnS_2	Nanoparticles	Hydrothermal	-	Cr	50	-	VL	0.3	58.0	80	[25]
SnO_2/ SnS_2	Nanoparticles	Thermal oxidation	-	Cr	50	-	VL	0.3	40.0	10	[25]
									60.0	20	
									80.0	40	
									95.0	60	
									100.0	80	
TiO_2/SnS_2	Hexagonal nanoplates	Hydrothermal	-	Cr	50	-	VL	1.0	87.0	150	[83]
ZrO_2	Nanotube	Anodization	-	Cr	5	2	VL	1.0	100.0	300	[26]

Bi_2O_3-ZrO_2	Nanoparticles	-	-	Cr	0.1mM	2	VL	0.8	40.0	180	[49]
								1.0	81.0		
								1.3	92.0		
								1.3	35.0	30	
									58.0	60	
									73.0	90	
									82.0	120	
									89.0	150	
$CoFe_2O_4/ZrO_2$	Nanoparticles	Hydrothermal	47	Cr	20	2	VL	20 mg	68.0	120	[76]
						4			54.0		
						6			47.0		
						8			39.0		
					20	2		20 mg	62.0		
								5 mg	59.0		
								10 mg	45.0		
					10	2		20 mg	82.0	180	
					20				75.0		
					30				62.0		
					40				54.0		
					50				29.0		
CdS	Nanoparticles	-	10–20	Cr	-	10	VL	30 mg	13.0	60	[27]

(Continued)

Table 4.3 (Continued)

Material	Shape	Method	Size(nm)	Metal removal	Initial concentration(mg/l)	pH	Light	Dosage (gl^{-1})	Removal efficiency (%)	Time(min)	Ref.
ZnO/CdS	Nanoparticle/ nanorod	Wet chemical	-	Cr	-	-	VL	-	85.0	60	[50]
rGO/CdS	-	-	-	Cr	100	-	Sunlight	20 $mg.ml^{-1}$	> 90.0	15	[84]
$ZnFe_2O_4$/CdS	Nanoparticle/ nanorod	Ultrasonic	-	Cr	-	2	VL	25 mg	20.0	120	[85]
$CuInS_2$/CdS	Irregular particles	Hydrothermal	80–100	Cr	-	10	VL	30 mg	100.0	60	[27]
WO_3	Grainy microparticles	Sol-gel	-	Cr	20	3	VL	30 mg	50.0	120	[40]
rGO/WO_3	Nanoplates/ sheets	Hydrothermal		Cr	20	-	VL	50 mgl^{-1}	90.0	360	[28]

UV: Ultraviolet, VL: Visible light, CNC: Cellulose nanocrystal

Effect of dosage: The dosage of photocatalyst also plays a vital role on the photocatalytic reduction of HM ions under visible light illumination. Generally, a higher dosage of photocatalyst provides greater surface area (i.e. more reactive sites). This enables more photons to be absorbed by the photocatalyst and subsequently generates more reactive species such as electrons, holes, and •OH radicals when illuminated. Consequently, this leads to an increase in the photodegradation rate of contaminants in water. For example, Mohamed et al. [24] synthesized a CNT/Pd/ZnO catalyst by the sol-gel method and employed a range of catalyst dosages (0.2, 0.4, 0.8, and 1.6 g L^{-1}) to remove Hg (100.0 mg L^{-1}). The Hg removal efficiency was enhanced by increasing catalyst dosage. Moreover, other studies also showed a similar trend in photocatalytic reduction of HM ions [23, 49, 76].

Effect of contact time: Another crucial factor in the photocatalysis process is the contact time, with a boost in the removal efficiency revealed at a longer contact time [79]. In the same study of green synthesized silicon nanocomposites, experiments were performed by varying contact time (0–120 min) at pH 4 while keeping all other parameters constant. The removal efficiency increased from 60.0% to 83.0% with time. By prolonging the UV irradiation time, more metal ions were reduced to metals and precipitated on the surface of the photocatalysts. The photodegradation efficiency was thus increased slowly until equilibrium was reached. A similar trend was also observed in other reported research works [25, 49].

Effect of light intensity: The photocatalytic reaction has a photonic nature in which the overall photocatalytic rate depends on the intensity of the light source. To achieve a high photocatalytic reaction rate for wastewater treatment, a relatively high light intensity is required to provide photocatalyst particles with enough photons to excite the electrons of the VB to the CB, and leave positive holes behind in the VB. These charge carriers further help in generating free radicals. Kumordzi et al. [80] investigated the impact of solar light intensity (between 25.0 and 100.0 mW/cm^2) on photoreduction of zinc ions by using hydrothermally synthesized graphene based TiO_2. It was observed that the zinc ion removal efficiency was reduced with the decrease of light intensity.

Effect of dopants: Dopants are the impurities that are introduced into the photocatalyst to improve its photocatalytic efficiency. Dopants help in photocatalytic reactions in the following manner: (a) by altering the bandgap of photocatalyst to avoid recombination of charge carriers [22], (b) by introducing impurity energy levels in the bandgap [77], (c) by trapping the electrons or holes [78], (d) by generating oxygen-deficient sites [44], and (e) by creating more active sites for the adsorption of pollutant molecules [23]. Introduction of extra energy levels through dopant addition helps in separating the charge carriers by trapping photo-induced electrons or holes. This also helps in modifying the bandgap of semiconductor photocatalysts. Additional energy levels, narrowing of the bandgap, and oxygen vacancies promote the adsorption of photons at visible regions to enhance the rate of photocatalytic reactions [49].

Alloy elements can be incorporated into the semiconductor metal oxide material either at interstitial sites or by substitution. For interstitial incorporation, the dopant radius should exhibit a smaller radius than the lattice space of the semiconductor. For substitutional mode incorporation, however, dopant substitutes the lattice oxide or ion. Optimization of dopant concentration should also be taken into consideration. Increasing the concentration of dopant beyond the optimum limit reduces the photocatalytic

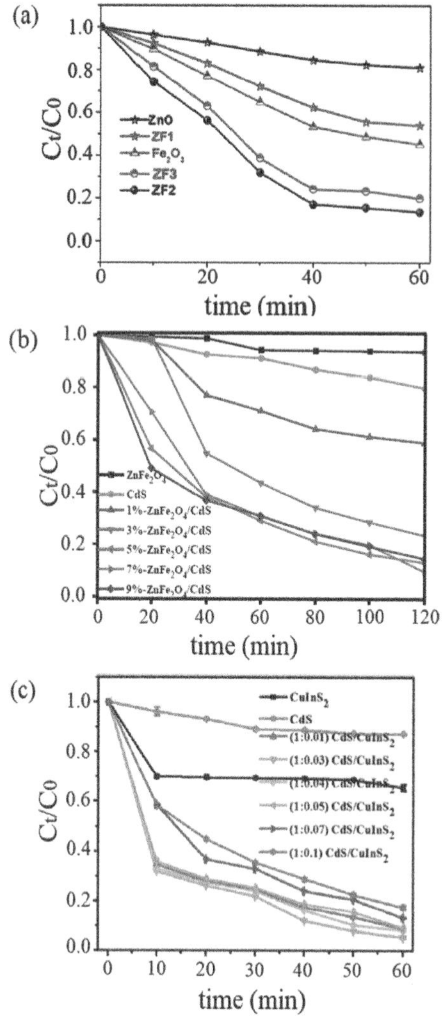

Figure 4.8 Effect of metal oxide doping on removal of Cr in visible light (a) ZnO/xFe_2O_3, (b) x$ZnFe_2O_4$/CdS, and (c) Cds/x$CuInS_2$ [27, 44, 85] / Elsevier. (Remark: ZF = ZnO/Fe_2O_3).

efficiency by (a) reducing the active surface area of the photocatalyst and (b) enhancing the photo-induced charge carrier recombination by narrowing the space charge region [48, 79].

For instance, Dhiman et al. [44] optimized the concentration of Fe_2O_3 (dopant) in ZnO for photocatalytic degradation of Cr. As shown in Figure 4.8, by changing the coupling concentration of Fe_2O_3 in ZnO, the degradation efficiency was improved. This was because Fe_2O_3 could act as charge trapping centers that prolonged the charge separation duration, which was needed for effective removal of metal ions. Furthermore, the coupling of Fe_2O_3 to ZnO was able to narrow the resultant bandgap of hybrid photocatalysts following the sequence of Fe_2O_3 (2.20 eV) > ZnO/ Fe_2O_3 (ratio 3:0.25) (ZF3) (2.24 eV) > ZnO/ 2Fe_2O_3 (ratio 3:0.5) (ZF2) (2.47 eV) > ZnO/ Fe_2O_3 (ratio 3:1) (ZF1) (2.82 eV) > ZnO (3.20 eV). Further increase of the dopant concentration, however, significantly reduced the removal efficiency of the photocatalyst. The dopant ions in the lattice spaces acted as charge carrier recombination centers, causing the deterioration of photocatalytic performance. The photoreduction from Cr under visible light without sacrificial agent at pH 4 for 60 min of exposure is as follows: ZF2 (87.5%) > ZF3 (80.1%) > Fe_2O_3 (54.8%) > ZF1 (46.2%) > ZnO (15.3%). The same trend was observed by other researchers [31, 53] as shown in Figure 4.8. Among factors affecting HM ion removal by photocatalysts, doping is an important effect due to the enhancement of electron movement by reduction of bandgap energy.

Effect of surface area: The last crucial factor for photocatalyst material is to have large surface area. It is preferred in removal of HMs as large surface area allows better light adsorption and provides more adsorption sites for photocatalytic reduction of HM ions. For example, Vignesh et al. [49] showed the effect of surface area on Cr removal in the presence of Bi_2O_3-ZrO_2 nanocomposite. The surface area of the Bi_2O_3-ZrO_2 nanocomposite (37.5 m^2/g) was larger than that of bare metal oxides, i.e. ZrO_2 (19.6 m^2/g) and Bi_2O_3 (18.2 m^2/g). The increase in surface area was due to the existence of a mixed metal oxide (Bi_2O_3 and

ZrO_2) phase. They recorded a 92.3% Cr removal, higher than Bi_2O_3 (65.6%) and ZrO_2 (40.0%) after 180 min visible light illumination. Fang et al. [85] reported the same trend for photoreduction of Cr based $ZnFe_2O_4$/CdS nanocomposites. The surface area of $ZnFe_2O_4$/CdS nanocomposites was 106.8 m^2/g, which was larger than CdS (32.6 m^2/g) and $ZnFe_2O_4$ (29.7 m^2/g). Consequently, the highest Cr photoreduction (90.0%) was obtained after 120 min of visible light illumination.

4.4 Selective Heavy Metal Ion Removal by Semiconductor Photocatalysts

Semiconductor photocatalysts display selective HM removal, depending on the band edge positions of the semiconductor photocatalysts in the NHE diagram, the types of HMs, and the excitation light sources.

4.4.1 ZnO

ZnO is one of the semiconductor photocatalysts showing high potential for environmental pollutant removal. ZnO provides numerous surface sites that are active for the adsorption of HM ions from aqueous solutions. There are plenty of studies reported on the use of porous micro or nanosized-ZnO with different morphologies including microspheres, nanoassemblies, nanoplates, nanorods, and nanosheets as the adsorbents for HM ion removal [86–90].

For example, porous ZnO nanoassemblies have been used by Singh et al. [87] for the removal of several toxic metal ions (i.e. Co, Ni, Cu, Cd, Pb, Hg, and As) from wastewater. In comparison to other HM ions, Hg, Pb, and As demonstrated stronger attraction toward ZnO nanoassemblies because of their high electronegativity, thereby showing better removal efficiency (i.e., 63.5% Hg, 100% Pb, and 100% As). Meanwhile, Thi et al. [34] synthesized ZnO submicron rods via the sol solid precipitation approach and investigated the removal efficiency of HM ions using ZnO particles in the presence of UV light (Figure 4.9). Based on Figure 4.9a, ZnO particles were found capable of removing the HM ions at different efficiencies with respect to time. The results showed 100.00% removal efficiency for Cu^{2+} ions, whereas it was about 97.9% and 85.1% removal efficiencies for Pb^{2+} and Cr^{4+} ions respectively, with the use of ZnO particles. However, the removal efficiencies of Ag, Mn, Cd, and Ni ions were found to be < 10.0 %, thereby indicating poor efficiency of ZnO particles in removing those particular HM ions.

Mesoporous spherical ZnO nanoassemblies fabricated via a soft chemical method was utilized by Singh et al. [87] to explore the removal of HM ions (Figure 9c). It was reported that with the use of 200 mg of ZnO nanoassemblies, the removal efficiencies of Cd, Co, Ni, Cu, and Hg ions were 7.0%, 15.5%, 17.8%, 25.0%, and 63.5%, respectively. At a similar dosage, the removal efficiencies of Pb and As ions were 100.0% using ZnO nanoassemblies. Interestingly, the preference of general hydrous solids for metals was related to the electronegativity of metals [91]. The values of electronegativity were noted to be 1.68, 1.88, 1.91, 1.95, 2.00, 2.33, and 2.18 for Cd, Co, Ni, Cu, Hg, Pd, and As ions, respectively. Based on the high electronegativity values of Hg, Pd, and As ions, these metal ions were

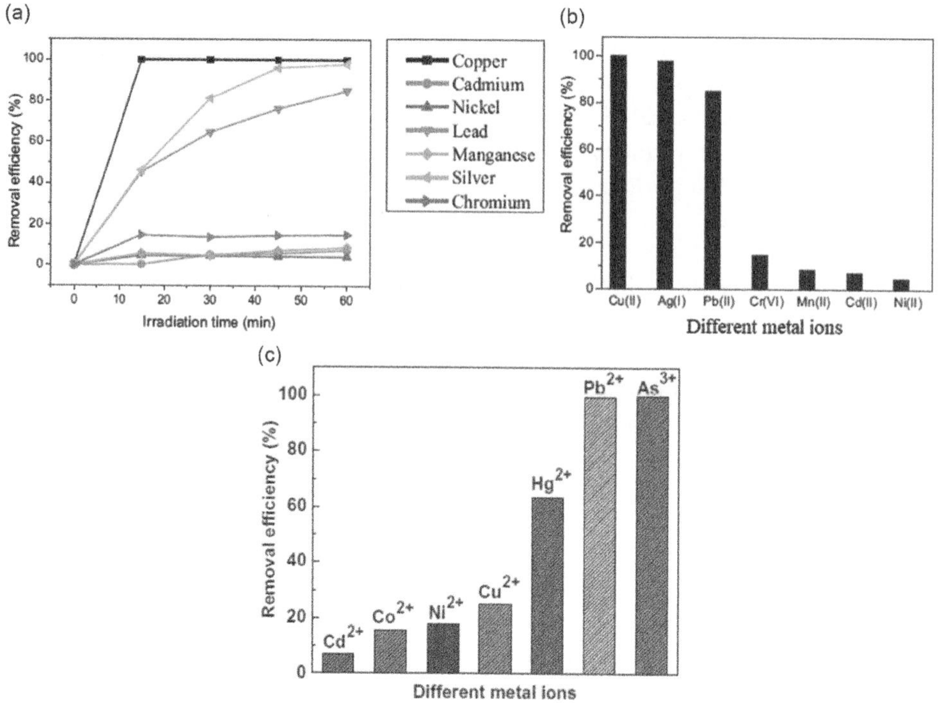

Figure 4.9 The removal efficiency of metal ions in an aqueous solution of ZnO rod-shaped structures (a) with respect to irradiation time [34] / Elsevier; (b) in 1 h; Adapted from [34] / Elsevier; and (c) HM ion removal of mesoporous spherical ZnO nanoassemblies; Adapted from [89] / Elsevier.

strongly attracted toward ZnO nanoassemblies, thereby demonstrating better removal efficiency [87].

4.4.2 TiO_2

TiO_2 is also extensively utilized for the elimination of Cr [92], Cd, Cu [93], As [94, 95], Pb, Cd, Cu, Zn, and Ni [96]. For example, Parida et al. [92] employed TiO_2-immobilized mesoporous MCM-41 to remove Cr from a solution containing 100 mgL^{-1} of Cr metal ions. It was reported that 91.0% removal of Cr was achieved in 80 min at pH ~ 5.5 and 323 K. In another work, Visa et al. [93] studied the influence of a fly ash-TiO_2 substrate on the elimination of Cu and Cd ions from synthetic wastewater. The removal efficiency of Cu ions from the synthetic wastewater was found to be greater than the removal of Cd ions with the utilization of fly ash-TiO_2 substrate. Meanwhile, Engates et al. [96] investigated and compared the removal efficiency of Pb, Cd, and Ni ions using TiO_2 nanoparticles and bulk particles. For TiO_2 nanoparticles, 100.0% removal of Pb, Cd, and Ni ions from solution (0.1 g L^{-1}) was observed within 120 min; however, only 20.0% of these HM ions were removed from solution (0.1 g L^{-1}) when TiO_2 bulk particles were used under the same test conditions. Moreover, Ramesh et al. [81] observed a poor removal efficiency for Cr (10.0%), Pr (17.0%), and Zn (25.0%) ions by TiO_2 nanocomposite under sunlight irradiation.

4.4.3 MnO_2

The photocatalytic removal of HM ions requires balanced absorption and photoredox reactions. A study by Chiam et al. reported that the removal of Ag^+ ions in a solution by a MnO_2 photocatalyst was optimum at pH 4 [97]. The removal of Ag^+ ions could be explained from the view of electrostatic attraction (adsorption) between the MnO_2 photocatalyst and Ag^+ ions, combined with the inhibition of electron–hole recombination in acidic environment due to the presence of excess H^+ ions. From the explanation provided, the Mn^{2+} ions were formed in the acidic condition. They then reacted with the photogenerated holes on the surface of β-MnO_2 particles, forming Mn^{3+} ions [98]. Since the holes were trapped by Mn^{2+} ions, the recombination of the electrons and holes in narrow bandgap MnO_2 was suppressed. As a result, the photo-induced electrons could freely react with the Ag^+ ions in the aqueous solution, forming Ag particles. The reactions involved are designated in Eqs. (4.4)–(4.7).

$$Ag^+(aq) + MnO_2(s) \rightarrow Ag^+ / MnO_2(s) \tag{4.4}$$

$$MnO_2(s)(\text{surface species}) + 4H^+ + 2e^- \rightarrow Mn^{2+}(s)(\text{surface species}) + 2H_2O \tag{4.5}$$

$$Mn^{2+}(s)(\text{surface species}) + h^+(MnO_2) \rightarrow Mn^{3+}(s)(\text{surface species}) \tag{4.6}$$

$$Ag^+(aq) + e^-(MnO_2) \rightarrow Ag^0(S)(E_0 = +0.79\,\text{V vs.NHE}) \tag{4.7}$$

Based on these results, the selectivity of HM ions by different photocatalysts is summarized in Table 4.4. As presented in Table 4.4, surface functionalized Fe_3O_4 shows a good removal efficiency to most of the metal ions such as Cr^{3+}, Co^{2+}, Ni^{2+}, Cu^{2+}, Cd^{2+}, Pb^{2+}, and As^{3+}. It is a potential material to be used in wastewater treatment for metal ion removal.

4.5 Major Drawbacks and Key Mitigation Strategies

Various photocatalysts have been developed throughout the years. Most of them were reported to possess high removal efficiency, reusability, and stability as an HM removal photocatalyst at laboratory-scale testing stages. The continuity of this work with industrial application, however, is small and performance in large-scale applications is in doubt. Furthermore, the technical concerns about the recycling of photocatalysts, which is one of the selling points of this technology, have not been solved completely. Nanopowder photocatalysts that wash away with wastewater could cause secondary pollution, but supported photocatalysts are still in the developmental stage. In addition, the potential safety threats that arise from the use of photocatalysts are not certain, and limited studies have been performed on the toxicity toward the human body and living creatures of the intermediates formed during the photocatalysis reaction. The eventual goal of this photocatalytic technology is large scale application for wastewater treatment, thereby relieving the issues of energy and water shortages for the benefit of human beings and the environment. To realize this goal, there are still lots of challenges to be addressed by the researchers.

Table 4.4 Selectivity of HM ion removal by semiconductor photocatalysts.

Types of photo-catalysts	Optical excitation source	Types of metal ions	Removal efficiency(%)	Removal selectivity	References
ZnO	UV light	Cd^{2+}	7.0	Poor	[87]
		Co^{2+}	15.5	Poor	
		Ni^{2+}	17.8	Poor	
		Cu^{2+}	25.0	Poor	
		Hg^{2+}	63.5	Moderate	
		Pb^{2+}	100.0	Good	
		As^{3+}	100.0	Good	
ZnO	UV light	Co^{2+}	<9.0	Poor	[34]
		Cd^{2+}	<9.0	Poor	
		Ni^{2+}	<9.0	Poor	
		Pb^{2+}	85.2	Good	
		Mn^{2+}	<9.0	Poor	
		Ag^{+}	97.9	Good	
		Cr^{6+}	43.3	Moderate	
MCM-41-doped TiO_2	-	Cr^{6+}	91.0	Good	[92]
TiO_2	UV light	Pb^{2+}	96.0	Good	[96]
		Cd^{2+}	99.0	Good	
		Ni^{2+}	71.0	Moderate	
Ga_2O_3-doped TiO_2	Sunlight	Cr^{6+}	10.0	Poor	[81]
		Pr^{3+}	17.0	Poor	
		Zn^{2+}	25.0	Poor	
MnO_2	Visible light	Cr^{6+}	14.8	Poor	[97]
		Ni^{2+}	17.1	Poor	
		Cu^{2+}	14.3	Poor	
		Ag^{+}	51.1	Moderate	
		Cd^{2+}	6.2	Poor	

4.6 Summary and Future Directions

Water is one of the crucial natural resources for all living beings. Global concerns have thus risen about water pollution and its impact to the ecosystem. Among all the water pollutants, HMs have received paramount attention due to their toxic nature. Although HMs are usually present in small amounts in water sources, many of them are toxic even at very low concentration. HM concentration in our water sources is currently an area of great concern. This is particularly vital during this industrialization era where many industries produce metal-containing effluents and discharge them into fresh water without adequate and proper treatment. The development and utilization of promising nanotechnology and nanomaterials have attracted the interest of researchers globally, particularly for wastewater treatment. Some wastewater containing HMs from both industry and domestic processes has been illegally discharged into the water streams. This could bring great havoc to the health of all living beings and cause destruction of the environment.

Future researches on wastewater treatment must look holistically into the balance of cost, efficiency, and toxicity to ecosystems. This chapter has critically covered several important photocatalyst materials, their limitations, and accomplishments toward the elimination of HM contaminants from water. Nevertheless, there are still many open issues and opportunities for upcoming research:

a) Cost. A variety of photocatalysts have been developed. Yet, a large number of photocatalysts require the use of chemicals that are costly and complex to synthesize. Although outstanding efficacy has been achieved, these photocatalysts are not practical for large-scale synthesis due to high production cost. The future development of photocatalysts should therefore focus on reduction of production cost to an economical level, simplification of process, scalability, and sustainability. These studies should focus moreover on photocatalysts triggered by sunlight instead of human-made UV or visible light sources, as sunlight is readily available and represents relatively economic green technology.
b) Toxicity. The safety of any new technology is the prime concern for consumers. The toxicity of intermediate products formed during the photocatalytic reactions needs to be well studied to avoid any undesirable side effects to humans and living organisms. If there are side effects, reasonable precautions and countermeasures should be established.
c) Development of new photocatalysts. The effectiveness of a photocatalytic reaction is greatly affected by the photocatalyst properties. It is believed that with today's technology and knowledge, an efficient photocatalyst can be modified / customized according to wastewater matrices and HM concentration. Nevertheless, efforts are always required to explore and come out with more efficient photocatalyst systems to cope with the increase in amounts of pollutants and the ever-changing environment.
d) Design of photocatalysts. The issue of complete recovery of the nanosize photocatalyst after water treatment is still waiting to be addressed. More competent and promising engineering design on the reactor system should be established. An appropriate nanostructure model should be designed moreover in accordance with the requirements and reactor system. Most importantly, the complete recovery of the catalyst is needed to avoid secondary contamination. Cost should also be taken into consideration so that it is economical and practical for large-scale applications.
e) Practical tests. Photocatalytic tests at the laboratory stage should be performed using the wastewater effluent from related industry rather than a laboratory-prepared mixed metal ion solution. Mostly, in the literature, only one specific type of metal ion at a time is targeted for study for ease of investigation. Researchers should be aware, however, that the industry effluents have complex matrices made up of thousands of complex

chemicals and impurities. The effect of those impurities on the efficacy of the photocatalyst system should be taken into consideration as soon as possible during the developmental stage of the photocatalyst. This is critical to ensure the developed materials are usable for industrial effluents treatment.

f) Modelling work. Computational modelling of the use of various photocatalytic systems for HM removal is required for a better and more solid understanding of the mechanisms involved during the reaction. This is required to understand and optimize the photocatalytic reaction conditions, removal efficiency, and kinetics of reactions.

References

1 Kjellstrom, T., Lodh, M., McMichael, T., Jamison, D.T., Breman, J.G., and Measham, A.R. (2006). Air and water pollution: burden and strategies for control. In: *The International Bank for Reconstruction and Development/The World Bank*, 2e. Washington, DC (Chapter 43).

2 Ajiboye, T.O., Oyewo, O.A., and Onwudiwe, D.C. (2021). Simultaneous removal of organics and heavy metals from industrial wastewater: a review. *Chemosphere* 262: 128379.

3 Bazrafshan, E., Mohammadi, L., Ansari-Moghaddam, A., and Mahvi, A.H. (2015). Heavy metals removal from aqueous environments by electrocoagulation process a systematic review. *Journal of Environmental Health Science & Engineering* 13: 74.

4 O'Connell, D.W., Birkinshaw, C., and O'Dwyer, T.F. (2008). Heavy metal adsorbents prepared from the modification of cellulose: a review. *Bioresource Technology* 99: 6709.

5 Callender, E. (2003). Heavy metals in the environment-historical trends. *Treatise on Geochemistry* 9: 67.

6 UNESCO. (2017). UN World Water Development Report, Wastewater, The Untapped Resource.

7 Sankpal, S.T. and Naikwade, P.V. (2012). Heavy metal concentration in effluent discharge of pharmaceutical industries. *Science Research Reporter* 2: 88–90.

8 Kosolapov, D.B., Kuschk, P., Vainshtein, M.B., Vatsourina, A.V., Wiebner, A., Kastner, M., and Muller, R.A. (2004). Microbial processes of heavy metal removal from carbon deficient effluents in constructed wetlands. *Engineering in Life Sciences* 4: 403.

9 Sud, D., Mahajan, G., and Kaur, M.P. (2008). Agricultural waste material as potential adsorbent for sequestering heavy metal ions from aqueous solutions-A review. *Bioresource Technology* 99: 6017.

10 Ngah, W.S.W. and Hanafiah, M.A.K.M. (2008). Removal of heavy metal ions from wastewater by chemically modified plant wastes as adsorbents: a review. *Bioresource Technology* 99: 3935.

11 Ali, H. and Khan, E. (2019). Trophic transfer, bioaccumulation, and biomagnification of non-essential hazardous heavy metals and metalloids in food chains/webs – concepts and implications for wildlife and human health. *Human and Ecological Risk Assessment: An International Journal* 25: 1353–1376.

12 Yi, Y., Yang, Z., and Zhang, S. (2011). Ecological risk assessment of heavy metals in sediment and human health risk assessment of heavy metals in fishes in the middle and lower reaches of the Yangtze River basin. *Environmental Pollution* 159: 2575–2585.

13 Litter, M.I. (2009). Treatment of chromium, mercury, lead, uranium, and arsenic in water by heterogeneous photocatalysis. *Advances in Chemical Engineering* 36: 37–67.

14 Ravindra, K., Bencs, L., and Grieken, R.V. (2004). Platinum group elements in the environment and their health risk. *Science of the Total Environment* 318: 1.

15 Hansima, M.A.C.K., Makehelwala, M., Jinadasad, K.B.S.N., Wei, Y., Nanayakkara, K.G.N., Herath, A.C., and Weerasooriya, R. (2021). Fouling of ion exchange membranes used in the electrodialysis reversal advanced water treatment: a review. *Chemosphere* 263: 127951.

16 Lyubimenko, R., Gutierrez Cardenas, O.I., Turshatov, A., Richards, B.S., and Schäfer, A.I. (2021). Photodegradation of steroid-hormone micropollutants in a flow-through membrane reactor coated with Pd(II)-porphyrin. *Applied Catalysis B: Environmental* 291: 120097.

17 Kangralkar, M.V., Kangralkar, V.A., and Manjanna, J. (2021). Adsorption of Cr (VI) and photodegradation of Rhodamine b, rose bengal and methyl red on Cu_2O nanoparticles. *Environmental Nanotechnology Monitoring & Management* 15: 100417.

18 Liu, M., Yin, W., Zhao, T.L., Yao, Q.Z., Fu, S.Q., and Zhou, G.T. (2021). High-efficient removal of organic dyes from model wastewater using $Mg(OH)_2$-MnO_2 nanocomposite: synergistic effects of adsorption, precipitation, and photodegradation. *Separation and Purification Technology* 272: 118901.

19 Gomes Júnior, O., Santos, M.G.B., Nossol, A.B.S., Clar, M., Starling, V.M., and Trovó, A.G. (2021). Decontamination and toxicity removal of an industrial effluent containing pesticides via multistage treatment: coagulation-flocculation-settling and photo-Fenton process. *Process Safety and Environmental Protection* 147: 674–683.

20 He, H., Huang, B., Fu, G., Xiong, D., Xu, Z., Wu, X., and Pan, X. (2018). Electrochemically modified dissolved organic matter accelerates the combining photodegradation and biodegradation of 17α-ethinylestradiol in natural aquatic environment. *Water Research* 137: 251–261.

21 Fu, F. and Wang, Q. (2011). Removal of heavy metal ions from wastewaters: a review. *Journal of Environmental Management* 92: 407.

22 Herrmann, J.M. (1999). Heterogeneous photocatalysis: fundamentals and applications to the removal of various types of aqueous pollutants. *Catalysis Today* 53: 115.

23 Li, Y., Zhang, J., Zhan, C., Kong, F., Lia, W., Yanga, C., and Hsiao, B.S. (2020). Facile synthesis of TiO_2/CNC nanocomposites for enhanced Cr(VI) photoreduction: synergistic roles of cellulose nanocrystals. *Carbohydrate Polymers* 233: 115838.

24 Mohamed, R.M. and Salam, M.A. (2014). Photocatalytic reduction of aqueous mercury (II) using multi-walled carbon nanotubes/Pd-ZnO nanocomposite. *Materials Research Bulletin* 50: 85–90.

25 Yao, L., Zhang, Y.C., Li, J., and Chen, Y. (2014). Photocatalytic properties of SnS_2/SnO_2 nanocomposite prepared by thermal oxidation of SnS_2 nanoparticles in air. *Separation and Purification Technology* 122: 1–5.

26 Bashiroma, N., Razak, K.A., and Lockman, Z. (2017). Synthesis of freestanding amorphous ZrO_2 nanotubes by anodization and their application in photoreduction of Cr (VI) under visible light. *Surface and Coatings Technology* 320: 371–376.

27 Denga, F., Lua, X., Luo, Y., Wang, J., Che, W., Yang, R., Luo, X., Luo, S., and Dionysiou, D.D. (2019). Novel visible-light-driven direct Z-scheme CdS/$CuInS_2$ nanoplates for excellent photocatalytic degradation performance and highly-efficient Cr (VI) reduction. *Chemical Engineering Journal* 361: 1451–1461.

28 Kumar, K.V.A., Chandana, L., Ghosal, P., and Subrahmanyam, C. (2018). Simultaneous photocatalytic degradation of p-cresol and Cr(VI) by metal oxides supported reduced graphene oxide. *Molecular Catalysis* 451: 87–95.

29 Mehrotra, K., Yablonsky, G.S., and Ray, A.K. (2003). Kinetic studies of photocatalytic degradation in a TiO_2 slurry system: distinguishing working regimes and determining rate dependences. *Industria & Engineering Chemistry Research* 42: 2273.

30 Chong, M.N., Jin, B., Chow, C.W.K., and Saint, C. (2010). Recent developments in photocatalytic water treatment technology: a review. *Water Research* 44: 2997.

31 S.L., Pung, S.Y., and Yeoh, F.Y. (2020). Recent development in MnO_2-based photocaralyats fororganic dyw removal: a review. *Environmental Science and Pollution Research* 27: 5759–5778.

32 Wu, W., Jiang, C., and Roy, V.A. (2015). Recent progress in magnetic iron oxide-semiconductor composite nanomaterials as promising photocatalysts. *Nanoscale* 7: 38–58.

33 Le, A.T. and Pung, S.Y. (2020). Reusability of metals/metal oxide coupled zinc oxide nanorods in degradation of Rhodamine B dye. *Pigment & Resin Technology* 50: 10–18.

34 Le, A.T., Pung, S.Y., Sreekantan, S., Matsuda, A., and PhuHuynh, D. (2019). Mechanisms of removal of heavy metal ions by ZnO particles. *Heliyon* 5: 01440.

35 Ruvarac-Bugarčić, I.A., Šaponjić, Z.V., Zec, S., Rajh, T., and Nedeljković, J.M. (2005). Photocatalytic reduction of cadmium on TiO_2 nanoparticles modified with amino acids. *Chemical Physics Letters* 407: 110–113.

36 Chenthamarakshan, C., Rajeshwar, K., and Wolfrum, E.J. (2000). Heterogeneous photocatalytic reduction of Cr (VI) in UV-irradiated titania suspensions: effect of protons, ammonium ions, and other interfacial aspects. *Langmuir* 16: 2715–2721.

37 Wu, Q., Zhao, J., Qin, G., Wang, C., Tong, X., and Xu, S. (2013). Photocatalytic reduction of Cr(VI) with TiO_2 film under visible light. *Applied Catalysis B: Environmental* 142–143: 142–148.

38 Fontana, K.B., Lenzi, G.G., Seára, E.C.R., and Chaves, E.S. (2018). Comparision of photocatalysis and photolysis processes for arsenic oxidation in water. *Ecotoxicology and Environmental Safety* 151: 127–131.

39 Zhang, G., Sun, M., Liu, Y., Lang, X., Liu, L., Liu, H., Qu, J., and Li, J. (2015). Visible-light induced photocatalytic activity of electrospun-TiO_2 in arsenic(III) oxidation. *ACS Applied Materials & Interfaces* 7: 511–518.

40 Nagarjuna, R., Challagulla, S., Sahu, P., Roy, S., and Ganesan, R. (2017). Polymerizable sol-gel synthesis of nano-crystalline WO_3 and its photocatalytic Cr(VI) reduction under visible light. *Advanced Powder Technology* 28: 3265–3273.

41 Malato, S., FernandezIbañez, P., Maldonado, M.I., Blanco, J., and Gernjak, W. (2009). Decontamination and disinfection of water by solar photocatalysis: recent overview and trends. *Catalysis Today* 147: 1–59.

42 Li, Y., Wang, W., Qiu, X., Song, L., and Meyer, H.M. (2011). Comparing Cr, and N only doping with (Cr, N)-codoping for enhancing visible light reactivity of TiO_2. *Applied Catalysis B: Environmental* 110: 148–153.

43 Savinkina, E., Obolenskaya, L., and Kuzmicheva, G. (2015). Efficiency of sensitizing nano-titania with organic dyes and peroxo complexes. *Applied Nanoscience* 5: 125–133.

44 Dhimana, P., Sharma, S., Kumar, A., Shekh, M., Sharma, G., and Naushad, M. (2020). Rapid visible and solar photocatalytic Cr(VI) reduction and electrochemical sensing of dopamine using solution combustion synthesized ZnO-Fe_2O_3 nano heterojunctions: mechanism elucidation. *Ceramics International* 46: 12255–12268.

45 Daghrir, R., Drogui, P., and Robert, D. (2013). Modified TiO_2 for environmental photocatalytic applications: a review. *Industrial & Engineering Chemistry Research* 52: 3581–3599.

46 Vaiano, V., Sacco, O., Sannino, D., and Ciambelli, P. (2015). Photocatalytic removal of spiramycin from wastewater under visible light with N-doped TiO_2 photocatalysts. *Chemical Engineering Journal* 261: 3–8.

47 Chowdhury, P. (2012). Visible-solar-light-driven photocatalytic degradation of phenol with dye-sensitized TiO_2: parametric and kinetic study. *Industrial & Engineering Chemistry Research* 51: 4523–4532.

48 Dozzi, M.V., Marzorati, S., Longhi, M., Coduri, M., Artiglia, L., and Sellia, E. (2016). Photocatalytic activity of TiO_2-WO_3 mixed oxides in relation to electron transfer efficiency. *Applied Catalysis B: Environmental* 186: 157–165.

49 Vignesh, K., Priyanka, R., Rajarajanc, M., and Suganthi, A. (2013). Photoreduction of Cr(VI) in water using Bi_2O_3-ZrO_2 nanocomposite under visible light irradiation. *Materials Science and Engineering : B* 78: 149–157.

50 Bai, L., Li, S., Ding, Z., and Wang, X. (2020). Wet chemical synthesis of CdS/ZnO nanoparticle/nanorod hetero-structure for enhanced visible light disposal of Cr(VI) and methylene blue. *Colloid Surface A* 607: 125489.

51 Zhang, L. and Wu, B. (2020). Dye-sensitization enhances photoelectrochemical performance of halide perovskite $CH_3NH_3PbI_3$ photoanode in aqueous solution. *Dyes and Pigments* 173: 108006.

52 Keşir, M.K., Dilber, G., Sökmen, M., and Durmuş, M. (2020). Use of new quaternized water soluble zinc phthalocyanin derivatives for effective dye sensitization of TiO_2. *Journal of Sol-Gel Science and Technology* 93: 687–694.

53 Chatterjee, D. and Mahata, A. (2001). Photoassisted detoxification of organic pollutants on the surface modified TiO_2 semiconductor particulate system. *Catalysis Communications* 2: 1–3.

54 Chatterjee, D. and Mahata, A. (2002). Visible light induced photodegradation of organic pollutants on dye adsorbed TiO_2 surface. *Journal of Photochemistry and Photobiology A: Chemistry* 153: 199–204.

55 Moon, J., Yun, C.Y., Chung, K.W., Kang, M.S., and Yi, J. (2003). Photocatalytic activation of TiO_2 under visible light using Acid Red 44. *Catalysis Today* 87: 77–86.

56 Kaur, S. and Singh, V. (2007). Visible light induced photocatalytic degradation of Reactive Red dye 198 using dye sensitized TiO_2. *Ultrasonics Sonochemistry* 14: 531–537.

57 Abe, R., Hara, K., Sayamaa, K., Domen, K., and Arakawa, H. (2000). Steady hydrogen evolution from water on eosin Y-fixed TiO_2 photocatalyst using a silane-coupling reagent under visible light irradiation. *Journal of Photochemistry and Photobiology A: Chemistry* 137: 63–69.

58 Rehman, S., Ullah, R., Butt, A.M., and Gohar, N.D. (2009). Strategies of making TiO_2 and ZnO visible light active. *Journal of Hazardous Materials* 170: 60–569.

59 Li, D. and Haneda, H. (2003). Photocatalysis of sprayed nitrogen-containing Fe_2O_3-ZnO and WO_3-ZnO composite powders in gas-phase acetaldehyde decomposition. *Journal of Photochemistry and Photobiology A: Chemistry* 160: 203–212.

60 Lin, H.F., Liao, S.C., and Hung, S.W. (2005). The dc thermal plasma synthesis of ZnO nanoparticles for visible-light photocatalyst. *Journal of Photochemistry and Photobiology A: Chemistry* 174: 82–87.

61 Asahi, R. and Morikawa, T. (2007). Nitrogen complex species and its chemical nature in TiO_2 for visible-light sensitized photocatalysis. *Chemical Physics* 339: 57–63.

62 Valentin, C.D., Pacchioni, G., Selloni, A., Livraghi, S., and Giamello, E. (2005). Characterization of paramagnetic species in N-doped TiO_2 powders by EPR spectroscopy and DFT calculations. *The Journal of Physical Chemistry B* 109: 11414–11419.

63 Wu, G. and Chen, A. (2008). Direct growth of F-doped TiO_2 particulate thin films with high photocatalytic activity for environmental applications. *Journal of Photochemistry and Photobiology A: Chemistry* 195: 47–53.

64 Mingce, L., Weimin, C., Heng, C., and Jun, X. (2007). Preparation, characterization and pho- tocatalytic activity of visible light driven chlorine-doped TiO_2. *Frontiers of Chemistry in China* 2: 278–282.

65 Ohno, T., Akiyoshi, M., Umebayashi, T., Asai, K., Mitsui, T., and Matsumura, M. (2004). Preparation of S-doped TiO_2 photocatalysts and their photocatalytic activities under visible light. *Applied Catalysis A: General* 265: 115–121.

66 Sakai, Y.W., Obata, K., Hashimoto, K., and Irie, H. (2009). Enhancement of visible light-induced hydrophilicity on nitrogen and sulfur-codoped TiO_2 thin films. *Vacuum* 83: 683–687.

67 In, S., Orlov, A., Berg, R., Garcıa, F., Pedrosa-Jimenez, S., Tikhov, M.S., Wright, D.S., and Lambert, R.M. (2007). Effective visible light-activated B-doped and B, N-codoped TiO_2 photocatalysts. *Journal of the American Chemical Society* 129: 13790–13791.

68 Zhang, X. and Liu, Q. (2008). Preparation and characterization of titania photocatalyst co-doped with boron, nickel, and cerium. *Materials Letters* 62: 2589–2592.

69 Xu, C., Killmeyer, R., Gray, M.L., and Khan, S.U.M. (2006). Photocatalytic effect of carbon- modified n-TiO_2 nanoparticles under visible light illumination. *Applied Catalysis B: Environmental* 64: 312–317.

70 Ren, W., Ai, Z., Jia, F., Zhang, L., Fan, X., and Zou, Z. (2007). Low temperature preparation and visible light photocatalytic activity of mesoporous carbon-doped crystalline TiO_2. *Applied Catalysis B: Environmental* 69: 138–144.

71 Naeem, M., Hasanain, S.K., and Mumtaz, A. (2008). Electrical transport and optical studies of ferromagnetic cobalt doped ZnO nanoparticles exhibiting a metal–insulator transition. *Journal of Physics: Condensed Matter* 20: 7.

72 Shinde, V.R., Gujar, T.P., Lokhande, C.D., Mane, R.S., and Han, S.H. (2006). Mn doped and undoped ZnO films: a comparative structural, optical and electrical properties study. *Materials Chemistry and Physics* 96: 326–330.

73 Wang, Y.S., Thomas, P.J., and O'Brien, P. (2006). Optical properties of ZnO nanocrystals doped with Cd, Mg, Mn, and Fe ions. *The Journal of Physical Chemistry B* 110: 21412–21415.

74 Kanade, K.G., Kale, B.B., Baeg, J.O., Lee, S.M., Lee, C.W., Moon, S.J., and Chang, H. (2007). Self-assembled aligned Cu doped ZnO nanoparticles for photocatalytic hydrogen production under visible light irradiation. *Materials Chemistry and Physics* 102: 98–104.

75 Marschall, R. and Wang, L. (2014). Non-metal doping of transition metal oxides for visible-light photocatalysis. *Catalysis Today* 225: 111–135.

76 Emadian, S.S., Ghorbani, M., and Bakeri, G. (2020). Magnetically separable $CoFe_2O_4/ZrO_2$ nanocomposite for the photocatalytic reduction of hexavalent chromium under visible light irradiation. *Synthetic Metals* 267: 116470.

77 Daneshvar, N., Salari, D., and Khataee, A.R. (2003). Photocatalytic degradation of azo dye acid red 14 in water: investigation of the effect of operational parameters. *Journal of Photochemistry and Photobiology A: Chemistry* 157: 111–116.

78 Kiriakidou, F., Kondarides, D.I., and Verykios, X.E. (2009). The effect of operational parameters and TiO_2-doping on the photocatalytic degradation of azo-dyes. *Catalysis Today* 54: 119–130.

79 Ren, H.T., Han, J., Li, T.T., Suna, F., Lin, J.H., and Lou, C.W. (2019). Visible light-induced oxidation of aqueous arsenite using facile Ag_2O/TiO_2 composites: performance and mechanism. *Journal of Photochemistry and Photobiology A: Chemistry* 377: 260–267.

80 Kumordzi, G., Malekshoar, G., Yanful, E.K., and Ray, A.K. (2016). Solar photocatalytic degradation of Zn^{2+} using graphene based TiO_2. *Separation and Purification Technology* 168: 294–301.

81 Ramesh, A.M., Gangadhar, A., Chikkamadaiah, M., and Shivanna, S. (2021). Hydrothermal synthesis of Ga_2O_3 /TiO_2 nanocomposites with highly enhanced solar photocatalysis and their biological interest. *Journal of Photochemistry and Photobiology A: Chemistry* 6: 100020.

82 Zhang, J., Ye, J., Chen, H., Qu, Y., Deng, Q., and Lin, Z. (2007). One-pot synthesis of echinus-like Fe-doped SnO_2 with enhanced photocatalytic activity under simulated sunlight. *Journal of Alloys and Compounds* 695: 3318–3323.

83 Deng, L., Liun, H., Gao, X., Su, X., and Zhun, Z. (2016). SnS_2/TiO_2 nanocomposites with enhanced visible light-driven photoreduction of aqueous Cr(VI). *Ceramics International* 42: 3808–3815.

84 Palve, A.M. (2021). Ultra-fast photoreduction of toxic Cr(VI) by CdS-rGO synthesized using single source precursor. *Journal of Alloys and Compounds* 868: 159143.

85 Fang, S., Zhou, Y., Zhou, M., Lia, Z., Xu, S., and Yao, C. (2018). Facile synthesis of novel $ZnFe_2O_4$/CdS nanorods composites and its efficient photocatalytic reduction of Cr(VI) under visible-light irradiation. *Journal of Industrial and Engineering Chemistry* 58: 64–73.

86 Wang, X., Cai, W., Liua, S., and Wang, G. (2013). Wu Z and Zhao H. ZnO hollow microspheres with exposed porous nanosheets surface: structurally enhanced adsorption toward heavy metal ions. *Colloids and Surfaces A: Physicochemical and Engineering Aspects* 422: 199–205.

87 Singh, S. (2011). Barick K C and Bahadur D. Novel and efficient three dimensional mesoporous ZnO nanoassemblies for environmental remediation. *International Journal of Nanoscience* 10: 1001–1005.

88 Wang, X.B., Cai, W.P., Lin, Y.X., Wang, G.Z., and Liang, C.H. (2010). Mass production of micro/nanostructured porous ZnO plates and their strong structurally enhanced and selective adsorption performance for environmental remediation. *Journal of Materials Chemistry* 20: 8582–8590.

89 Kumar, K.Y., Muralidhara, H.B., Arthoba, N.Y., Balasubramanyam, J., and Hanumanthappa, H. (2013). Hierarchically assembled mesoporous ZnO nanorods for the removal of lead and cadmium by using differential pulse anodic stripping voltammetric method. *Powder Technology* 239: 208–216.

90 Sheela, T., Nayaka, Y.A., Viswanatha, R., Basavanna, S., and Venkatesha, T.G. (2012). Kinetics and thermodynamics studies on the adsorption of Zn(II), Cd(II) and Hg(II) from aqueous solution using zinc oxide nanoparticles. *Powder Technology* 217: 163–170.

91 Hu, J., Chen, G., and Lo, I.M.C. (2006). Selective removal of heavy metals from industrial wastewater using maghemite nanoparticle: performance and mechanisms. *Journal of Environmental Engineering* 132: 709.

92 Parida, K., Krushna Mishra, G., and Dash, S.K. (2012). Adsorption of toxic metal ion Cr(VI) from aqueous state by TiO_2-MCM-41: equilibrium and kinetic studies. *Journal of Hazardous Materials* 241–242: 395–403.

93 Visa, M. and Duta, A. (2013). TiO_2/fly ash novel substrate for simultaneous removal of heavy metals and surfactants. *Chemical Engineering Journal* 223: 860–868.

94 Jing, C., Meng, X., Calvache, E., and Jiang, G. (2009). Remediation of organic and inorganic arsenic contaminated groundwater using a nanocrystalline TiO_2 based adsorbent. *Environmental Pollution* 157: 2514–2519.

95 Luo, T., Cui, J., Hu, S., Huang, Y., and Jing, C. (2010). Arsenic removal and recovery from Copper Smelting Wastewater Using TiO_2. *Environmental Science & Technology* 44: 9094–9098.

96 Engates, K.E. and Shipley, H.J. (2011). Adsorption of Pb, Cd, Cu, Zn, and Ni to titanium dioxide nanoparticles: effect of particle size, solid concentration, and exhaustion. *Environmental Scienc and Pollution Research* 18: 386–395.

97 Chiam, S.L., Le, A.T., Pung, S., and Yeoh, F.Y. (2021). Effect of pH on the photocatalytic removal of silver ions by β-MnO_2 particles. *International Journal of Minerals, Metallurgy and Materials* 28: 325–334.

98 Devi, L.G., Kottam, N., Narasimha Murthy, B., and Girish Kumar, S. (2010). Enhanced photocatalytic activity of transition metal ions Mn^{2+}, Ni^{2+} and Zn^{2+} doped polycrystalline titania for the degradation of Aniline Blue under UV/solar light. *Journal of Molecular Catalysis A: Chemical* 328: 44–52.

5

Smart Photocatalysts in Water Remediation

Albert Serrà

Grup d'electrodeposició de capes primes i nanoestructures (Ge-CPN), Department of Materials Science and Physical Chemistry, Universitat de Barcelona, C/ Martí i Franquès, Barcelona, Catalonia, Spain
Institute of Nanoscience and Nanotechnology (IN^2UB), Universitat de Barcelona, Barcelona, Catalonia, Spain

5.1 Introduction

This chapter is titled "Smart Photocatalysts in Water Decontamination," but what exactly do we mean when we say *smart photocatalysts*? The term *smart*, meaning fast, active, and intelligent, is often used as an adjective to emphasize a feature with some advantage over its counterparts. Researchers often misuse such adjectives in scientific publications, however, as if all new materials being developed were somehow smarter. In this chapter, we seek to holistically explain the state of the art in the field of smart photocatalysts by considering photocatalysts with some advantage or particularity that gives them added value for photocatalytic activity. In this section, we attempt to clarify why water decontamination is important, why photocatalysis is promising for water decontamination, and why smart photocatalysts are necessary [1].

- *Why is water decontamination important?*

In September 2015, the United Nations hosted a summit on the sustainable development of Planet Earth. At the gathering, representatives of the social, political, scientific, and private sectors drafted a 15-year agenda with 17 Sustainable Development Goals (SDGs), which entered into force on January 2016 [2, 3]. One SDG is dedicated to water and sanitation (SDG6), namely with the aim to "ensure the availability and sustainable management of water and sanitation for all." Achieving that goal is crucial, for more than 40% of the world's population is currently affected by water scarcity, which is expected to worsen due to global warming and indiscriminate anthropogenic activities [4, 5].

Throughout the 21st century, water decontamination will be a critical challenge, one that exceeds local, national, and state interests due to the extremely harmful effects of contaminated water on ecosystems, human life, and human health. The adverse situation is further aggravated in developing countries, especially by rapid industrialization, which generally increases energy demands and promotes lenient environmental regulations that can exacerbate pollution. As a consequence of anthropological and industrial activities, approximately 1.8 billion people around the world use a source of drinking water that is contaminated. Currently more than 80% of China's rivers have been deemed unusable for supplying drinking water, and in the United States, health authorities have prohibited fishing and/or bathing in 40% of all rivers [3, 6, 7].

Photocatalysts and Electrocatalysts in Water Remediation: From Fundamentals to Full Scale Applications, First Edition. Edited by Prasenjit Bhunia, Kingshuk Dutta, and S. Vadivel.

The principal pollutants in water – pathogens, inorganic materials, organic materials, and micro and nanopollutants – cannot be effectively eliminated by traditional methods of decontaminating water, especially when it contains emerging and persistent organic pollutants such as toxins, drugs, antibiotics, hormones, and personal hygiene products [8, 9]. Thus, to mitigate the current crisis of water pollution and water shortages facing Earth's societies, the protection and recovery of aqueous ecosystems and the implementation of efficient water treatment technologies and strategies are urgently needed. Ideally, those processes should be characterized by simplicity, sustainability, scalability, low cost, and easy worldwide implementation [10].

- *Why is photocatalysis promising for water decontamination?*

Photocatalysis exploits the intrinsic response of semiconductors when they interact with photons that have an energy equal to or greater than the bandgap energy (E_g) [11, 12]. The fundamental process is based on the photoexcitation of an electron (e_{CB}^-) from the valence band (VB) to the empty conduction band (CB), as shown in Scheme 5.1. The photo-driven promotion of an electron to the CB causes the formation of a positively charged hole or vacancy at the VB (h_{VB}^+), as first reported by Fujishima and Honda in 1972 [13]. The photogenerated species e_{CB}^- and h_{VB}^+ are usually called *charge carriers*, for they are responsible for the electrical conductivity of semiconductors [12, 14] and are highly reactive species with important implications for water decontamination [15]. On the one hand, h_{VB}^+ is a strong oxidizing agent capable of oxidizing organic matter, if not also even mineralizing it via its conversion to carbon dioxide and another inorganic species. In aqueous media, when h_{VB}^+ reacts with water, hydroxyl ($^{\bullet}OH$) radicals form as a result, which are the strongest oxidizing agents known except for fluorine. On the other hand, e_{CB}^- reacts with oxygen to form the weak oxidant superoxide radical $O_2^{\bullet-}$ as well as $^{\bullet}OH$ or other scavengers (e.g. H_2O_2) to form more $^{\bullet}OH$ or to reduce inorganic ions [9, 11, 12]. Those

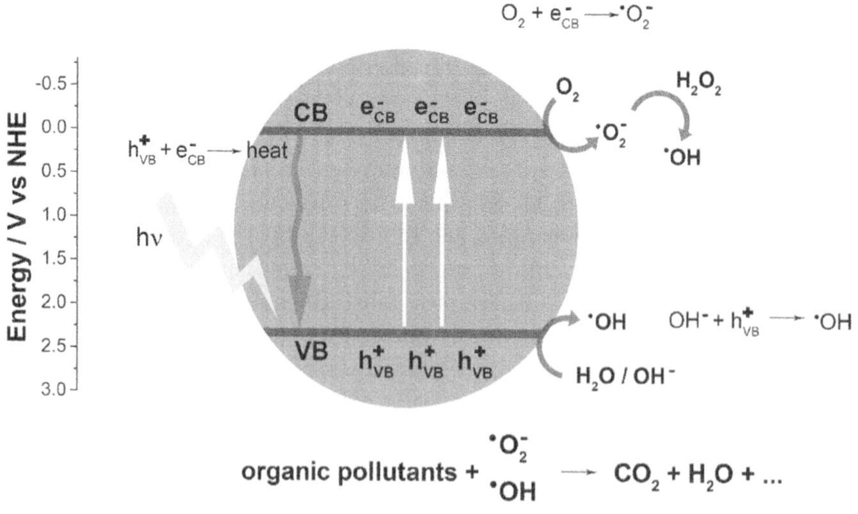

Scheme 5.1 Schematic representation of a photocatalysis process.

species, which linger close to the photocatalyst's surface, are responsible for the various photocatalytic processes in water decontamination. Photocatalysis is thus a superficial process in which the adsorption and/or interaction between the catalyst's surface and the pollutant are pivotal, as are the number of photons delivered to the photocatalyst's surface per second [16].

Solar light and solar light-driven photocatalysts rank among the most promising solutions for alleviating energy and environmental problems. Effectively using sunlight, however, requires the development of smart, highly efficient photocatalysts, for most of the potential catalysts, including TiO_2 and ZnO, have relatively large bandgaps and are thus photoexcitable only with UV light [17]. Because the spectrum of solar radiation has only a slight amount of UV light (<4.0%), researchers have recently focused on designing, synthesizing, and characterizing solar or visible light-driven photocatalysts based on doped or modified semiconductors, and also investigating the synthesis of other photocatalysts with smaller bandgap energies [14, 18, 19].

- *Why are smart photocatalysts necessary?*

Smart catalysts need to be developed for improved specificity and improved competitiveness. On the one hand, the use and extension of new technologies at the micro and nanoscale, including micro and nanofluidics, requires the development of intelligent systems capable of performing multiple tasks at once [20–23]. For example, (photo)catalysts with not only excellent photocatalytic activity but also locomotor capacity benefit from a certain amount of directionality and mobility when provided with an internal and/or external chemical, physical, and/or biological stimulus and, as a result, have the specificity to selectively execute a specific task. In such development efforts, micro and nanocatalysts are particularly important [23–25].

On the other, smart photocatalysts need to be designed to become more competitive – due to improved catalytic activity, for instance, especially when using sunlight. To that purpose, researchers have focused on developing simple pathways of synthesis that produce photocatalysts with excellent photocatalytic activity using sunlight. The reasons for the complexity of using solar energy are not only the nature of the catalyst but also the architecture, shape, and/or engineering of the reactor design to maximize the use of sunlight, especially because photocatalytic water decontamination has to be competitive in developing countries, where climates tend to facilitate the use of solar energy [20–22]. Improving the durability of the materials by improving their chemical stability and photocorrosion resistance is also an important effort. Intelligent photocatalysts can not only perform multiple tasks at once but also improve their capacity for decontamination via mechanisms that directly affect the production of reactive species or modify the kinetics or dynamics of the process. As a result, they facilitate the penetration of light, catalyst–pollutant interaction, and/or the transport of pollutants within the reaction medium. Lastly, competitiveness can be improved by designing catalysts that can be integrated into circular processes and become the raw material of a new material, and thus reduce the generation of waste [21].

This chapter describes the major smart photocatalysts used in water decontamination as well as possible pathways and strategies for improving photocatalysts [20, 21].

5.2 Advances in the Development of Visible-light Driven Photocatalysts

The photoexcitation of electrons via the absorption of photons is a fundamental mechanism of the photocatalytic process. However, pure TiO_2 and ZnO photocatalysts have energy bandgaps of 3.20 and 3.37 eV, respectively, which makes them photosensitive to UV light only [9, 26, 27]. For water decontamination achieved by processes applicable worldwide and especially in developing countries, it is important that photocatalysts can use solar irradiation as a free, renewable source of photons. Unfortunately for that need, the low intensity of UV light in the solar spectrum, less than 5% (Figure 5.1a), substantially inhibits the use of pure photocatalysts. At the same time, using conventional artificial light sources (Figure 5.1b) such as Hg or Xe lamps is unsustainable, for most of the photons lack sufficient energy to promote the photoexcitation of electrons [27–29]. Recently, the advent of light-emitting diode (LED) technology offers light sources with better prospects in terms of energy consumption for the production of UV light (Figure 5.1c), thereby making classic photocatalysts more viable. Even so, the priority in water decontamination should be harnessing solar energy as the most environmentally and economically sustainable option worldwide [27].

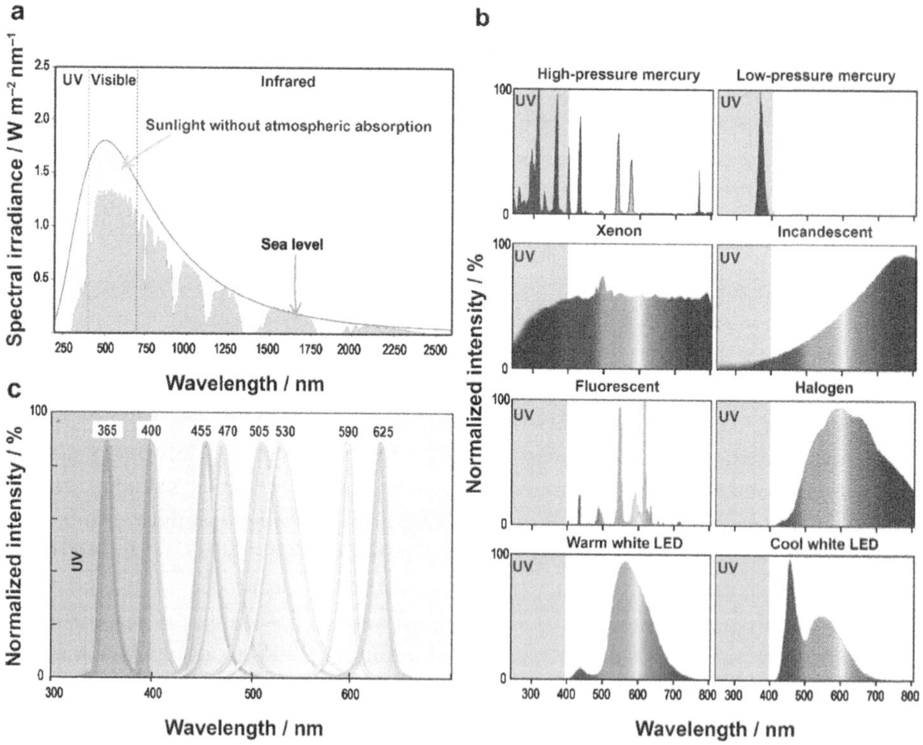

Figure 5.1 (a) Irradiance spectra of sunlight in the atmosphere above Earth's surface, with normalized irradiance spectra emitted by (b) high-pressure mercury, low-pressure mercury, xenon, incandescent, fluorescent, halogen, warm white LEDs, and cool white LEDs; and (c) currently available monochromatic LEDs [9].

Current research on water decontamination focuses on developing photocatalysts driven by visible light that harness solar energy as a benign strategy to promote the application of photocatalysis without harming the environment. Despite the wide bandgaps of most semiconductor materials, single-component semiconductors are not ideal owing to the high recombination of photogeneration charge carriers, which results in poor quantum and photocatalytic efficiencies [30–33]. Numerous methods and strategies are available, however, to shift the bandgaps of semiconductor materials to the visible light region (i.e. 44% of the entire solar spectrum), if not reduce recombination losses or photocorrosion activity; they include doping, the formation of heterojunctions, the generation of oxygen vacancies, morphology tuning, surface modification, quantum confinement effects, and dye sensitization [4, 31, 34–39]. Sections 5.2.1–5.2.5 briefly describe the primary strategies used to synthesize smart photocatalysts for water decontamination.

5.2.1 Doping

Doping involves introducing low levels of impurities to tune bandgaps and inducing the prevalence of free e^- and h^+, which considerably broaden light absorption and significantly accelerate the separation of electron–hole pairs [35, 40, 41]. Doped semiconductors with a prevalence of free e^- are known as n-type semiconductors, whereas p-type semiconductors possess h^+ as their primary charge carrier [42, 43]. Typically, metal doping, nonmetal doping, and codoping are investigated as strategies for improving the photocatalytic performance of TiO_2 and ZnO. As shown in Figure 5.2a, dopants capable of giving an electron to the CB when irradiated with visible light and, in turn, generating a hole are known as *impurity donors*, whereas ones able to accept an electron from the VB when irradiated with visible light are known as *impurity acceptors*. The narrowing of the bandgap is a consequence of the interaction between levels of impurity and the valence or conduction bands due to the creation of interbandgap levels [11, 34].

5.2.2 Heterojunction Formation

Heterojunctions are interfaces built by joining two different semiconductors, either with two different phases of one material, sometimes called *homojunctions* (e.g. TiO_2 anatase–brookite) or two different materials (e.g. ZnO–ZnS) with different bandgap energies [10, 36, 37, 44, 45]. As shown in Figure 5.2b, depending on the bandgap energies and the electronic affinity of semiconductors, three different band alignments are possible. The most promising, most studied heterojunction in smart photocatalysis for water decontamination is the fabrication of type II heterostructures [46, 47]. Two prerequisites should be satisfied to prepare efficient photocatalysts driven by visible light: (i) the outer shell material should allow strong visible light absorption and act as a photosensitizer, and (ii) the band alignment at the interface should satisfy the conditions of type II heterostructures. The joining of an internal UV-excited semiconductor with a semiconductor excited by visible light (i.e. narrow bandgap) such as CdSe and C_3N_4 enhances quantum and photocatalytic efficiencies under irradiation via solar light. Heterojunctions formed by different UV-excited semiconductors, however, such as ZnO and ZnS, also possess visible-light photocatalytic activity. In the past decade, various type II heterostructured photocatalysts such as

ZnSe–ZnO, ZnS–ZnO, SnO_2–CdS, CdS–TiO_2, and CdS–ZnO have been shown to exhibit higher visible-light photocatalytic efficiency [34, 48–50]. As shown in Figure 5.2b, the CB position of one material, A, is more negative than the other, B, whereas the VB position of the B semiconductor is more positive than that of A. In that system, holes and electrons transfer in opposite directions, and such a heterojunction improves the separation of photogenerated electron–hole pairs, inhibits recombination processes, and enhances the photocatalytic activity of visible light [46, 51–53].

5.2.3 Surface Modification: Localized Surface Plasmon Resonance

Although modifying a surface entails several phenomena, this section focuses exclusively on the decoration of a semiconductor's surface with a metal nanostructure, which is especially relevant for using metal nanoparticles. This strategy is based on exploiting the localized surface plasmon resonance effect. As shown in Figure 5.2c, visible light radiation is captured by metal nanoparticles (e.g. Ag nanoparticles), and an increased local electromagnetic field is generated. The electromagnetic field subsequently induces the formation of charge carriers near the surface of the semiconductor. This phenomenon enhances electron–hole separation, reduces recombination loses, and can enhance the photocatalytic efficiency of photocatalysts with visible light [4, 54–57].

The quantum confinement effect occurs when extremely small semiconductor crystals (i.e. at the nanometer scale) are used. While the confinement of electrons and/or electron-hole pairs (i.e. excitons) in nanocrystals significantly depends on the properties of the material, the optoelectronic properties of semiconductors are functions of their size and shape. As shown in Figure 5.2d, in the case of nanoparticles, the bandgap energy increases with the increase of the diameter of the nanoparticle due to the quantum size effect. That phenomenon is frequently exploited to tune the bandgap of nanometric photocatalysts [54–57].

5.2.4 Dye Sensitization

Dye sensitization is widely used for photocatalytic applications driven by visible light, especially for the production of hydrogen. The strategy is not commonly used in water decontamination because the sensitizer (e.g. erythrosin B and thionine) can also become degraded and mineralized, which hinders its realistic application for water treatment. The efficient degradation of various organic pollutants by using dye sensitization, however, is reported in the literature from the past decade [39, 58]. As shown in Figure 5.2e, during visible light irradiation, photogenerated electrons from organic dyes are transferred to the CB of semiconductors and thus form cationic radicals and active electrons. This dynamic results in the enhanced photocatalytic activity of photocatalysts [39, 58].

5.2.5 Photocatalyst Alternatives to TiO_2 and ZnO

A wide variety of visible-light active semiconductor alternatives to TiO_2-based and ZnO-based photocatalysts have recently been investigated, most of which are based on oxides and sulfides [19]. In the context of water remediation, the most important are oxides and sulfides,

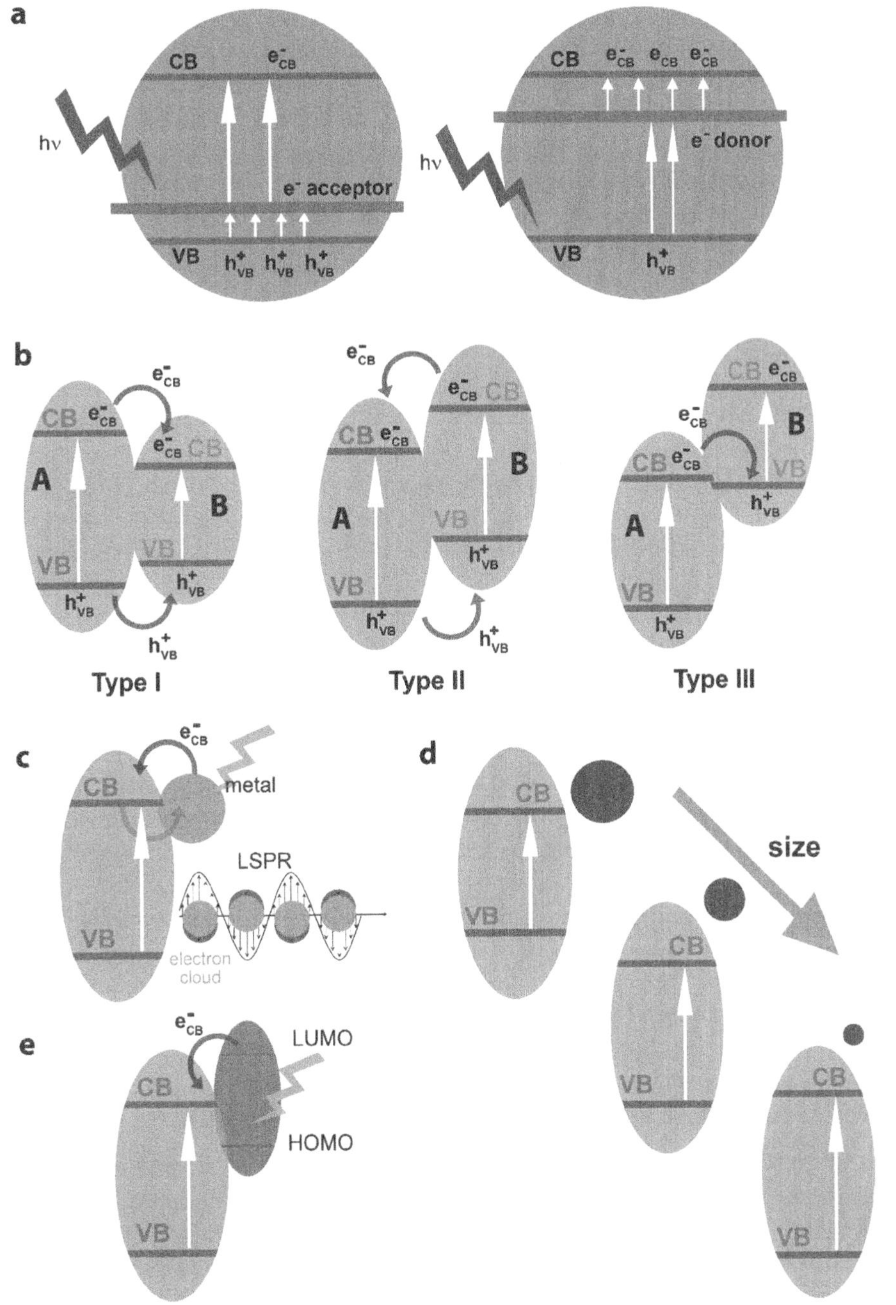

Figure 5.2 (a) Schematic representation of the energy levels of semiconductor doped with dopants with electron acceptor and electron donor nature. (b) Schematic representation of the various types of heterojunctions, where A and B correspond to semiconductor A and B, respectively. (c) Localized surface plasmon resonance effect of metal nanoparticles for semiconductor photocatalysis under light irradiation. (d) Effect of quantum confinement effects on electronic structure of semiconductors. (e) Schematic representation of the mechanism of the dye sensitization process. The organic dye is adsorbed on the surface of a semiconductor.

not only CeO_2 given its high stability and strong absorption of visible light, but also WO_3 and SnO_3, which are frequently doped or modified with other elements, as well as ZnS. Various metal sulfides exhibit high photocatalytic activity driven by visible light, but also high photocorrosion. Other important examples are metallates, particularly $BiVO_4$, but also members of the Aurivillius-related oxide family ($Bi_2A_{n-1}B_nO_{3n+3}$) such as Bi_2WO_6 or Bi_2MoO_6, $ZnWO_4$, $PbWO_4$, alkali earth metal indates (e.g. $CaIn_2O_4$, $SrIn_2O_4$, and $BaIn_2O_4$), titanates, stannates, ferrites, and even titanoniobates. Bismuth oxyhalides, especially BiOCl, are also important photocatalysts that have recently been examined [19, 38, 41].

5.3 Advances in Photocatalyst Immobilization and Supports

Photoreactors are critical elements in the design of photocatalysts, which have to integrate the photocatalysts while maximizing the use of light, especially sunlight. Divided into fluidized bed reactors and fixed bed reactors [59–62], photoreactors are not typically scalable to the industrial level, despite significant research in the design of new reactors that seeks to improve their fixation and use of supports. The immobilization of photocatalysts within fixed-bed reactors should guarantee efficient light penetration, the mass transport of pollutants, and/or the interaction between the semiconductor's surface and pollutants. In turn, immobilized photocatalysts should ideally increase or maintain the active surface area, light absorption, and photocatalytic performance of bare photocatalysts [59, 63, 64]. Because immobilizing a bare photocatalyst involves creating a new photocatalyst that may have a different level of efficiency and different properties from the bare photocatalyst, the process has to be done intelligently in order to maximize the new catalyst's photocatalytic activity. The photocatalyst immobilizer should thus possess (i) a highly active, highly accessible surface area before and after the photocatalyst's immobilization, (ii) excellent chemical stability in working conditions and photochemical stability under UV light and visible light irradiation against highly oxidative photogenerated radicals, (iii) the capacity to adsorb or interact with pollutants, (iv) strong interaction with the photocatalyst permanently immobilized in the entrapment medium, (v) the ability to minimally reduce the bare catalyst's photocatalytic performance, and (vi) optical transparency [9, 65]. Among the most promising supports and immobilizers are glass, activated carbons, silica, polymers, alumina, mineral clays, quartz, zeolites, and biodegradable materials such as cellulose and optical fibers [65, 66]. The smartest photocatalysts listed here are based on various immobilization strategies:

- **Photocatalytic membranes**: Photocatalytic membranes have attracted enormous interest in both industry and academia due to their excellent antifouling and antimicrobial properties, superhydrophilicity, concurrent photocatalytic activity and separation properties, and innovative design [18, 67–69]. That photocatalytic membranes do not require photocatalyst separation ranks among their most important advantages in the practical application of photocatalytic water treatment. Although metallic, ceramic, and polymeric membranes are the most commonly used materials for fabricating photocatalytic membranes, polymeric membranes may be the most promising. Various methods of synthesis, including hydrothermal synthesis, anodization, liquid phase deposition, electrospinning, and sputtering, are frequently used for fabricating photocatalytic membranes. However, dry–wet cospinning and sputtering techniques have been revealed as the most promising for immobilizing a uniformly distributed photocatalyst within the polymeric membrane. Typically, two strategies are used to fabricate photocatalytic membranes based on the immobilization of photocatalysts into and onto polymeric membranes, which result in a mixed-matrix membrane or a coated membrane, respectively (Figure 5.3a) [69–71]. To date, research on the immobilization of photocatalysts into and onto membranes has shown that because TiO_2-based photocatalysts, especially nanoparticles, are the most widely used semiconductors in the field, photocatalytic membrane systems have typically been studied under UV light irradiation. Another finding is that mixed-matrix membranes prevent the leaching of photocatalysts during water treatment, thanks to the strong attachment of the photocatalyst, but demand an excellent distribution of semiconductors; otherwise, photocatalyst agglomeration occurs, and the photocatalytic performance becomes significantly reduced as a result of the reduced active and accessible surface area and decreased light penetration. Lastly, coated membranes suffer from considerable photocatalyst leaching and photocorrosion compared with mixed-matrix membranes but are more often used for metallic and ceramic membranes. In all, photocatalytic membranes have outstanding potential despite their significant shortcomings in durability. Important challenges in immobilizing active photocatalysts under visible light and in improving the distribution of photocatalysts will mark research in the field in the coming years [18, 70, 72].

- **Porous supports**: Porous materials, especially nano and mesoporous ones, are basic elements in heterogeneous catalysis owing to their large ratios of surface area to volume. Most porous materials, however, especially those with a volumetric porosity, are not suitable for application in photocatalysis, because they need to be transparent such that light may penetrate and reach the surface of the immobilized semiconductor [73]. Porous materials, despite being normally inactivated by photocatalytic processes, can improve the photocatalytic activity of immobilized photocatalysts thanks to their remarkable ability to adsorb and/or interact with pollutants, facilitate light absorption, and reduce recombination losses (Figure 5.3b). Natural clays, including bentonite, sepiolite, zeolite, and kaolinite, have been extensively explored and proven to be effective as immobilizers for various photocatalysts for water remediation [59]. Recently, covalent organic frameworks (COFs) and conjugated microporous polymers (CMPs) have shown potential for designing more efficient photocatalytic systems, including Fe_2O_3@COFs, CFT-3D-BiOBr, COF-PD–AgI, Py-BF-CMP, Fc-TEB-CMP, and UPC-CMP-1 [74, 75].
- **Fibers, rods, wires, and cables**: Glass, carbon, ceramic, and metallic optical fibers, rods, wires, and cables have also been explored as support materials for designing immobilized photocatalytic systems (Figure 5.3c). Using such supports is especially relevant in manufacturing photocatalytic membranes and ultrafiltration systems [77–79]. The direct growth of photocatalysts on such structures via different physical and chemical methods such as chemical deposition and electroless processes allows creating more efficient, more complex photocatalytic systems that, more importantly, are scalable at the industrial level. The primary disadvantages of these supports are their low durability, especially when based on glass, woven, or natural fibers, and the increased pressure drop in the system when incorporated in photoreactors. The support material can benefit photocatalysis by reducing the processes of recombination and photocorrosion while facilitating mass transfer and catalyst–pollutant interaction [78, 80].

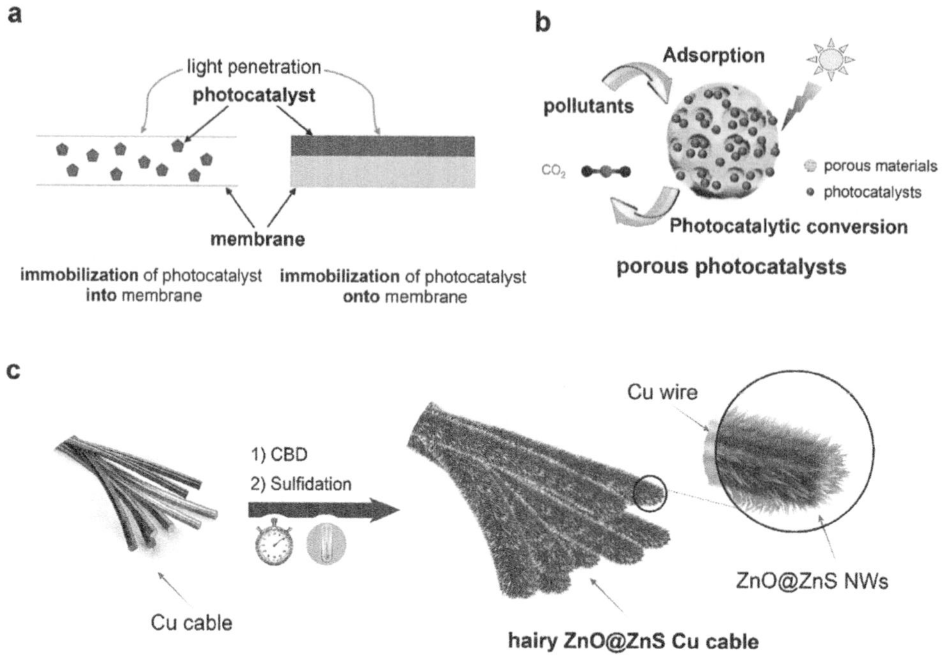

Figure 5.3 (a) Schematic representation of the immobilization of photocatalysts into and onto membranes. With permission of [70]. (b) Schematic representation of porous photocatalysts. [76]. (c) Schematic representation of hairy ZnO@ZnS core@shell Cu cables. CBD means chemical bath deposition. [77] / Elsevier / CC BY 4.0.

Of all photocatalysts, immobilized photocatalysts clearly have the greatest applicability in water decontamination because they lower costs by simplifying the later stages of the separation and recovery of the photocatalysts. They typically exhibit, however, fewer competitive photocatalytic activities than the free catalysts.

5.4 Advances in Nonimmobilized Smart Photocatalysts

The photocatalytic decontamination of water is based on the use of light as an external stimulus to promote the photogeneration of reactive species able to oxidize, reduce, and/or destroy organic and/or inorganic matter or microorganisms present in bodies of water. While light induces the formation of reactive species, it can also facilitate other parallel processes that improve the photocatalytic activity of photocatalysts.

Nonimmobilized smart photocatalysts with additional locomotive functionality are usually categorized as micro or nanomotors, micro or nanorobots, micro or nanoswimmers, micro or nanomachines, and micro or nanocleaners [22, 81–84]. Regardless of category, their primary feature is that they can move or even be directed toward specific places via an external stimulus such as magnetic or electric fields or an internal stimulus such as the concentration gradient, pH, or pollutant concentration (Figure 5.4) [85, 86]. In using nonimmobilized smart photocatalysts, the goal is fabricating smart

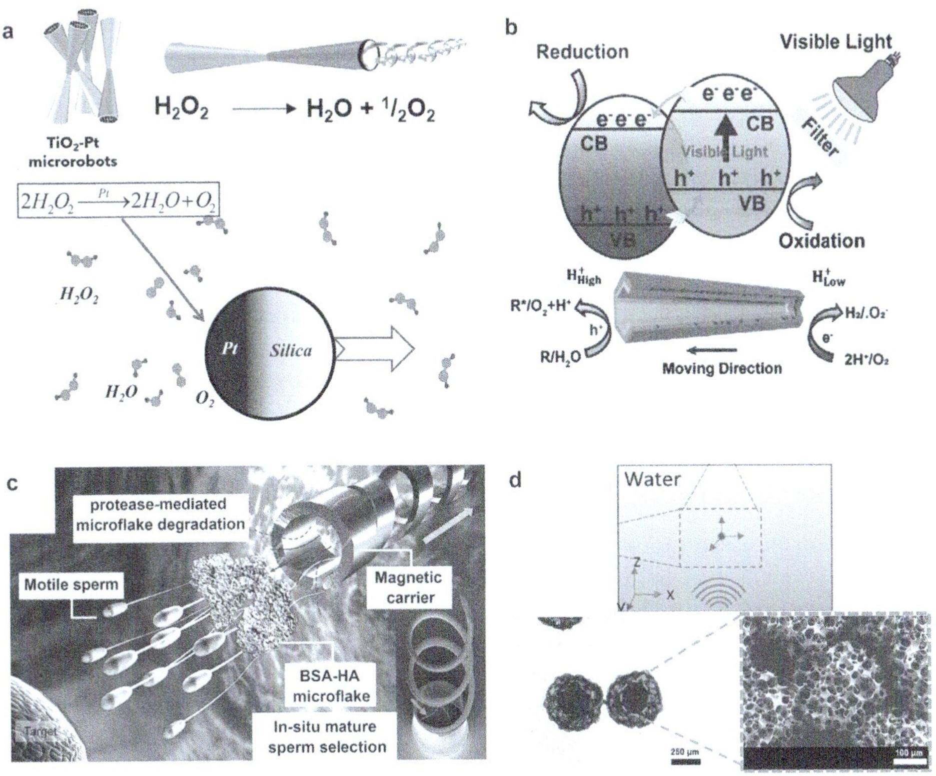

Figure 5.4 Four common locomotion mechanisms of micro and nanomotors. (a) Chemically powered micromotors. Schematic illustration of TiO_2/Pt microrobots. Reproduced with permission from [88]. Self-propulsion of Pt-SiO_2 Janus micromotors. [89] / MDPI / CC BY 4.0. (b) Light-driven micromotors. Motion mechanism under visible light irradiation of TiO_2/Fe_3O_4/CdS microsubmarines. With permission of [90]. (c) Magnetic field-controlled micromotors. Schematic representation of magnetic actuation of Fe-TiO_2 helical micromotor. With permission of [91]. (d) Acoustic field-actuated micromotors. PGLA micromotors controlled by ultrasonic fields. Adapted from [92].

systems able to integrate photocatalytic, catalytic, optical, or acoustic properties and their combination in order to make them more competitive. The first micromotors were based on a mechanism of self-propulsion afforded by the catalytic properties of metals such as Ni and Pt to reduce hydrogen peroxide to oxygen. Recently, smart photocatalytic micro and nanomotors have been investigated given their promising applicability in water remediation [85–87].

Li et al. fabricated TiO_2–Au–Mg microspheres that propel themselves autonomously in natural water (Figure 5.5a), thereby facilitating the fluid transport and dispersion of the photogenerated reactive oxidative species while improving the interaction of photocatalysts with agents of chemical and biological warfare. The system has proven to be highly effective in destroying the cell membrane of the anthrax simulant *Bacillus globigii* spore and mineralizing highly persistent organophosphate nerve agents [72]. As shown in Figure 5.5a, the motion of the smart photocatalyst is driven by hydrogen bubbles formed via the oxidation of the Mg surface when immersed in chloride-rich environments such as seawater. The system's durability is relatively low, for it loses mobility as the Mg is consumed.

Pané and colleagues have developed 3D hybrid CoNi–Bi_2O_3–BiOCl segmented microcylinders and microhelices via a two-step electrodeposition process using 3D photolithographed membranes (Figure 5.5b). The system is formed by a short ferromagnetic CoNi segment that confers magnetic control over locomotion and a Bi_2O_3–BiOCl photocatalytic segment that oxidizes organic pollutants. Under UV light irradiation, the hybrid photocatalysts can degrade Rhodamine B with 90% efficiency within 6 h [41]. Using a simple electrodeposition process, these authors also created TiO_2–PtPd–Ni tubular structures able to photodegrade 100% of Rhodamine B, methyl orange, and methylene blue under sunlight in only 30 min. In that process, propulsion was based on the catalyzed decomposition of hydrogen peroxide added as a fuel to the working environment [93].

In other recent work, Pumera et al. have synthesized various self-propelled smart photocatalysts, some of which are light controlled [94, 95]. For example, mesoporous ZnO-based or Pt-based Janus microparticles, both locomoted via the decomposition of hydrogen peroxide fuel by a Pt layer, have shown excellent efficiency in removing nitroaromatic explosive and dye pollutants under visible light irradiation (Figure 5.5c). Those systems were fabricated by simple water-based ZnO precipitation followed by the thermal treatment and physical vapor deposition of Pt on the nanosheets of the hierarchical and mesoporous ZnO microparticles [94]. The same authors have also fabricated metal-free tubular micromotors based only on graphitic carbon, which proved effective for simultaneously removing and monitoring heavy metal pollutants. The motion and photocatalytic activity of both types of hybrid photocatalysts are controlled by visible light irradiation [95]. The authors additionally developed ZnO micromotors doped with Ag in order to photocatalytically remove bacterial biofilms. All of these smart photocatalysts can be fabricated at a low cost via a simple, scalable hydrothermal route. Under visible light irradiation and via a self-electrophoretic mechanism, the self-propelled micromotors exhibited excellent locomotion control as well as efficient removal of both gram-positive and gram-negative bacteria biofilms (Figure 5.5d) [96]. Lastly, they developed light-powered magnetic hematite–metal (i.e. Au, Au–Pd, Pt, and Pt–Pd) Janus micromotors able to move, capture, and degrade polymers and plastics using light (Figure 5.5e). The speed of those micromotors depended on the nature of the metal fragment and was faster for bimetallic architectures than single ones. Such materials are attractive for their low-cost, large-scale chemical synthesis based on asymmetrical deposition [97].

The photocatalytic micromotors developed by Ren and colleagues [98, 99] derived from studies on the synthesis of self-propelled Au-WO_3@C [98] and TiO_2–Au [99] Janus micromotors driven by UV light. In both cases, the synthetic routes were simple and scalable and achieved magnetic functionalization without affecting the photocatalytic performance via the deposition of paramagnetic Ni. The Au–WO_3@C micromotors proved to be efficient in mineralizing various organic pollutants such as Rhodamine B and sodium-2,6-dichloroindophenol at extremely low concentrations, whereas the TiO_2–Au micromotors were useful for degrading methyl blue, cresol red, and methyl orange [98, 99].

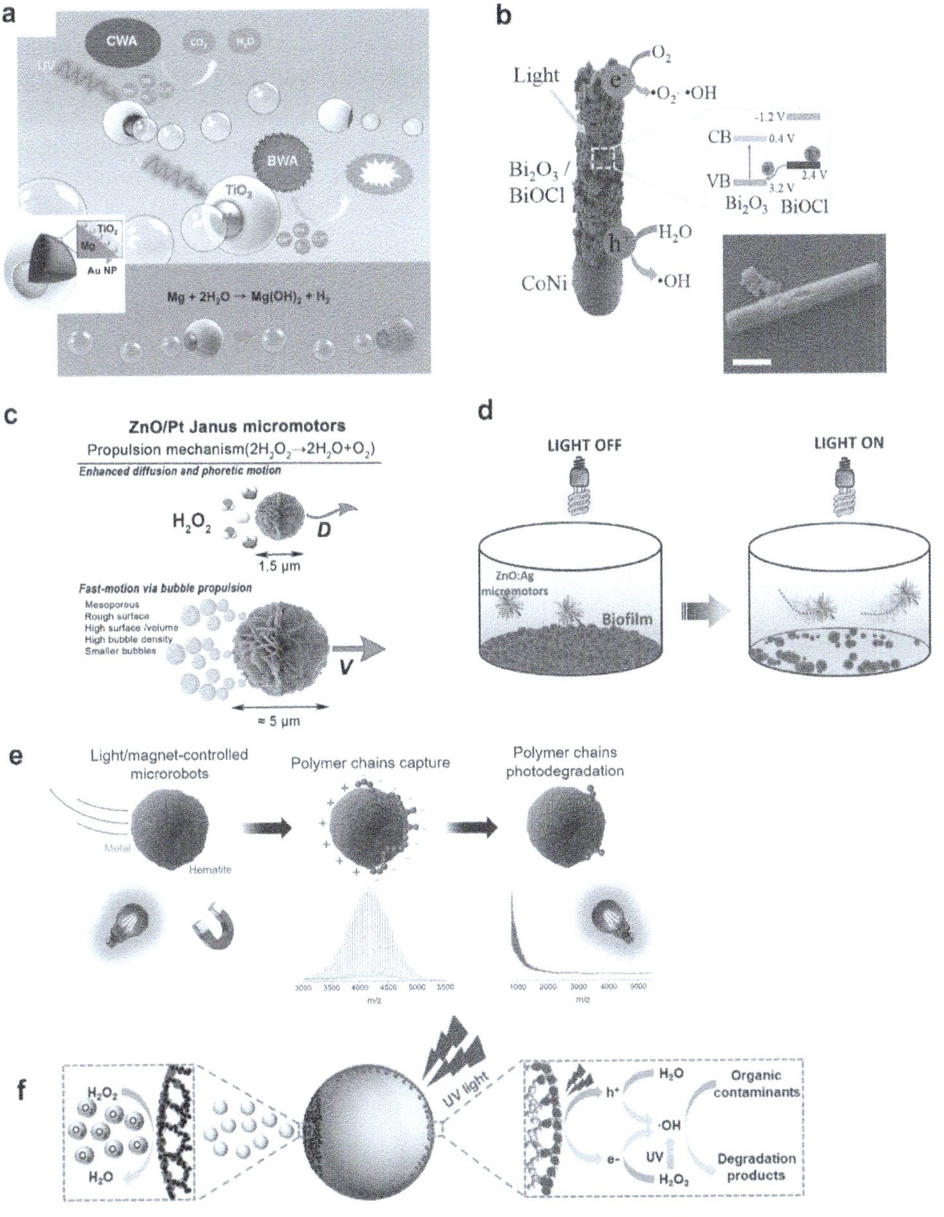

Figure 5.5 (a) Schematic representation of the self-propulsion and photocatalytic degradation of biological and chemical warfare pollutants by water-driven photoactive TiO_2/Au/Mg micromotors. With permission of [72]. (b) Schematic representation photocatalytic mechanism in magnetically driven Bi_2O_3/BiOCl-based hybrid micromotors for degrading organic pollutants. Scale bar: 2 μm. [41]. (c) Schematic representation of propulsion mechanism of ZnO/Pt Janus micromotors with hydrogen peroxide. With permission of [94]. (d) Schematic representation of light-powered antibiofilm ZnO micromotors for bacterial eradication. Reproduced with permission from [96]. (e) Schematic representation of self-propelled light-powered magnetic hematite/metal Janus micromotor for photodegrading polymeric materials in wastewater. Reproduced with permission from [97]. (f) Schematic illustration of photocatalytic mechanism of Fe_3O_4@Ag nanoparticles and zeolitic imidazolate framework-8@ZnO nanoparticles composite. Reproduced with permission from [20].

Metal–organic frameworks (MOF) have additionally shown potential for the fabrication of photocatalytic micromotors [20]. By using a simple one-step strategy based on template synthesis with emulsion microdroplets from microfluidics, uniform porous micromotors based on zeolitic imidazolate framework-8@ZnO nanoparticles (ZIF-8@ZnONPs) and Fe_3O_4@Ag nanoparticles (Figure 5.5f) can be continuously fabricated. In turn, this simple strategy can be flexibly extended to other materials used in MOF. Such hybrids exhibit excellent locomotive properties via the decomposition of hydrogen peroxide, easy magnetic recycling, and excellent photocatalytic activity in degrading organic pollutants such as Rhodamine B [20].

In recent research on the development of iron-based micromotors for water treatment [100, 101], Dong and colleagues have designed ecofriendly micromotors based in iron phthalocyanine and gelatin that exhibit rapid movement via self-diffusiophoresis thanks to self-propulsion driven by visible light using water as fuel. The hybrid catalysts exhibited excellent photocatalytic performance in degrading organic pollutants such as Rhodamine B [100]. Using iron-based micromotors thus seems quite promising given its general usefulness in combining photocatalytic and magnetic functionalization with only one material.

In still other work, Serrà and colleagues have described a highly effective, sustainable, inexpensive, holistic circular process that integrates solar photocatalytic water decontamination, bioethanol production, and CO_2 fixation. Central to that process is a novel hybrid Ni@ZnO@ZnS-plated microalgal (*Spirulina platensis*) micromotor that acts as the photocatalyst for water decontamination. Driven by visible light, the hybrid micromotor could remove toxic cyanobacteria blooms, cyanotoxins, and various organic pollutants such as methylene blue, Rhodamine B, and 4-nitrophenol in the presence of magnetic hybrid ZnO-based photocatalysts. The magnetic functionalization facilitated the micromotor's locomotion as well as the reusability of the photocatalysts. At the end of the photocatalyst's lifetime, the micromotors can be used as feedstock for producing bioethanol and fabricating biopellets, while the clean water and CO_2 gas produced by water decontamination and CO_2 gas generated during bioethanol production at the final stage of the cycle can be used to cultivate new microalgae. Pivotal to the outstanding performance of that process are the excellent photocatalytic performance of the hybrid Ni@ZnO@ZnS–*Spirulina* structure and the exceptional yield of *Spirulina platensis* algae in generating ethanol and microalgal biopellets. Given those findings, the authors' work marks an important step forward in circular chemistry and the circular economy. Because the residues produced in each step of the process can be reused or recycled in the next step, the process mobilizes cleaner production as a way to increase efficiency in the use of energy, water, and natural resources [10, 102].

Although numerous micromotors can be classified as smart photocatalysts, this chapter's purpose is not to describe all such structures that have been manufactured in recent years but only to illustrate the primary strategies being investigated. Most of the synthetic processes used to manufacture those multitasking photocatalysts are economical and easily scalable. Despite their high photocatalytic activity and relative competitiveness at present, these processes currently seem to be unlikely candidates for decontaminating large volumes of water. Nevertheless, they can be crucial for eliminating waste in small spaces and in new applications in microfluidics.

5.5 Advances in Improving the Efficiency of Light Delivery

At present, the study of photocatalysts and photoreactors does not consider light transport to be a fundamental aspect of light–matter interaction. At its base, photocatalytic water treatment relies on the photoexcitation of semiconductor materials able to produce charge carriers that directly react with pollutants and/or originate reactive species that achieve water remediation. Although the efficient harvesting of photons at the semiconductor's surface is crucial to ensuring a high quantum yield, current semiconductor architectures have normally disregarded that fact [14, 103].

Bioinspired, biomimetic, or biomimicry thinking emulates natural models, systems, and elements to overcome obsolete conceptual frameworks in the face of complex global challenges. To be sure, the scientific community has always been fascinated by the incredible, sophisticated systems, architectures, strategies, and materials found in nature, and nature is frequently identified as a source for the design of more competitive materials. Leveraging the experience of nature, especially of photosynthetic organisms, is also an intelligent way to prepare more efficient, smarter photocatalysts for improved light capture [104].

The major tasks for improving light harvesting are maximizing light coupling at the surface, improving light trapping in weakly absorbed wavelength regions, and eliminating optical losses. Those three tasks can be performed on nonreflective surfaces found in nature depending on surface architecture and morphology. The most efficient way of achieving nonreflective materials is based on developing subwavelength surface structures (Figure 5.6a) or hierarchical micro and/or nanoscale architectures (Figure 5.6b), in which light losses are minimal or nonexistent unlike with disorganized macrostructures (Figure 5.6c) [14, 104]. At the same time, the shape of the material, especially in the case of micro and nanophotocatalysts, is especially relevant to intensifying the interaction between light and the material's surface. In adjacent work, the direct use of microorganisms has been explored for designing new photocatalysts that allow replicating the microorganism's architecture and morphology while at once recycling them for new applications at the end of the photocatalyst's useful life [10, 105]. Such processes have offered important opportunities for proposing simple, inexpensive, environmentally friendly, scalable synthetic pathways that can be integrated into circular processes. Sections 5.5.1–5.5.3 describe the primary systems currently inspiring the manufacture of more efficient photocatalysts; some of these have not yet been explored for water decontamination, the most relevant being algae and plants as major photosynthetic systems, and insects [103].

5.5.1 Inspiration from Algae

Algae are prokaryotic or eukaryotic unicellular or multicellular organisms that normally live in saline or freshwater environments, where they perform photosynthesis as photosynthetic autotrophs. Algae, especially microalgae, rank among the most efficient photosynthetic organisms. Microalgae are prokaryotic or eukaryotic photosynthetic microorganisms able to convert sunlight, water, carbon dioxide, and some organic and inorganic compounds into algal biomasses. Able to grow rapidly and live in harsh environmental

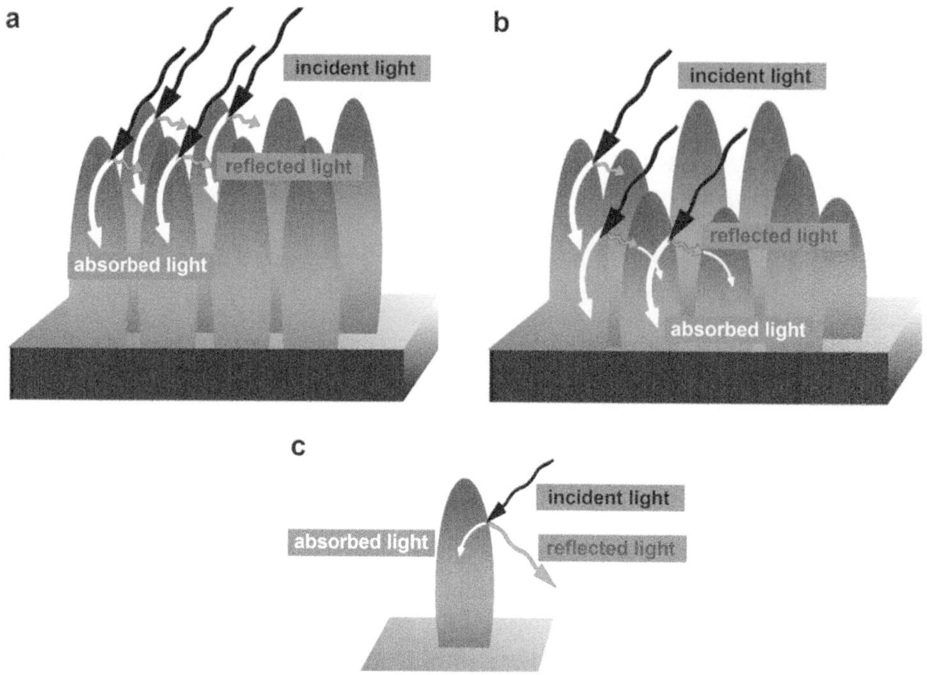

Figure 5.6 Schematic representation of light interaction with (a) subwavelength surface structures, (b) hierarchical micro and/or nanoscale architectures, and (c) disorganized macrostructures. Reproduced with permission from [14] / Elsevier.

conditions in all of Earth's ecosystems, microalgae tolerate an array of temperatures, salinities, pH levels, degrees of light intensity, and other conditions. In recent decades, microalgae and cyanobacteria have emerged as promising raw materials for multiple applications in fields such as human health, cosmetics, wastewater purification, the development of pharmaceuticals, aquaculture, and the production of pigments and antibiotics, among others. Microalgae also demonstrate a wide variety of morphologies, architectures, and shapes, some of which favor interaction with their environments while facilitating the absorption of sunlight. The primary microalgal morphologies are spherical, ellipsoidal, triangular, helical, sigmoidal, and rectangular [106–109].

Microalgae, especially spirulina, have thus emerged as microbiotemplates for different deposition processes, especially for chemical and electroless deposition [83, 110–112]. The coating of those microorganisms with different metals or semiconductors supports an easy, scalable way of manufacturing large quantities of highly homogeneous microstructures with complex shapes. Spirulina is often used due to its 3D helical architecture, whose diameter, helical pitch, and/or length can be easily tuned by controlling the growing environment [83].

Zhang and colleagues have synthesized hollow metallic Cu microhelices via simple electroless deposition [110], and, following the same fabrication strategy, Kamata et al. have investigated the electromagnetic response of spirulina-templated Cu microcoils [111]. In turn, Xia et al. have examined the use of spirulina for fabricating hierarchical $LiFePO_4/C$

microstructures, which exhibit excellent electrochemical performance as cathodes for lithium ion batteries [112]. Zheng and colleagues have also used spirulina as a biotemplate for manufacturing porous hollow carbon@magnetite core@shell magnetic helical micro and nanoswimmers (Figure 5.7a), which showed potential for photothermal antibacterial therapy, biological detoxification, and/or the targeted delivery of drugs [83].

In other work, Serrà et al. have integrated principles of green chemistry and circular chemistry by combining electroless deposition and biotemplating to fabricate efficient, sunlight driven, hybrid helicoidal Ni@ZnO@ZnS–microalgae and Cu@Cu_2O@CuO–microalgae photocatalysts for the photoremoval of persistent organic pollutants, cyanobacteria, and cyanotoxins (Figure 5.7b) [10]. The biomimetic architecture of the microalgae-based photocatalyst was designed to enhance its sunlight-trapping capability and improve the adsorption of pollutants and pollutant–photocatalyst interaction. Beyond that, the hybrid microalgae system can be easily recycled to produce biofuels such as bioethanol, biodiesel, and biopellets for combustion after its effective lifetime, thereby offering a holistic, green, scalable, inexpensive circular process with no residues. Cultivating microalgal biotemplates can also benefit the environment when used to fixate carbon dioxide and release atmospheric oxygen. The integrated strategy could be easily integrated into industrial processes to simultaneously achieve water purification and biofuel production in a simple, smart way.

5.5.2 Inspiration from Insects

In the case of insects, their microstructure and micro or nanoscale architecture or organization, not their shape or morphology, have inspired the design of nonreflective coatings. In the last decade, various structures based on moth eyes and butterfly wings have been extensively explored as means to enhance light harvesting [104, 113].

The moth-eye architecture is a typical example of efficient, bioinspired, nonreflective structures due to the hexagonal arrays of nanometric cones. In the past 10 years, the deposition of coating inspired by moth eyes for different types of solar cells has been widely investigated [113–115], while the moth-eye configuration has been explored in various spatial arrangements with different nanometric shapes, including nanowires and nanorods. For example, Guldin and colleagues have proposed the fabrication of low-cost, nonreflective coatings based on a moth-eye structure with self-cleaning properties [114]. Knowing that the adsorption of various contaminants such as organic pollutants can significantly reduce the efficacy of solar cells, the authors proposed a simple method to manufacture self-cleaning surfaces via the self-assembly of a block copolymer in combination with a silica-based sol-gel process and the incorporation of TiO_2 nanocrystals, the latter of which translated into self-cleaning properties. Ultimately, that antireflective coating could effectively decompose stearic acid, as part of a trend of other fabricated self-cleaning coatings extensively explored in the last decade.

Zheng and colleagues have fabricated ZnO–Si moth-eye structures (Figure 5.8a) with enhanced charge-separating, nonreflective, and photocatalytic properties based on a pyramidal silicon substrate covered with ZnO nanorods [115], a configuration that can be especially attractive for solar cells and sensors as well as photocatalysis. Those authors also investigated the photodegradation of Rhodamine B using the moth-eye structure, which

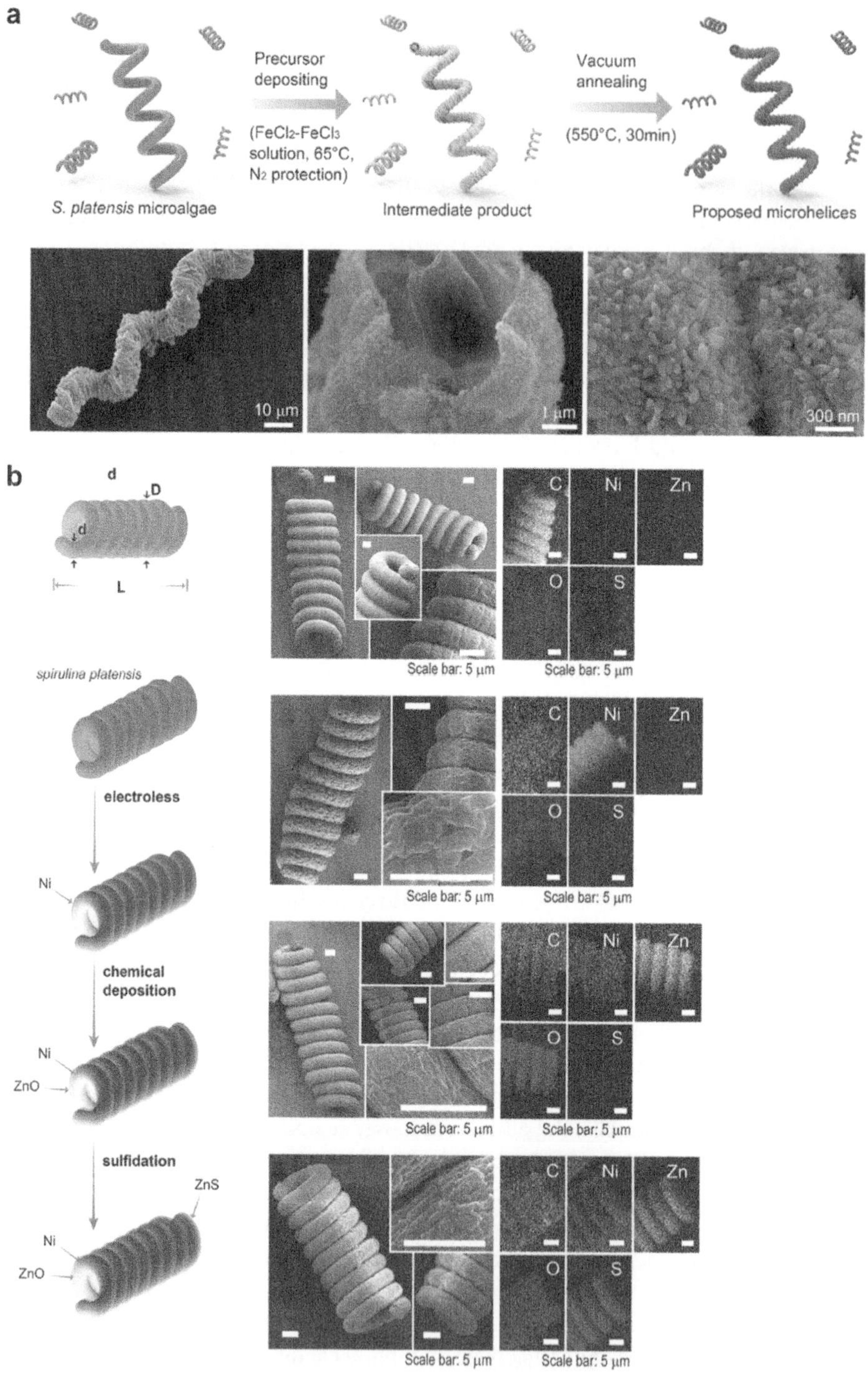

Figure 5.7 Schematic representation of the biotemplating process and electron microscopy images of (a) porous hollow carbon@magnetite core@shell magnetic helical micro and nanoswimmers. Reproduced with permission from [83] / Elsevier; and (b) Ni@ZnO@ZnS-microalgae biohybrid photocatalyst. [10] / John Wiley & Sons / CC BY 4.0.

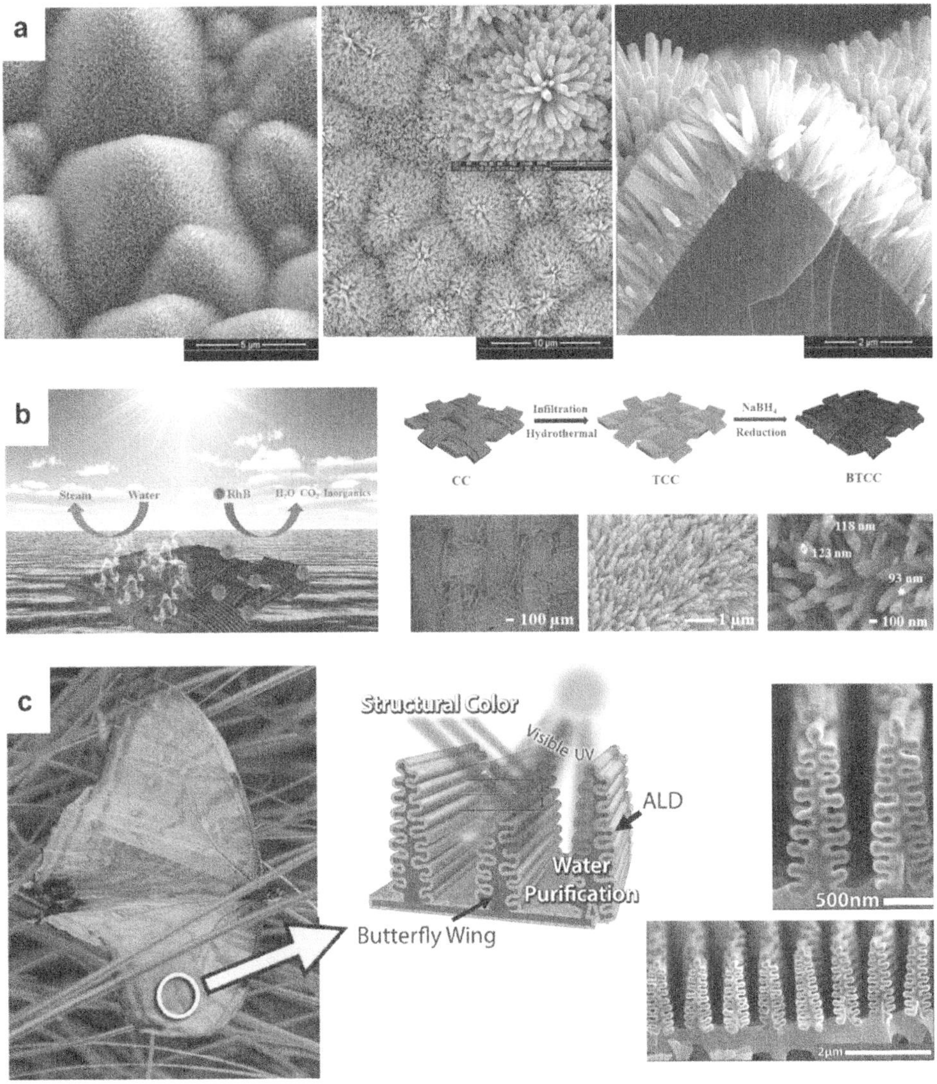

Figure 5.8 (a) Field-emission scanning electron microscopy images of p-n heterostructure ZnO/Si moth-eye structures. Adapted with permission from [115]. (b) Schematic representation and field-emission scanning electron microscopy images of moth-eye-like nanostructured black titania nanocomposites for solar steam generation and photocatalytic water decontamination. Adapted from [113]. (c) Picture of *Morpho sulkowskyi* butterfly, schematic illustration of *Morpho sulkowskyi* butterfly wings, and field-emission scanning electron microscopy images of biotemplated ZnO films on *Morpho* nanostructures. With permission of [117].

proved to be significantly greater than with Si-p or ZnO seeds and Si-p due to improved light-trapping characteristics [115]. By contrast, Liu et al. investigated the development of bioinspired moth-eye photocatalysts for solar-driven water decontamination. The authors reported the fabrication of bioinspired black titania nanocomposites (Figure 5.8b) with outstanding light absorption, approximately 96%, in the full solar spectrum due to effective propagation via absorbing light and enhancing light scattering promoted by the moth-eye architecture [113]. When those nanocomposites were explored in the solar steam generation and photocatalytic degradation of organic pollutants for water remediation, solar steam generation was effective in seawater, whereas photocatalytic degradation was achieved with Rhodamine B [113]. Lastly, Jin and colleagues have designed a novel composite based on TiO_2 pyramids and polyaniline nanoparticles via soft imprinting. Their nanocomposite effectively separated the photoinduced charges, enhanced the antireflective properties, and effectively photodegraded methylene blue [116].

Along with moth eyes, butterfly wings also frequently inspire the design of smart antireflective coatings, especially the wings of black *Morpho* and glasswing butterflies [117, 118]. Among researchers who have explored this bioinspired strategy in the context of photocatalytic water decontamination, Rodríguez and colleagues directly deposited ZnO onto *Morpho sulkowskyi* butterfly wings via low-temperature deposition on the atomic layer (Figure 5.8c). Such a hybrid photocatalyst maximized the photocatalytic activity due to the improved light harvesting capacity [117]. Photocatalysts driven by visible light and based on C-doped $BiVO_4$ using *Papilio paris* butterfly wings as a biotemplate as part of a simple one-step sol-gel process have also been synthesized. Those hybrid photocatalysts exhibited enhanced photocatalytic activity in mineralizing methylene blue and the evolution of oxygen from splitting water. The improved photocatalytic performance was clearly caused by the unique quasihoneycomb architecture and carbon doping, both of which improved visible light harvesting and absorption and facilitated the separation of photogenerated charge carriers [118].

Structures bioinspired by moth eyes and butterfly wings have tremendous potential for the design of more efficient solar cells but are less commonly used in designing smart photocatalysts. Even so, the fundamentals of those structures can be perfectly integrated into the design of photocatalytic systems for more competitive processes of water decontamination.

5.5.3 Inspiration from Plants

Plants and trees as photosynthetic systems also represent an important source for the design of more efficient, sophisticated systems that maximize light absorption. As in the case of algae and microalgae, the light-harvesting complexes of such organisms are primarily responsible for capturing photons of sunlight by acting as antennae and channeling the energy absorbed by the photosynthetic mechanism. At the same time, the shape of those organisms has also evolved to facilitate sunlight absorption.

In the synthesis of intelligent photocatalysts, many researchers have sought to mimic the shape and architecture of various flowers [119–122] and, as a result, generated complex structures with large surface areas that can nevertheless efficiently maximize the absorption of sunlight. In multiple works, the catalysts have reproduced the shape of flowers such as cauliflower [123] and rose petals [124] at the milli, micro, and nanoscales. Various materials and synthetic pathways have additionally been explored to fabricate flower-shaped photocatalysts, including SnS_2 [125], bimetallic (Fe–Bi)-povidone-iodine [126], rGO-functionalized ZnO-TiO_2 [127], BiOI-MoS_2 [128], Ag-AgCl-Bi_2MoO_6 [122], Bi_2WO_6 [129], $BiOCl_xBr_{1-x}$ [130], Ag–AgCl–BiOCl [131], BiOBr [132–135], Ag_3VO_4–BiOBr [136], Bi_2SeO_5 [137], Ag–ZnO [138], and N-TiO_{2-x}@MoS_2 [139]. Examples in Figure 5.9a showcase such sophisticated photocatalysts that have been synthesized – namely UiO-66–$CdIn_2S_4$ flower-like 3D microspheres and Fe-(8-hydroxyquinoline-7-carboxilic)–TiO_2 flower-like composites with large surface areas and enhanced light-harvesting properties. On the one hand, the bioinspired UiO-66–$CdIn_2S_4$ photocatalyst, developed via an easy hydrothermal route, demonstrated improved photocatalytic efficiency in degrading triclosan [140, 141]. In particular, the 3D microflower architecture proved effective in enhancing reactive sites on the surface, facilitating the channelization of charge carriers, and hindering the recombination process. On the other hand, the Fe-(8-hydroxyquinoline-7-carboxilic)–TiO_2 composite was highly effective in eliminating antibiotics such as ciprofloxacin, tetracycline, amoxicillin, and erythromycin. The photocatalyst's flower morphology facilitated light harvesting and accelerated the separation and transmission of photogenerated electrons and holes, thereby preventing recombination and providing many porous structures with large active surface areas [140, 141].

Another potential strategy for improving the light harvesting of smart photocatalysts is mimicking the phototropic mechanism of sunflowers, which self-orient toward sunlight throughout the day. Although no smart photocatalysts incorporating phototropic characteristics have been developed for photocatalytic water decontamination, the strategy should be considered given that human-made omnidirectional trackers with phototropic capacity have shown up to a 400%-greater energy harvesting capacity than nonphototropic ones [14, 142]. As shown in Figure 5.9b, the nanostructured polymers are oriented toward

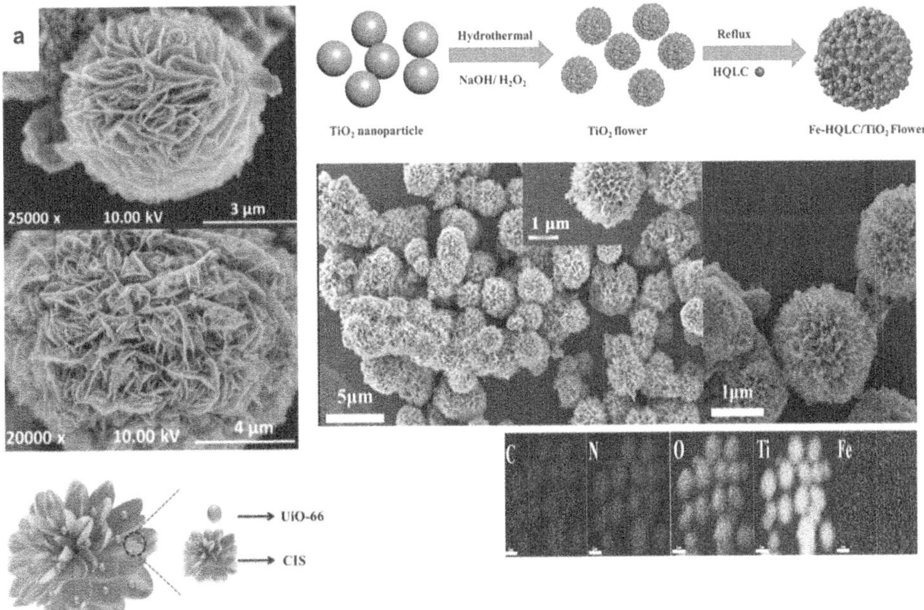

Figure 5.9 (a) Field-emission scanning electron microcopy images of $CdIn_2S_4$/UiO-66 nanocomposites with flower-like 3D microspheres. Adapted from [140]. Field-emission scanning electron microcopy images and schematic representation of the synthesis of Fe-(8-hydroxyquinoline-7-carboxilic)–TiO_2 composite flowers. Adapted from [141].

(Continued)

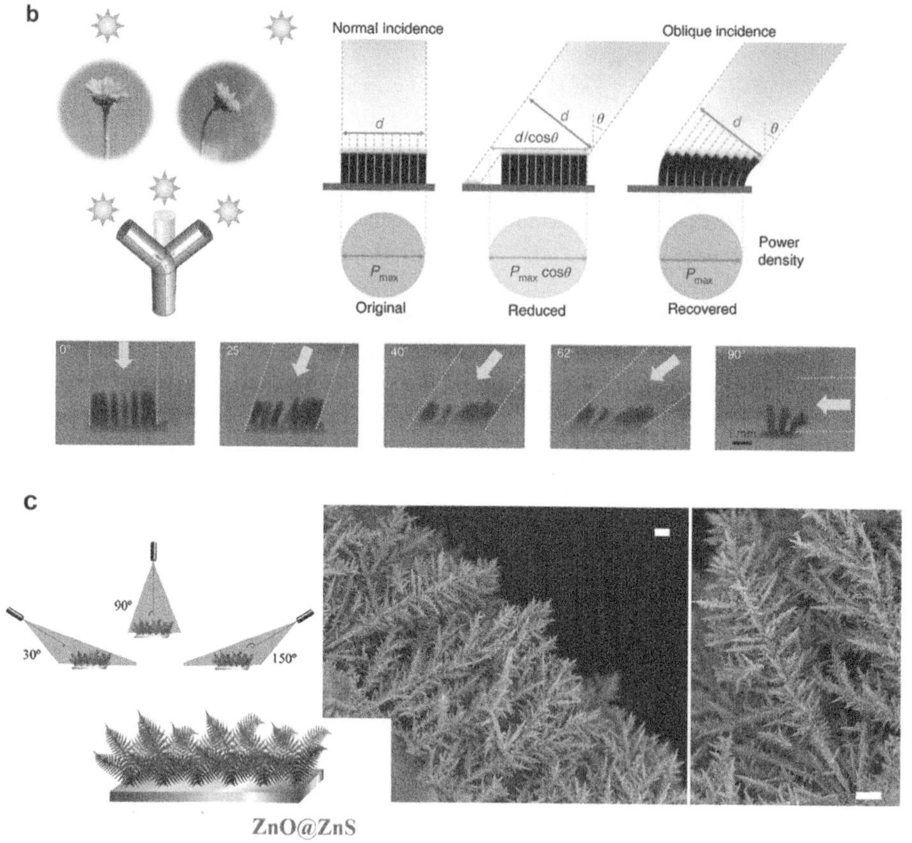

Figure 5.9 (Cont'd) (b) Schematic representation of phototropism mechanism of sunflowers of human-made omnidirectional trackers and optical micrographs of human-made omnidirectional tracker(s) in water. Adapted with permission from [142]. (c) Schematic representation and field-emission scanning electron microcopy images of ZnO@ZnS core@shell biomimetic fern-like microleaves. Adapted from [4].

light in all three dimensions, which improved light capture while maintaining the same maximum angle of incidence. The strategy may be especially relevant for constructing smart photocatalysts [14, 142].

Another component of plants and trees used in the design of new photocatalysts are leaves. For example, Serrà and colleagues have proposed a simple, green, low-cost electrosynthesis of biomimetic ZnO-based microferns for the rapid photodegradation and photoremediation of three types of persistent organic pollutants: methylene blue, 4-nitrophenol, and Rhodamine B. The fractal architecture of the prepared ZnO micro and nanofern photocatalysts showed a central trunk and different levels of ramifications. BET surface area measurements of the microfern-shaped photocatalysts confirmed the large increase in surface area compared with nonbranched nanowires, which was caused by their fractal and dendritical architecture. Moreover, photocurrent experiments conducted by irradiating the photocatalysts with different angles of irradiation have demonstrated the largely enhanced current photogenerated by visible light in modified ZnO micro or nanoferns, especially ZnO@ZnS heterostructures. That trend supports the independence of the photocurrent on the irradiation angle in micro and nanofern fractal architectures and, by extension, their improved ability to trap light (Figure 5.9c) [14, 143].

By contrast, other researchers have used leaves as templates. For example, highly active carbon-supported nano-CdS photocatalysts have been synthesized via the carbonization of lotus leaves [144]. This strategy efficiently yielded photocatalysts with incredibly high surface areas able to greatly facilitate the absorption of both light and pollutants and consequently enhance the photocatalytic degradation of organic pollutants. The highly ordered porous carbonaceous support derived from lotus leaves synergistically improves light harvesting, enhances electron–hole separation, and facilitates pollutant–photocatalyst interaction [144].

Plants and trees have been another important source of inspiration for reproducing shapes, architectures, and strategies at the milli, micro, and nanoscales or for use as biotemplates.

5.6 Advances in Biomaterials for Designing Smart Photocatalysts

Research on the use of new materials with superior properties is a primary field within materials science. Recently, such research has increasingly focused on using greener bioinspired materials that promote sustainable development based on the UN's Sustainable Development Goals for 2030. These goals have encouraged many research groups to focus on using biomaterials within photocatalysts to improve their stability, light-harvesting capacity, biocompatibility, generation of reactive species, reusability, and even recyclability at the end of their lifetimes [32, 145, 146].

Biomaterials have important potential for the synthesis of smart photocatalysts thanks to the following properties and characteristics:

- **Biodegradability**: Sustainability and biodegradability have become virtually nonnegotiable requirements in the development of new materials. Driving that trend have been new environmental policies and social demands requiring the design of synthetic strategies and materials that can be easily recycled. The use of biomaterials implies the possibility of easily eliminating waste generated at the end of the lifetime of catalysts, and opens the door to circular processes. In particular, hybrid photocatalysts based on biomaterials can be used as raw materials for other processes and/or applications [105, 108]. As illustrated in Section 5.5.1, hybrid photocatalysts based on a microalgae matrix have been used to produce biofuels such as bioethanol and biopellets [32, 147].
- **Biocompatibility and low toxicity**: Because metal-based photocatalysts generally exhibit high cytotoxicity, they may ultimately become incorporated into the environment as new pollutants. By contrast, biomaterials, due to their nature, structure, and shape, have a lower impact on aquatic biota and the environment in general in the event that they end up in aquatic systems [32].
- **Antimicrobial properties**: Several biomaterials have an inherent antibacterial activity that can be added to the similar activity of semiconductors used to produce reactive species

and facilitate water remediation. The antimicrobial nature of photocatalysts can be explained in light of various phenomena that include the formation of reactive species, the production of hydrogen peroxide, and the attack and rupture of cell membranes, among others. The incorporation of natural biopolymers (e.g. chitosan, pectin, and alginate), natural agents (e.g. nisin and thymol), and enzymes can be especially useful for manufacturing hybrid smart photocatalysts with superior antibacterial properties [32, 147].

- **Structural, textural, superficial, and optical properties**: Most biomaterials possess improved structural, textural, surface, and optical properties that promote the absorption of light while improving the photocatalytic process via synergistic effects achieved by their shape and chemical nature, effects that facilitate the separation of load carriers that form on semiconductors and, in turn, prevent recombination losses. Section 5.5 illustrates the primary strategies and materials that have been used or are currently being studied to produce smart photocatalysts [9, 32].

The following briefly illustrates the state of the art in the use of biomaterials for synthesizing smart photocatalysts:

- **Photocatalysts based on biopolymers:** During the past two decades, synthetic polymers have been widely used as components or base materials for the development of photocatalysts. More recently, biopolymers such as cellulose, pectin, collagen dextran, cyclodextrins, gelatin, chitosan, alginate, agar, and chitin have been explored to design new materials with multiple applications that are especially relevant in water treatment [145, 147, 148]. Natural biopolymers can be used to synthesize smart photocatalysts as surface coating agents, support matrices, and immobilizing agents and to prepare composites and/or core@shell structures. Various composite materials based on biopolymers such as TiO_2–microcrystalline cellulose [145], chitosan coated with ZnO [147], TiO_2-loaded *Moringa oleifera* gum-activated carbon [146], Fe^0@Guar gum-crosslinked-soya lecithin nanocomposite hydrogel [148], and $CoFe_2O_4$/PVA–SiO_2 [149] photocatalysts with excellent photocatalytic performance for water remediation have recently been reported.
- **Photocatalysts based on immobilized biomolecules:** The immobilization of biomolecules such as enzymes, peptides, and microbes also shows outstanding potential for the synthesis of new materials, especially for multitasking systems. Its application, however, in water decontamination via photocatalytic processes has been limited due to the complexity of those systems and the associated costs. The immobilization of biomolecules on photocatalysts affords the photocatalysts greater stability, thereby increasing their applicability in extreme pH conditions in which naked photocatalysts would be unstable. Enzymes such as lactases and peroxidases are frequently used in industrial bioremediation processes as a means to remove organic pollutants. In that context, researchers have developed photocatalytic systems based on enzymes, including polymeric films prepared via the dispersion of titania nanoparticles and the enzyme soybean peroxidase within a polymer matrix [150], or composites based on horseradish peroxidase enzyme and titania [66] as well as on porphyrins, including hemin combined with g-C_3N_4 and titania, which has been shown to be effective for mineralizing organic pollutants in Fenton-based processes [151, 152]. The primary challenges in using these molecules for

designing smart photocatalysts relate to the photocatalysts' deactivation during irradiation, including the negative effect of photogenerated reactive oxygen species. Given these challenges, this approach is not the best strategy for developing realistic photocatalysts for water decontamination in the near future.

- **Photocatalysts based on microorganisms, plants, and natural products:** This broad group of biomaterials involves the direct use of microorganisms, plants, or other natural products as biotemplates for the synthesis of new materials (see Section 5.5 for examples), as well as the use of greener synthetic pathways to develop more sustainable processes and/or more complex architectures. These clean synthetic pathways are especially relevant in cleaner production of new materials, because they usually prevent the use of potential secondary contaminants as aggressive reducing agents, which is necessary for many classical synthetic pathways, and therefore allow avoidance of the incorporation of new secondary contaminants in waters. In view of such a broad group, this section focuses only on the use of microorganisms, plants, or other natural products for the synthesis of new photocatalysts that introduce the concept of cleaner production. Cleaner production has the final aim of preventing the generation of waste as well as increasing the efficient use of energy, water, and resources [108, 153]. In that context, during the last two decades, researchers have investigated the synthesis of various photocatalysts using extracts such as *Catharanthus roseus* extract for $CuSnO_3$ nanoparticles [154], *Moringa oleifera* peel extract for CeO_2 nanoparticles [155], *Ulva lactuca* seaweed extract for ZnO nanoparticles [156], *Cystoseira trinodis* extracts for CuO nanoparticles [157], *Parkia speciosa* Hassk pod extract for SnO_2 quantum dots [158], oils such as flaxseed oil for fluorescent onion-like carbon nanoparticles [159] and *Persea americana* (avocado) oil for Au nanoparticles [160], and leaves such as *Jatropha curcas* leaves for titania nanoparticles [161].
- **Photocatalysts based on biochar:** Biochar is charcoal obtained from plant debris and biomass residues. Biochar-based photocatalysts have significant potential, because they allow the use of residues generated from other processes. Biochar is highly beneficial for photocatalytic water decontamination, because it can act as electron donor or acceptor, sink, and facilitator. Several studies have revealed that biochar can also generate hydroxyl and superoxide anion radicals and hydrogen peroxide [153]. During the last two decades, various smart photocatalysts based on biochar, including Fe_3O_4–BiOBr–biochar, $CaWO_4$–biochar, TiO_2–biochar, TiO_2-Fe-Fe_3C-biochar, and ZnO–bone char, have been identified as efficient photocatalysts for water decontamination, especially for removal of organic pollutants and heavy metals [162–167].

5.7 Advances Toward Improving Photocatalytic Activity via External Stimuli

The action of an external stimulus can be used to provide materials with new functionalities, including locomotor capacity, and to improve the photocatalytic activity of photocatalysts. This section introduces the effects of different external agents on photocatalytic activity.

5.7.1 Sonophotocatalysis

Ultrasound waves are sound waves with a certain frequency above human hearing (i.e. 20–1000 kHz). As shown in Figure 5.10a, the effect of ultrasound is the collapse of cavitation bubbles when water is subjected to a series of high-power compression and rarefaction waves. At high power, cavitation-related phenomena, including nucleation, formation, growth, and the violent collapse of cavities, generate highly localized temperatures of approximately 4000–5000 K and pressures exceeding 1000 atm that decompose water into extremely highly reactive radicals [108, 168, 169], as captured in the following equations:

$$H_2O + \text{ultrasound} \rightarrow H^{\bullet} + {}^{\bullet}OH$$

$$2\,{}^{\bullet}OH \rightarrow H_2O_2$$

$$H^{\bullet} + O_2 \rightarrow HO_2^{\bullet}$$

$$2HO_2^{\bullet} \rightarrow H_2O_2 + O_2$$

$$H_2O + {}^{\bullet}OH \rightarrow H_2O_2 + H^{\bullet}$$

The generation of highly reactive species depends on the intensity of the cavity implosion, which in turn is a function of various factors, including the sound wave's amplitude and frequency [170–172]. Although ultrasonically generated radicals can oxidize organic pollutants, massive energy loss in thermal dissipation (i.e. >50%) hinders the use of sonocatalysis for water decontamination despite rapid degradation [171, 172]. The use of ultrasound, however, seems more viable as a complementary stimulus. In that context, sonophotocatalysis, even with lower-power ultrasound waves, has been shown to significantly improve the photocatalytic activity of many semiconductor photocatalysts [170, 173–175]. Typically, sonophotocatalysis involves the simultaneous combination of ultrasonic sound waves, light, and the use of a semiconductor photocatalyst to enhance the formation of highly reactive species. In sonophotocatalysis, the homogeneous and heterogeneous nucleation of bubbles occurs, while, according to the literature, three phenomena govern the generation of radical species [168, 172]. As shown in Figure 5.10b, homogeneous nucleation produces highly reactive species due to a pyrolytic reaction of water under extreme conditions. Homogeneous nucleation always exists, with greater or lesser intensity, when ultrasound is applied to water. By contrast, the presence of a photocatalyst, usually in the form of a micro or nanostructure, when subjected to an ultrasonic field causes three effects:

i) *An increase in cavitation-related nucleation*: The solid surface or phase boundaries increase the formation of cavitation bubbles and, consequently, the number of radical species. The favored nucleation on the surface of the photocatalyst depends on the nature of the catalyst (i.e. size, shape, composition, and physicochemical properties) and is more favorable on hydrophobic surfaces. It also strongly depends on the temperature, sonication parameters, and concentration of electrolytes and/or other chemicals present in the aqueous medium [168, 172, 176].
ii) *Sonoluminescence effects*: The collapse of cavitation bubbles emits light with a wide wavelength range (i.e. 200–700 nm) and relatively high intensity that is absorbed by

semiconductor photocatalysts. Light with energy equal or greater than the bandgap of the photocatalysts can excite electrons and, in turn, promote the formation of electron–hole pairs. Sonoluminescence, therefore, offers an extra source of light capable of photoexciting photocatalysts [168].

iii) *Thermal effects*: High temperatures can induce the thermal excitation of semiconductors, a phenomenon that translates into the significantly enhanced production of electron–hole pairs and consequently improves the photocatalytic performance of photocatalysts [168, 169].

Additional phenomena may explain the improved photocatalytic activity of photocatalysts when subjected to ultrasound. According to the literature, however, the three mentioned effects are the most plausible [169].

In parallel with these three effects, ultrasound waves are also useful for improving the transport of pollutants to catalyst surfaces while cleaning and preventing the poisoning and inactivation of photocatalyst surfaces [169]. In recent years, various materials such as NiO [177], Ca-doped ZnO nanoparticles [169], graphene-based ZnCr-layered double hydroxide nanocomposites [178], Bi_2WO_6 [179], N/Ti^{3+} codoping biphasic TiO_2–Bi_2WO_6 [180], TiO_2 [181], and FeCuMg and CrCuMg-layered double hydroxides [174] have proven to be efficient as sonophotocatalysts and shown significant improvement over the use of light alone.

5.7.2 Photoelectrocatalysis

In photoelectrocatalysis, a constant anodic potential or constant current density is simultaneously applied to light in order to prevent the recombination of electron–hole pairs (Figure 5.11). The photoanode, based on a semiconductor photocatalyst supported on a conductive substrate, is irradiated with light, and electrons from the VB of the semiconductor are promoted

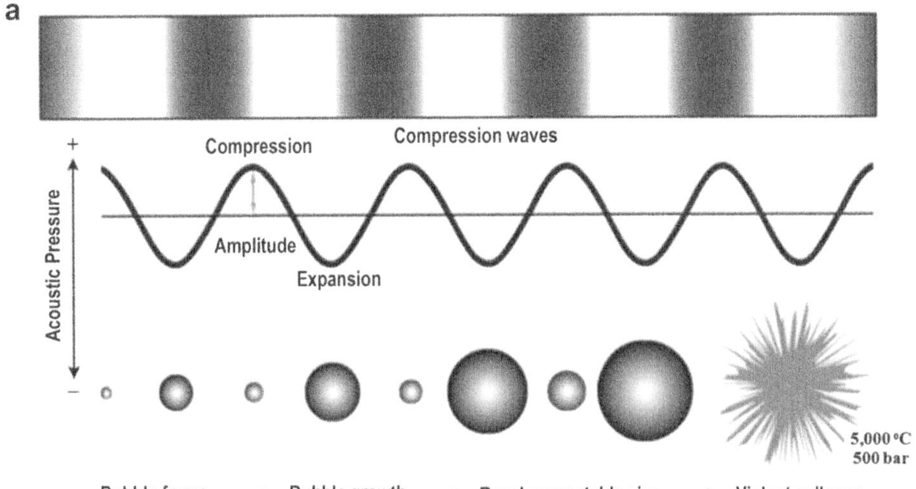

Figure 5.10 (a) Ultrasound-assisted generation of a cavitation bubble. (b) Schematic representation of homogenous and heterogeneous mechanism in sonolytic and sonophotocatalytic mechanisms. Reproduced with permission from [168] / Elsevier.

(Continued)

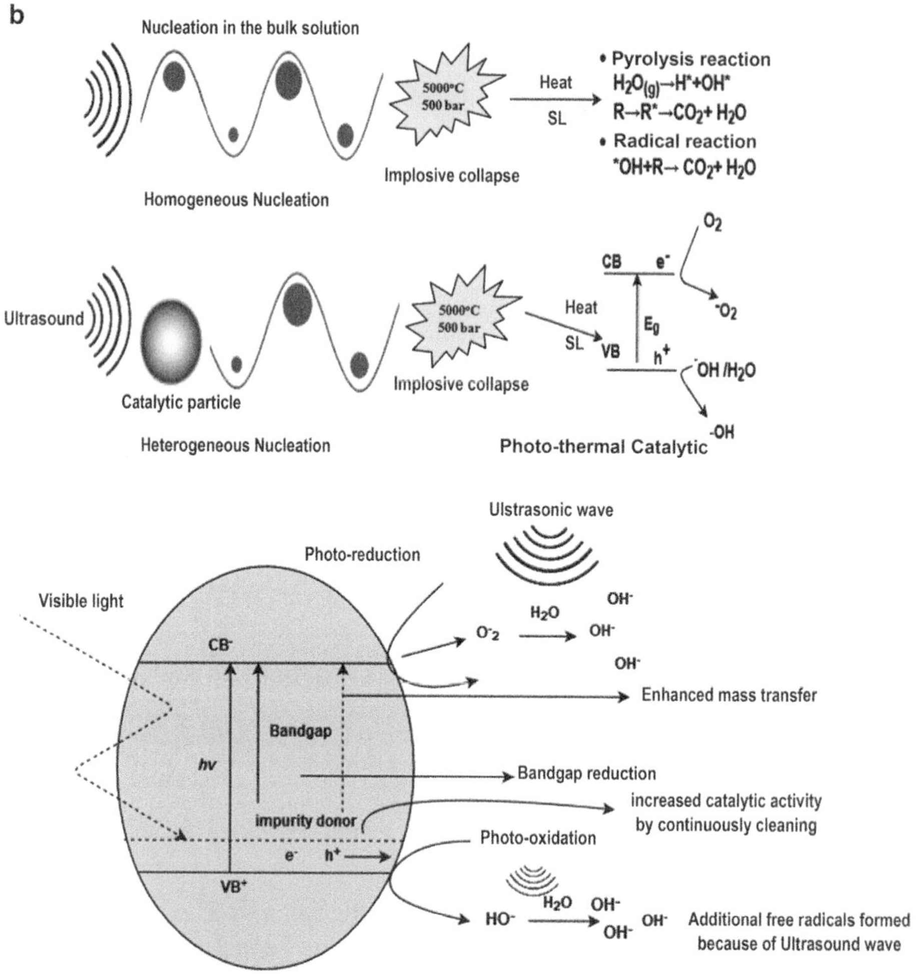

Figure 5.10 (Cont'd) (b) Schematic representation of homogenous and heterogeneous mechanism in sonolytic and sonophotocatalytic mechanisms. Reproduced with permission from [168] / Elsevier.

to its CB, thereby generating a positive vacancy in the VB (h^+_{VB}). As the photogenerated hole reacts with water, reactive oxygen species form such as hydroxyl radicals while electrons (e^-_{CB}) react with oxygen to form superoxide radicals [182–186]. This behavior is identical to that described in Sec. 5.1 for photocatalysis; however, to avoid the fast recombination losses that hinder the use of various photocatalysts and to improve the effectiveness of the photocatalytic process, an external potential can be applied to the semiconductor (i.e. the photoanode). The application of constant anodic potential or constant current density selectively ejects electrons (e^-_{CB}) and consequently decreases the rate of recombination. The photogenerated electrons also can induce reduction reactions at the cathode, e.g. formation of hydrogen peroxide, formation of hydrogen, formation of ammonia of hydrazine via the reduction of nitrogen present in water, reduction of heavy metals, and/or reduction of carbon dioxide into carbon monoxide, methane, or longer chain carbon species [15, 187, 188].

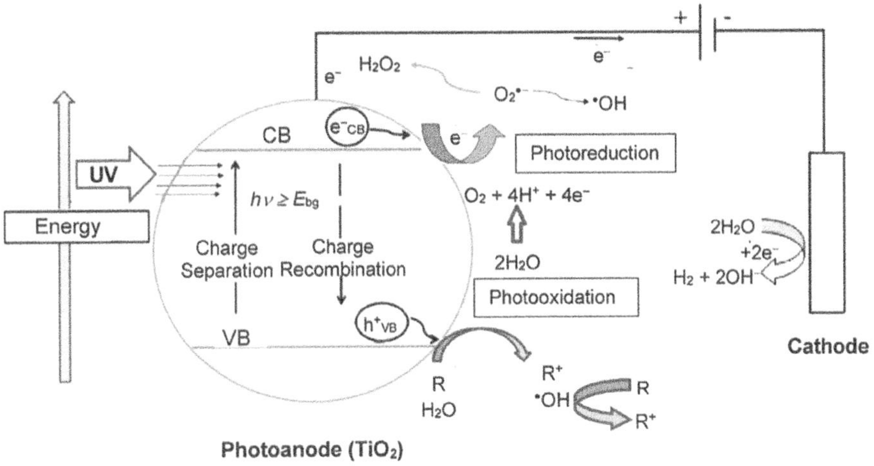

Figure 5.11 Schematic representation of photoelectrochemical process for degrading organic pollutants. [214] / with permission of Elsevier.

The photoelectrochemical process shows remarkable promise for water decontamination technology. In that field, the development of intelligent photocatalysts focuses on the selection of more sustainable synthetic pathways and greener materials, the increased durability and useful life of electrodes, improved photocatalytic activity using photocatalysts with large surface areas and a greater ability to harvest light (e.g. biomimetic architectures), and photocatalysts powered by sunlight or visible light [14]. Examples of photoelectrodes with significant potential for water decontamination, namely for the elimination of persistent organic pollutants, include perovskite oxide-based photoelectrodes [188], TiO_2-based photoanodes [183, 184], WO_3, ZnO, CdS, Fe_2O_3, and SnO_2 [186].

5.7.3 Thermophotocatalysis

Thermophotocatalysis takes advantage of thermocatalytic and photocatalytic effects in synergistic ways. Thermocatalysis works in extreme conditions (e.g. in high pressure and at high temperatures) and often generates significant amounts of harmful byproducts that end up affecting the life of catalysts. By comparison, photocatalysis, especially using sunlight, wastes most of the visible and infrared radiation – that is, most of the solar spectrum (see Figure 5.1). In the field of solar photocatalysis, a pressing challenge has been using the entire solar spectrum from UV to infrared light, but to date, no materials have been developed that allow all solar radiation to be harnessed in the photocatalytic process [189]. Against this background, thermophotocatalysis emerges as an intelligent strategy that allows the use of the whole solar spectrum, including UV, visible, and infrared light, without the external inputs of high temperature and pressure required in industrial thermocatalytic processes. Despite growing interest in this topic, knowledge of the physicochemical grounds for thermophotocatalytic processes remains limited [190–192].

Researchers are currently devoting their efforts to developing efficient thermophotocatalysts, for most state-of-the-art thermocatalysts exhibit low photocatalytic performance, whereas most photocatalysts are only efficient under UV light or part of the visible light spectrum. Nair and collaborators have reported four effective strategies for developing smart photocatalysts for water decontamination that use thermal effects to maximize the photocatalytic performance [190]:

i) Incorporate narrow bandgap photocatalysts that can be excited by near-infrared light – for example, coupling TiO_2, which is photoactive only under UV irradiation, with oxygen-deficient Bi_2WO_6, which is photoactive in near-infrared light [190, 191, 193].
ii) Fabricate nanocomposites of UV and/or visible-light driven photocatalysts with materials such as rare earth metal ions or carbon quantum dots that can convert infrared light to UV and/or visible light, which subsequently photoexcites the photocatalyst and generates reactive species that can oxidize, reduce, or attack pollutants [190, 192, 194].
iii) Fabricate visible and/or near-infrared plasmonic photocatalysts (e.g. oxides, sulfides, and metal-based materials) that utilize plasmonic hot electrons for the photocatalytic decontamination of water [195].
iv) Tune the photocatalyst by defect chemistry, which is mostly used in nonplasmonic oxide materials [190, 196, 197].

In the context of water treatment, despite limited testing to date, the most commonly investigated materials for thermophotocatalysis are Mn_3O_4–$MnCO_3$, CeO_2–Ce_2O_3, Er^{3+}-doped $BiVO_4/TiO_2$, and Er^{3+}-doped $BiVO_4$ [190, 197, 198].

5.8 Advances in Inhibiting the Photocorrosion of Semiconductor-based Photocatalysts

Under light irradiation, semiconductors undergo photocorrosion, a complex process that is often ignored when evaluating the potential use of photocatalysts. However, an in-depth understanding of the mechanisms of photocorrosion is essential to designing more efficient smart photocatalysts [199, 200].

The mechanisms of photocorrosion can be divided into two general categories. On the one hand, the mechanism of photogenerated hole-induced destabilization typically governs the photocorrosion of metal sulfides, oxynitrides, and various metal oxides, including ZnO and Cu_2O. In metal sulfides, hole-driven oxidation converts surface sulfide ions (S^{2-} to sulfur (S) and/or sulfate (SO_4^{2-}). Over time, the accumulation of photogenerated holes causes the photodissolution of the photocatalyst's surface, thereby restricting the use of those photocatalysts and releasing metal ions into the water [199, 201–204]. In ZnO-based photocatalysts, the photoinduced holes are trapped on the ZnO's surface, where they form oxygen molecules and release Zn^{2+} from the surface. As a result, Zn(II) ions could be released into bodies of water during the irradiation of ZnO-based photocatalysts [205]. In the case of Cu_2O-based photocatalysts, Cu_2O can be transformed into CuO [206, 207], whereas in oxynitride-based photocatalysts, nitrogen ions are oxidized to N_2 by photogenerated holes [199, 208].

On the other hand, the mechanism of photogenerated electron-induced destabilization has been described in silver-based and copper-based photocatalysts such as silver orthophosphate, silver halides, and copper (II) oxide. Silver-containing photocatalysts suffer from silver's reduction into metallic Ag due to the action of photogenerated electrons. In the case of CuO, the photocatalysts can be reduced into Cu_2O by photoinduced electrons [199].

Efficiently exporting photogenerated charge carriers is crucial to improving the stability of photocatalysts and hindering their photocorrosion. Table 5.1 summarizes the major strategies that have been developed to design photocatalysts that show low photocorrosion and superior photocatalytic activity.

Because inhibiting photocatalysts' photocorrosive activity is essential before the use of photocatalysts in water decontamination technology can be proposed, smart photocatalysts need to be designed to reduce their photodecomposition.

Table 5.1 Major strategies for hindering the photocorrosion of photocatalysts.

Strategy	Effect	Examples
Modifying the structure, size, and morphology	• High crystallinity generally promotes efficient charge separation and reduces the number of recombination centers, thereby hampering photocorrosion [199]. • Photocorrosion is more prominent in photocatalysts with high surface areas and more unstable, reactive surfaces, both of which circumstances also benefit photocatalytic activity. The size, architecture, and morphology of photocatalysts thus need to be optimized in order to maximize activity and photostability [199]. • Porous morphologies favor charge carrier separation and improve the adsorption of organic pollutants, both of which processes hinder photocorrosion [199, 209, 210].	CdS, ZnO, and Ag_3PO_4, etc.
Doping semiconductors with heteroatoms	• Doping generally tunes the bandgap of semiconductors by introducing impurities or forming solid solutions. Those dynamics extend light absorption toward the visible light domain, which improves the separation of photogenerated charge carriers and, in turn, reduces recombination and improves both photostability and photocatalytic activity [30, 35, 199] (see Section 5.2).	Ni-doped ZnS, Bi-doped Ag_3PO_4, Dy-doped ZnO, N-doped ZnS, S-doped ZnO, etc.
Fabricating homo and heterojunctions	• Heterojunctions are especially useful in promoting the separation of photogenerated charge carriers (see Section 5.2). The combination of tailored individual components can facilitate charge separation and inhibit recombination losses, which consequently hampers photoinduced corrosion [199, 202, 204].	ZnS/ZnO, ZnCdS/CdS, CQD/Ag/Ag_3PO_4, CdS – ZnS, TiO_2/CuO, etc.
Hybridizing semiconductors with cocatalysts and carbon materials	• Adding cocatalysts induces the consumption and/or improves the transport of photogenerated charge carriers, both of which enhance photostability [199, 203].	Ag–Ag_2O, Ag/AgBr, Ag/AgI, Ag-decorated ZnO, Pt/ZnO, RGO – ZnCdS, etc.
Tuning reaction conditions	• Although reaction conditions are often not considered, the pH, electrolyte content, number of sacrificial agents, and other environmental conditions during reactions dramatically affect the chemical and photochemical stability of photocatalysts. Organic pollutants generally consume photogenerated holes and reduce the photocorrosion activity [169].	

5.9 Advances in Recycling Photocatalysts: Assessing the Photocatalyst Life Cycle

To date, the synthesis of photocatalysts has been based primarily on linear processes (e.g. take–make–dispose) that often allow evaluation of the useful life of catalysts but rarely of their recycling. Awareness, however, of the finite nature of many resources and of the climate emergency facing our generation, coupled with the obvious anthropological effects on our environment, can be an important driver of developing circular pathways for synthesizing new materials, especially new photocatalysts. The ultimate goal of circular processes is to make the management of natural resources as efficient as possible, generally by repurposing the waste of one phase as the raw material of a new phase and thereby achieving a virtually waste-free production cycle [10, 105, 211, 212].

Researchers have recently begun investigating holistic, circular, self-sustainable processes at the water–energy nexus and even at the water–energy–food–pharmaceuticals nexus if the nutritional and pharmaceutical properties of hybrids are considered. For example, Serrà and collaborators have proposed contributing to sustainable circular economies by developing recyclable biomimetic hybrid photocatalysts using microalgae as a template in a simple, economical, scalable synthetic process (see Figure 5.12) [10, 105, 211–213]. Their proposed synthesis pathway integrates the principles of green chemistry and circular chemistry by combining the use of electrochemical deposition (i.e. electrodeposition or electroless deposition) and biotemplates to manufacture more efficient micro and nanometric photocatalysts, all powered by sunlight and based on various semiconductors. After their effective lifetime is spent eliminating emerging pollutants or microplastics, the photocatalysts can be transformed into a source of energy for the

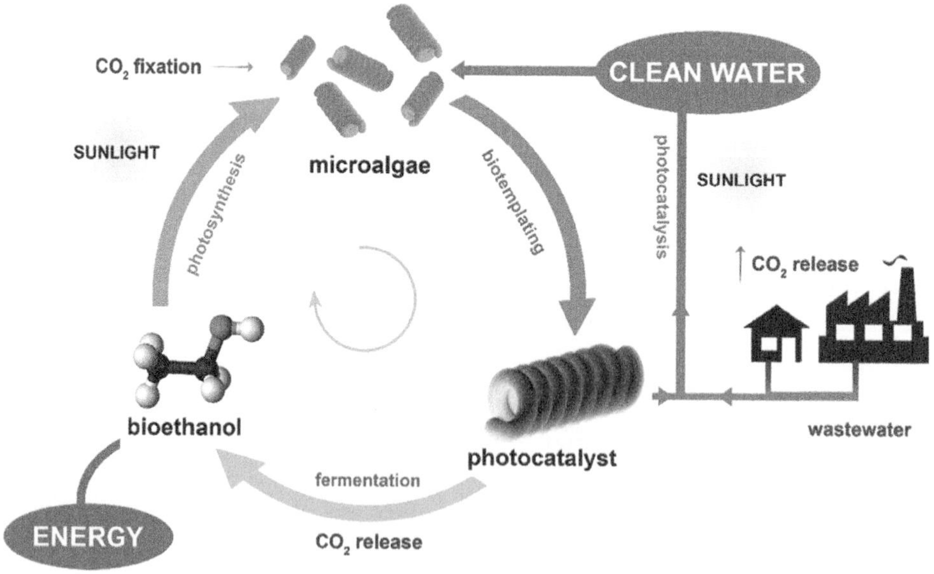

Figure 5.12 Schematic representation of the circular process based on hybrid microalgae photocatalysts. [10] / John Wiley & Sons / CC BY 4.0.

production of biofuel by being recycled – namely, as biopellets. Beyond that, their strategy allows all waste generated during the process to be reused in the production of new catalysts, biofuels, and/or supplements for growing plants.

A major part of the potential of the process proposed in Figure 5.12 is that it places sustainability at the heart of the synthesis of new smart photocatalysts. The development of such circular processes can promote not only environmental but also economical sustainability in the use of photocatalysts for water decontamination.

5.10 Summary, Future Challenges, and Prospects for Further Research

In September 2015, the United Nations held a summit to address the sustainable development of Planet Earth. Representatives of the social, political, scientific, and private sectors drafted a 15-year agenda containing 17 Sustainable Development Goals (SDGs), which entered into force in January 2016. Among its other goals, the agenda contains an SDG dedicated to water and sanitation (i.e. SDG6) with the aim to "ensure the availability and sustainable management of water and sanitation for all." SDG6 responds to the fact that, to date, more than 40% of the world's population is affected by water scarcity, a highly concerning trend that is expected to worsen due to global warming and indiscriminate anthropogenic activities. To mitigate the current dual crises of water pollution and water shortages, the protection and recovery of aqueous ecosystems, as well as the implementation of efficient water treatment technologies, are urgently needed.

Micro and nanotechnologies currently play an important role in the development of novel water treatment approaches thanks to significant advances in micro and nanostructured materials able to, among other things, degrade and/or capture pollutants. The high surface area of micro and nanostructured materials ranks among the primary features exploited in heterogeneous catalysis or adsorption chemistry, two disciplines with major relevance in environmental chemistry. Water cleaning and remediation involve the application of physical, chemical, thermal, and/or biological means to target specific types of contaminants. To date, however, most current treatments are either pollutant-specific or do not discriminate between harmful or innocuous species, the latter of which can be problematic if the technology is to be employed directly in aquatic ecosystems. A clear need thus exists to develop a technology capable of targeting several kinds of pollutants (i.e. organic compounds, heavy metals, and microorganisms) at once.

The water crisis, and more specifically the need to decontaminate bodies of water, currently drives the development of new technologies and strategies that should be scalable and applicable at a low cost worldwide. The problems of water scarcity and pollution are especially important in developing countries and states with significant social and economic inequalities. In that context, photocatalysis has emerged as a clean, sustainable pathway that can be exploited at low cost using sunlight as a source of energy across the globe. Significant improvements, however, in the design of photocatalysts, the design of photoreactors, and their industrial application need to be resolved before real, extendable applications can be proposed for worldwide use in the near future. For the design of catalysts, especially intelligent catalysts with multifunctionality or improved catalytic activity, the primary challenges are (summary in Table 5.2):

Table 5.2 Recent progress in smart photocatalysis for water remediation.

Challenge	Strategy	Examples
Improving the photosensitivity of photocatalysts by extending their use to the domain of visible light	• Doping • Homo/heterojunction formation • Surface modification • Dye sensitization. • Synthesis of alternative photocatalysts	• Anionic doping (e.g. C, N, and S), cationic doping (e.g. Al, Sb, Mn, Ni, Co, Bi, Fe, K, and Mg), rare earth doping (e.g. Ce, Dy, Er, Eu, Gd, Ho, and Nd), and codoping • TiO_2 anatase–brookite, ZnO–ZnS, ZnSe–ZnO, SnO_2–CdS, CdS–TiO_2, Bi_2MoO_6–Bi_2MoO_{6-x}, and CdS–ZnO • Core@shell architectures and nanoparticle decoration • Visible light photocatalysts such as Bi_2MoO_6 and Bi_2WO_6
Immobilizing photocatalysts without reducing their photocatalytic activity due to lower interaction with light and/or pollutants	• Photocatalytic membranes • Porous supports • Fibers, rods, wires and cables	• Fe_2O_3@COFs, CFT-3D–BiOBr, COF-PD–AgI, Py-BF-CMP, Fc-TEB-CMP, and UPC-CMP-1 • Cu@Cu_2O@CuO cables
Harnessing the advantages of nonimmobilized photocatalysts	• Micro and nanodevices	• TiO_2/Pt microrobots [88] • Pt-SiO_2 Janus micromotors [89] • Light-driven TiO_2/Fe_3O_4/CdS microsubmarines [90] • Magnetic Fe-TiO_2 helical micromotors [91] • TiO_2/Au/Mg micromotors [72] • Magnetically driven Bi_2O_3/BiOCl-based hybrid micromotors [41] • ZnO/Pt Janus micromotors [94] • ZnO micromotors [96]
Improving the absorption of light, especially sunlight	• Inspiration from algae • Inspiration from insects • Inspiration from plants	• Ni@ZnO@ZnS-microalgae biohybrid photocatalysts • ZnO/Si moth-eye structures [115] • $CdIn_2S_4$/UiO-66 nanocomposites with flower-like 3D-microspheres [140] • Fe-(8-hydroxyquinoline-7-carboxilic)–TiO_2 composite flowers [141] • Human-made omnidirectional trackers [14] • ZnO@ZnS core@shell biomimetic fern-like microleaves [4]

(Continued)

Table 5.2 (Continued)

Challenge	Strategy	Examples
Designing intelligent photocatalysts	• Biodegradability • Biocompatibility and low toxicity • Antimicrobial properties • Structural, textural, superficial, and optical properties	• Photocatalysts based on biopolymers • Photocatalysts based on immobilized molecules • Photocatalysts based on microorganisms, plants, and natural products • Photocatalysts based on biochar
Determining the effect of external variables that maximize the efficiency of photocatalysis	• Sonophotocatalysis • Photoelectrocatalysis • Thermophotocatalysis	• Ca-doped ZnO nanoparticles [169] • NiO [177] • Graphene-based ZnCr-layered double hydroxide nanocomposites [178], Bi_2WO_6 [179] • N/Ti^{3+} codoping biphasic TiO_2–Bi_2WO_6 [180]
Improving the chemical and photochemical stability of photocatalysts	• Modifying the structure, size, and morphology • Doping with heteroatoms • Fabricating homo and heterojunctions • Hybridizing semiconductors with cocatalysts and carbon materials • Tuning reaction conditions	• Ni@ZnO@ZnS-microalgae biohybrid photocatalysts [10] • Ca-doped ZnO nanoparticles [169]
Incorporating green chemistry thinking	• Circular processes • Sustainable synthetic pathways	• Ni@ZnO@ZnS-microalgae biohybrid photocatalysts [10]

- *Improving the photosensitivity of photocatalysts by extending their use to the domain of visible light.* Although LED technology significantly reduces energy consumption compared with conventional light sources and despite monochromatic LEDs in the domain of UV light, the photocatalytic decontamination of water should instead take advantage of sunlight to assure its economic and social sustainability and applicability in developing countries. Important advances in the design of photocatalysts have been driven by visible or solar light based on doping processes, the formation of homo and heterojunctions, surface modifications (e.g. decoration with metal nanoparticles), dye sensitization, and/or the search for new semiconductor materials with smaller bandgaps. Significant efforts should be made to modify the optoelectronic properties of photocatalysts to extend their use into the visible light domain, to improve the efficiency of the separation of charge carriers, and, in turn, to reduce recombination and maximize photocatalytic performance.

- *Immobilizing photocatalysts without reducing their photocatalytic activity due to lower interaction with light and/or pollutants.* Immobilizing photocatalysts without lowering their photocatalytic activity is an important challenge that needs to be overcome in order to design more efficient photocatalytic reactors. The subchallenges therein lie in the hassle of integrating the use of sunlight and designing photoreactors able to harvest sunlight, which requires holistic thinking that incorporates both the scientific method and engineering. Some of the primary pathways being investigated involve the use of photocatalytic membranes, porous supports, and structures such as cables, fibers, and threads.
- *Harnessing the advantages of nonimmobilized photocatalysts.* Nonimmobilized photocatalysts offer important advantages over immobilized ones, especially in relation to the efficiency of water decontamination, for they facilitate pollutant–photocatalyst interaction, light–photocatalyst interaction, and the incorporation of additional functionalities. Of particular importance in the field are robotic platforms at the micro and nanoscale, which, as tiny devices, can be propelled via fluids by means of external sources of energy (i.e. magnetic or electric fields and/or acoustic waves or light) or as result of chemical reactions occurring at the surface of the machine. Recently, several works have shown that micro and nanoswimmers can be used to capture heavy metals and oil droplets, degrade organic compounds, and/or kill bacteria. On top of that, micro and nanomotors offer several advantages over other micro and nanosystems for applications of environmental remediation. For one, because they exhibit locomotive capabilities, they can be recovered and recycled. For another, the motility of micro and nanoswimmers can enhance the rate of certain degradation reactions. Another advantage of micro and nanoswimmers is their large cargo-towing force that, combined with their functional components, can be exploited to target and isolate specific harmful compounds. Additionally, some micro and nanomachines can be designed to follow chemical gradients. Deepening knowledge of the mechanisms of locomotion as well as their effects on photocatalytic properties are an important challenge, for which the aim for the near future is twofold. First is to create highly sophisticated nanorobotic systems that can integrate components for multiple locomotion mechanisms and several building blocks for cleaning and detoxifying contaminants in water. Second is to develop scalable synthetic pathways for the mass production of such sophisticated, efficient nanorobotic platforms.
- *Improving the absorption of light, especially sunlight.* Improving the absorption of sunlight can help to maximize the efficiency of photocatalytic processes. The design of biomimetic photocatalysts is a good strategy that needs to be further investigated to take advantage of the greater interaction between light and photocatalysts. In that sense, nature serves as a school for the design of new intelligent photocatalysts.
- *Designing intelligent photocatalysts.* Intelligent photocatalysts need to be based on the principles of green chemistry and sustainable synthetic processes must always be prioritized. At the same time, photocatalysts need to have minimal impact on the generation of new waste at the end of their useful lives. The use of biodegradable, recyclable,

biocompatible base materials should also be incorporated into the synthesis of new photocatalysts in the coming years.

- *Determining the effect of external variables that maximize the efficiency of photocatalysis.* The cost associated with external stimuli cannot be high considering the need for the widespread application of such techniques of water decontamination. The application of ultrasound significantly improves photocatalytic activity but at once involves exceedingly high energy consumption and poses problems for scalability at the industrial level, both of which make it virtually infeasible. Albeit a technology currently applied in many processes, photoelectrocatalysis can be greatly improved considering the aspects that have been raised thus far. Thermophotocatalysis, though among the least known strategies, can have important advantages, starting with the use of sunlight. In that field, it is necessary to extend the design of photocatalysts and photoreactors for the benefit of the processes that maximize photocatalytic activity.
- *Improving the chemical and photochemical stability of photocatalysts.* The durability of photocatalysts has to be guaranteed. Whereas most research to date has evaluated photocatalyst activity over several laboratory-scale cycles, it has rarely done so over the course of a few weeks. It is therefore necessary to improve the stability of photocatalysts and ensure acceptable photocatalytic activity for a period that makes the synthesis of the material economically, humanly, and environmentally sustainable. At the same time, the photodecomposition of photocatalysts can preclude the new entry of pollutants into aqueous systems. Improving the resistance to photocorrosion, therefore, is a prerequisite for the potential application of smart photocatalysts. Although the primary strategies being investigated partly solve those problems, the long-term stability of photocatalysts needs to be analyzed.
- *Incorporating green chemistry thinking.* Although green chemistry thinking is clearly essential at present, incorporating the principles of circular chemistry can prevent the production of waste, particularly because products of the new process can contribute to a smart strategy for improving the overall efficiency of water treatment and facilitate the use of photocatalysis for the decontamination of water and/or other applications. In that sense, biomimicry and biotemplating strategies open doors to the design of circular processes.

References

1 Pol, R., Guerrero, M., García-Lecina, E., Altube, A., Rossinyol, E., Garroni, S. et al. (2016). Ni-, Pt- and (Ni/Pt)-doped TiO_2 nanophotocatalysts: a smart approach for sustainable degradation of Rhodamine B dye. *Applied Catalysis B: Environmental* 181: 270–278. http://dx.doi.org/10.1016/j.apcatb.2015.08.006.

2 Mara, D. (2016). The all-inclusive sustainable development goals: the WASH professional's guide (or should that be 'nightmare'?). *Journal of Water, Sanitation and Hygiene for Development* 6 (3): 349–352.

3 Bain, R., Johnston, R., Mitis, F., Chatterley, C., and Slaymaker, T. (2018). Establishing sustainable development goal baselines for household drinking water, sanitation and hygiene services. *Water (Switzerland)* 10: 12.

4 Serrà, A., Zhang, Y., Sepúlveda, B., Gómez, E., Nogués, J., Michler, J. et al. (2019). Highly active ZnO-based biomimetic fern-like microleaves for photocatalytic water decontamination using sunlight. *Applied Catalysis B: Environmental* 248: 129–146. https://doi.org/10.1016/j.apcatb.2019.02.017

5 Cumberlege, O. (1993). Making every drop count. *Waterlines* 11 (3): 23–26.

6 Jury, W.A. and Vaux, H.J. (2007). The emerging global water crisis: managing scarcity and conflict between water users. *Advances in Agronomy* 95 (07): 1–76.

7 Axon, S. and James, D. (2018). The UN sustainable development goals: how can sustainable chemistry contribute? A view from the chemical industry. *Current Opinion in Green and Sustainable Chemistry* 13: 140–145. https://doi.org/10.1016/j.cogsc.2018.04.010

8 Byrne, C., Subramanian, G., and Pillai, S.C. (2018). Recent advances in photocatalysis for environmental applications. *Journal of Environmental Chemical Engineering* 6 (3): 3531–3555. https://doi.org/10.1016/j.jece.2017.07.080

9 Serrà, A., Philippe, L., Perreault, F., and Garcia-Segura, S. (2021). Photocatalytic treatment of natural waters. Reality or hype? The case of cyanotoxins remediation. *Water Research* 188: 116543.

10 Serrà, A., Artal, R., García-Amorós, J., Sepúlveda, B., Gómez, E., Nogués, J. et al. (2020). Hybrid Ni@ZnO@ZnS-microalgae for circular economy: a smart route to the efficient integration of solar photocatalytic water decontamination and bioethanol production. *Advancement of Science* 7 (3): 1–9.

11 Etacheri, V., Di Valentin, C., Schneider, J., Bahnemann, D., and Pillai, S.C. (2015). Visible-light activation of TiO2 photocatalysts: advances in theory and experiments. *Journal of Photochemistry and Photobiology C: Photochemistry Reviews* 25: 1–29. https://doi.org/10.1016/j.jphotochemrev.2015.08.003

12 Hoffmann, M.R., Martin, S.T., Choi, W., and Bahnemann, D.W. (1995). Environmental applications of semiconductor photocatalysis. *Chemical Reviews* 95 (1): 69–96.

13 Fujishima, A. and Honda, K. (1972). Electrochemical photolysis of water at a semiconductor electrode one and two-dimensional structure of alpha-helix and beta-sheet forms of poly (L-Alanine) shown by specific heat measurements at low temperatures (1.5–20 K). *Nature* 238 (March): 37–38.

14 Brillas, E., Serrà, A., and Garcia-Segura, S. (2021). Biomimicry designs for photoelectrochemical systems: strategies to improve light delivery efficiency. *Current Opinion in Electrochemistry* 26 (2): 100660.

15 Garcia-Segura, S. and Brillas, E. (2017). Applied photoelectrocatalysis on the degradation of organic pollutants in wastewaters. *Journal of Photochemistry and Photobiology C: Photochemistry Reviews* 31: 1–35. https://doi.org/10.1016/j.jphotochemrev.2017.01.005

16 Xiong, Z., Lei, Z., Li, Y., Dong, L., Zhao, Y., and Zhang, J. (2018). A review on modification of facet-engineered TiO2 for photocatalytic CO2 reduction. *Journal of Photochemistry and Photobiology C: Photochemistry Reviews* 36: 24–47. https://doi.org/10.1016/j.jphotochemrev.2018.07.002

17 Manzano, C.V., Philippe, L., and Serrà, A. (2021). Recent progress in the electrochemical deposition of ZnO nanowires: synthesis approaches and applications. *Critical Reviews in Solid State and Materials Sciences* 0 (0): 1–34. https://doi.org/10.1080/10408436.2021.1989663

18 Zhang, X., Wang, D.K., and Diniz Da Costa, J.C. (2014). Recent progresses on fabrication of photocatalytic membranes for water treatment. *Catalysis Today* 230: 47–54. https://doi.org/10.1016/j.cattod.2013.11.019

19 Hernández-Alonso, M.D., Fresno, F., Suárez, S., and Coronado, J.M. (2009). Development of alternative photocatalysts to TiO2: challenges and opportunities. *Energy & Environmental Science* 2 (12): 1231–1257.

20 Chen, L., Zhang, M.J., Zhang, S.Y., Shi, L., Yang, Y.M., Liu, Z. et al. (2020). Simple and continuous fabrication of self-propelled micromotors with photocatalytic metal-organic frameworks for enhanced synergistic environmental remediation. *ACS Applied Materials & Interfaces* 12 (31): 35120–35131.

21 Zhang, Z., Zhao, A., Wang, F., Ren, J., and Qu, X. (2016). Design of a plasmonic micromotor for enhanced photo-remediation of polluted anaerobic stagnant waters. *Chemical Communications* 52 (32): 5550–5553. https://doi.org/10.1039/C6CC00910G

22 Wang, Q., Dong, R., Wang, C., Xu, S., Chen, D., Liang, Y. et al. (2019). Glucose-fueled micromotors with highly efficient visible-light photocatalytic propulsion. *ACS Applied Materials & Interfaces* 11 (6): 6201–6207.

23 Jang, B., Hong, A., Kang, H.E., Alcantara, C., Charreyron, S., Mushtaq, F. et al. (2017). Multiwavelength light-responsive Au/B-TiO2 janus micromotors. *ACS Nano* 11 (6): 6146–6154.

24 García-Torres, J., Serrà, A., Tierno, P., Alcobé, X., and Vallés, E. (2017). Magnetic propulsion of recyclable catalytic nanocleaners for pollutant degradation. *ACS Applied Materials & Interfaces* 9 (28): 23859–23868.

25 Serrà, A. and García-Torres, J. (2021). Electrochemistry: a basic and powerful tool for micro and nanomotor fabrication and characterization. *Applied Materials Today* 22: 100939.

26 Pirhashemi, M. and Habibi-Yangjeh, A. (2018). Facile fabrication of novel ZnO/CoMoO4nanocomposites: highly efficient visible-light-responsive photocatalysts in degradations of different contaminants. *Journal of Photochemistry and Photobiology A: Chemistry* 363 (January): 31–43.

27 Jo, W.K. and Tayade, R.J. (2014). New generation energy-efficient light source for photocatalysis: LEDs for environmental applications. *Industrial & Engineering Chemistry Research* 53 (6): 2073–2084.

28 Akbal, F. (2005). Photocatalytic degradation of organic dyes in the presence of titanium dioxide under UV and solar light: Effect of operational parameters. *Environmental Progress* 24 (3): 317–322.

29 Tayade, R.J., Surolia, P.K., Kulkarni, R.G., and Jasra, R.V. (2007). Photocatalytic degradation of dyes and organic contaminants in water using nanocrystalline anatase and rutile TiO2. *Science and Technology of Advanced Materials* 8 (6): 455–462.

30 Han, C., Pelaez, M., Likodimos, V., Kontos, A.G., Falaras, P., O'Shea, K. et al. (2011). Innovative visible light-activated sulfur doped TiO2 films for water treatment. *Applied Catalysis B: Environmental* 107 (1–2): 77–87. https://doi.org/10.1016/j.apcatb.2011.06.039

31 Xie, Z., Feng, Y., Wang, F., Chen, D., Zhang, Q., Zeng, Y. et al. (2018). Construction of carbon dots modified MoO3/g-C3N4 Z-scheme photocatalyst with enhanced visible-light photocatalytic activity for the degradation of tetracycline. *Applied Catalysis B: Environmental* 229 (November 2017): 96–104. https://doi.org/10.1016/j.apcatb.2018.02.011

32 Kumar, A., Sharma, G., Naushad, M., Al-Muhtaseb, A.H., García-Peñas, A., Mola, G.T. et al. (2020). Bio-inspired and biomaterials-based hybrid photocatalysts for environmental detoxification: a review. *Chemical Engineering Journal* 382 (April 2019).

33 Steplin Paul Selvin, S., Radhika, N., Borang, O., Sharmila, L.I., and Princy, M.J. (2017). Visible light driven photodegradation of Rhodamine B using cysteine capped ZnO/GO nanocomposite as photocatalyst. *Journal of Materials Science: Materials in Electronics* 28 (9): 6722–6730.

34 Ong, C.B., Ng, L.Y., and Mohammad, A.W. (2018). A review of ZnO nanoparticles as solar photocatalysts: synthesis, mechanisms and applications. *Renewable & Sustainable Energy Reviews* 81 (July 2016): 536–551. https://doi.org/10.1016/j.rser.2017.08.020

35 Vinesh, V., Shaheer, A.R.M., and Neppolian, B. (2019). Reduced graphene oxide (rGO) supported electron deficient B-doped TiO2 (Au/B-TiO2/rGO) nanocomposite: an efficient visible light sonophotocatalyst for the degradation of Tetracycline (TC). *Ultrasonics Sonochemistry* 50 (July 2018): 302–310.

36 Ranjith, K.S., Castillo, R.B., Sillanpaa, M., and Rajendra Kumar, R.T. (2018). Effective shell wall thickness of vertically aligned ZnO-ZnS core-shell nanorod arrays on visible photocatalytic and photo sensing properties. *Applied Catalysis B: Environmental* 237 (March): 128–139. https://doi.org/10.1016/j.apcatb.2018.03.099

37 Chen, F., Yang, Q., Sun, J., Yao, F., Wang, S., Wang, Y. et al. (2016). Enhanced photocatalytic degradation of tetracycline by AgI/BiVO4 heterojunction under visible-light irradiation: mineralization efficiency and mechanism. *ACS Applied Materials & Interfaces* 8 (48): 32887–32900.

38 Guo, J., Shi, L., Zhao, J., Wang, Y., Tang, K., Zhang, W. et al. (2018). Enhanced visible-light photocatalytic activity of Bi2MoO6 nanoplates with heterogeneous Bi2MoO6-x@Bi2MoO6 core-shell structure. *Applied Catalysis B: Environmental* 224 (November2017): 692–704. https://doi.org/10.1016/j.apcatb.2017.11.030

39 Chowdhury, P., Moreira, J., Gomaa, H., and Ray, A.K. (2012). Visible-solar-light-driven photocatalytic degradation of phenol with dye-sensitized TiO 2: parametric and kinetic study. *Industrial & Engineering Chemistry Research* 51 (12): 4523–4532.

40 Oliveros, A.N., Pimentel, J.A.I., de Luna, M.D.G., Garcia-Segura, S., Abarca, R.R.M., and Doong, R.A. (2021). Visible-light photocatalytic diclofenac removal by tunable vanadium pentoxide/boron-doped graphitic carbon nitride composite. *Chemical Engineering Journal* 403 (July 2020).

41 Mushtaq, F., Guerrero, M., Sakar, M.S., Hoop, M., Lindo, A.M., Sort, J. et al. (2015). Magnetically driven Bi2O3/BiOCl-based hybrid microrobots for photocatalytic water remediation. *Journal of Materials Chemistry A* 3 (47): 23670–23676.

42 El-Shaer, A., Dev, A., Richters, J.P., Waldvogel, S.R., Waltermann, J., Schade, W. et al. (2010). Hybrid LEDs based on ZnO-nanowire arrays. *Physica Status Solidi: Basic Research* 247 (6): 1564–1567.

43 Lifson, M.L., Levey, C.G., and Gibson, U.J. (2013). Diameter and location control of ZnO nanowires using electrodeposition and sodium citrate. *Applied Physics A: Materials Science & Processing* 113 (1): 243–247.

44 Shen, H., Fu, F., Xue, W., Yang, X., Ajmal, S., Zhen, Y. et al. (2021). In situ fabrication of Bi2MoO6/Bi2MoO6-x homojunction photocatalyst for simultaneous photocatalytic phenol degradation and Cr(VI) reduction. *Journal of Colloid and Interface Science* 599: 741–751. https://doi.org/10.1016/j.jcis.2021.04.122

45 Zhang, G., Nadagouda, M.N., O'Shea, K., El-Sheikh, S.M., Ismail, A.A., Likodimos, V. et al. (2014). Degradation of cylindrospermopsin by using polymorphic titanium dioxide under UV-Vis irradiation. *Catalysis Today* 224: 49–55. https://doi.org/10.1016/j.cattod.2013.10.072

46 Wang, Y., Wang, Q., Zhan, X., Wang, F., Safdar, M., and He, J. (2013). Visible light driven type II heterostructures and their enhanced photocatalysis properties: a review. *Nanoscale* 5 (18): 8326–8339.

47 Zhai, C., Sun, M., Zeng, L., Xue, M., Pan, J., Du, Y. et al. (2019). Construction of Pt/graphitic C3N4/MoS2 heterostructures on photo-enhanced electrocatalytic oxidation of small organic molecules. *Applied Catalysis B: Environmental* 243 (June 2018): 283–293.

48 Ghoul, M., Braiek, Z., Brayek, A., Ben Assaker, I., Khalifa, N., Ben Naceur, J. et al. (2015). Synthesis of core/shell ZnO/ZnSe nanowires using novel low cost two-steps electrochemical deposition technique. *Journal of Alloys and Compounds* 647: 660–664. https://doi.org/10.1016/j.jallcom.2015.06.100

49 Torabi, A. and Staroverov, V.N. (2015). Band gap reduction in ZnO and ZnS by creating layered ZnO/ZnS heterostructures. *The Journal of Physical Chemistry Letters* 6 (11): 2075–2080. https://doi.org/10.1021/acs.jpclett.5b00687

50 Pirhashemi, M., Habibi-Yangjeh, A., and Rahim Pouran, S. (2018). Review on the criteria anticipated for the fabrication of highly efficient ZnO-based visible-light-driven photocatalysts. *Journal of Industrial and Engineering Chemistry* 62: 1–25. https://doi.org/10.1016/j.jiec.2018.01.012

51 Brayek, A., Ghoul, M., Souissi, A., Ben Assaker, I., Lecoq, H., Nowak, S. et al. (2014). Structural and optical properties of ZnS/ZnO core/shell nanowires grown on ITO glass. *Materials Letters* 129: 142–145. https://doi.org/10.1016/j.matlet.2014.04.192

52 Braiek, Z., Brayek, A., Ghoul, M., Ben Taieb, S., Gannouni, M., Ben Assaker, I. et al. (2015). Electrochemical synthesis of ZnO/In2S3 core-shell nanowires for enhanced photoelectrochemical properties. *Journal of Alloys and Compounds* 653: 395–401. https://doi.org/10.1016/j.jallcom.2015.08.204

53 Hao, L., Ju, P., Zhang, Y., Zhai, X., Sun, C., Duan, J. et al. (2021). Fabrication of hierarchical flower-like BiOI/MoS2 heterostructures with highly enhanced visible-light photocatalytic activities. *Colloids and Surfaces A: Physicochemical and Engineering Aspects* 610 (August 2020): 125714. https://doi.org/10.1016/j.colsurfa.2020.125714

54 Naya, S.I., Yamauchi, J., Okubo, T., and Tada, H. (2017). Rapid removal and mineralization of bisphenol A by heterosupramolecular plasmonic photocatalyst consisting of gold nanoparticle-loaded titanium(IV) oxide and surfactant admicelle. *Langmuir* 33 (40): 10468–10472.

55 Zhu, Y., Wang, B., Deng, C., Wang, Y., and Wang, X. (2021). Photothermal-pyroelectric-plasmonic coupling for high performance and tunable band-selective photodetector. *Nano Energy* 83 (January).

56 Wang, C.C., Shieu, F.S., and Shih, H.C. (2020). Enhanced photodegradation by RGO/ZnO core-shell nanostructures. *Journal of Environmental Chemical Engineering* 8 (1): 103589.

57 Fang, J., Gu, J., Liu, Q., Zhang, W., Su, H., and Zhang, D. (2018). Three-dimensional CdS/Au butterfly wing scales with hierarchical rib structures for plasmon-enhanced photocatalytic hydrogen production. *ACS Applied Materials & Interfaces* 10 (23): 19649–19655.

58 Qin, G., Sun, Z., Wu, Q., Lin, L., Liang, M., and Xue, S. (2011). Dye-sensitized TiO2 film with bifunctionalized zones for photocatalytic degradation of 4-cholophenol. *Journal of Hazardous Materials* 192 (2): 599–604.

59 Chong, M.N., Jin, B., Chow, C.W.K., and Saint, C. (2010). Recent developments in photocatalytic water treatment technology: a review. *Water Research* 44 (10): 2997–3027. https://doi.org/10.1016/j.watres.2010.02.039

60 Liu, M., Hou, L., Yu, S., Xi, B., Zhao, Y., and Xia, X. (2013). MCM-41 impregnated with A zeolite precursor: synthesis, characterization and tetracycline antibiotics removal from aqueous solution. *Chemical Engineering Journal* 223: 678–687. https://doi.org/10.1016/j.cej.2013.02.088

61 McCullagh, C., Skillen, N., Adams, M., and Robertson, P.K.J. (2011). Photocatalytic reactors for environmental remediation: a review. *Journal of Chemical Technology and Biotechnology* 86 (8): 1002–1017.

62 Braham, R.J. and Harris, A.T. (2009). Review of major design and scale-up considerations for solar photocatalytic reactors. *Industrial & Engineering Chemistry Research* 48 (19): 8890–8905.

63 Chauhan, I., Aggrawal, S., and Mohanty, P. (2015). ZnO nanowire-immobilized paper matrices for visible light-induced antibacterial activity against Escherichia coli. *Environmental Science: Nano* 2 (3): 273–279. https://doi.org/10.1039/C5EN00006H

64 Rincón, G.J. and La Motta, E.J. (2019). A fluidized-bed reactor for the photocatalytic mineralization of phenol on TiO2-coated silica gel. *Heliyon* 5: 6.

65 Srikanth, B., Goutham, R., Badri Narayan, R., Ramprasath, A., Gopinath, K.P., and Sankaranarayanan, A.R. (2017). Recent advancements in supporting materials for immobilised photocatalytic applications in waste water treatment. *Journal of Environmental Management* 200: 60–78.

66 Cheng, H., Hu, M., Zhai, Q., Li, S., and Jiang, Y. (2018). Polydopamine tethered CPO/HRP-TiO2 nano-composites with high bio-catalytic activity, stability and reusability: Enzyme-photo bifunctional synergistic catalysis in water treatment. *Chemical Engineering Journal* 347 (December 2017): 703–710. https://doi.org/10.1016/j.cej.2018.04.083

67 Priya, A.K., Suresh, R., Kumar, P.S., Rajendran, S., Vo, D.V.N., and Soto-Moscoso, M. (2021). A review on recent advancements in photocatalytic remediation for harmful inorganic and organic gases. *Chemosphere* 284 (May): 131344.

68 Gonzalez-Perez, A. and Persson, K.M. (2016). Bioinspired materials for water purification. *Materials (Basel)* 9 (6): 447.

69 Yin, J. and Deng, B. (2015). Polymer-matrix nanocomposite membranes for water treatment. *Journal of Membrane Science* 479: 256–275. https://doi.org/10.1016/j.memsci.2014.11.019

70 Zakria, H.S., Othman, M.H.D., Kamaludin, R., Sheikh Abdul Kadir, S.H., Kurniawan, T.A., and Jilani, A. (2021). Immobilization techniques of a photocatalyst into and onto a polymer membrane for photocatalytic activity. *RSC Advances* 11 (12): 6985–7014.

71 Wang, C., Wu, Y., Lu, J., Zhao, J., Cui, J., Wu, X. et al. (2017). Bioinspired synthesis of photocatalytic nanocomposite membranes based on synergy of Au-TiO2 and polydopamine for degradation of tetracycline under visible light. *ACS Applied Materials & Interfaces* 9 (28): 23687–23697.

72 Li, J., Singh, V.V., Sattayasamitsathit, S., Orozco, J., Kaufmann, K., Dong, R. et al. (2014). Water-driven micromotors for rapid photocatalytic degradation of biological and chemical warfare agents. *ACS Nano* 8 (11): 11118–11125.

73 Serrà, A. and Vallés, E. (2018). Advanced electrochemical synthesis of multicomponent metallic nanorods and nanowires: fundamentals and applications. *Applied Materials Today* 12: 207–234.

74 Gan, J.S., Li, X.B., Rizwan, K., Adeel, M., Bilal, M., Rasheed, T. et al. (2022). Covalent organic frameworks-based smart materials for mitigation of pharmaceutical pollutants from aqueous solution. *Chemosphere* 286: 131710. https://doi.org/10.1016/j.chemosphere.2021.131710

75 Luo, S., Zeng, Z., Zeng, G., Liu, Z., Xiao, R., Xu, P. et al. (2020). Recent advances in conjugated microporous polymers for photocatalysis: designs, applications, and prospects. *Journal of Materials Chemistry A* 8 (14): 6434–6470. https://doi.org/10.1039/D0TA01102A

76 Ma, Y., Wang, Z., Xu, X., and Wang, J. (2017). Review on porous nanomaterials for adsorption and photocatalytic conversion of CO2. *Cuihua Xuebao/Chinese Journal of Catalysis* 38 (12): 1956–1969. https://doi.org/10.1016/S1872-2067(17)62955-3

77 Serrà, A. and Philippe, L. (2020). Simple and scalable fabrication of hairy ZnO@ZnS core@shell Cu cables for continuous sunlight-driven photocatalytic water remediation. *Chemical Engineering Journal* 401: 126164.

78 O'Neal Tugaoen, H., Garcia-Segura, S., Hristovski, K., and Westerhoff, P. (2018). Compact light-emitting diode optical fiber immobilized TiO2 reactor for photocatalytic water treatment. *Science of the Total Environment* 613–614: 1331–1338.

79 Wang, J.C., Li, Y., Li, H., Cui, Z.H., Hou, Y., Shi, W. et al. (2019). A novel synthesis of oleophylic Fe2O3/polystyrene fibers by Γ-Ray irradiation for the enhanced photocatalysis of 4-chlorophenol and 4-nitrophenol degradation. *Journal of Hazardous Materials* 379 (June): 120806. https://doi.org/10.1016/j.jhazmat.2019.120806

80 Malato, S., Fernández-Ibáñez, P., Maldonado, M.I., Blanco, J., and Gernjak, W. (2009). Decontamination and disinfection of water by solar photocatalysis: recent overview and trends. *Catalysis Today* 147 (1): 1–59.

81 Pané, S., Puigmartí-Luis, J., Bergeles, C., Chen, X.Z., Pellicer, E., Sort, J. et al. (2019). Imaging technologies for biomedical micro and nanoswimmers. *Advanced Materials Technologies* 4 (4): 1–16.

82 Li, T., Li, J., Zhang, H., Chang, X., Song, W., Hu, Y. et al. (2016). Magnetically propelled fish-like nanoswimmers. *Small* 12 (44): 6098–6105.

83 Zheng, C., Li, Z., Xu, T., Chen, L., Fang, F., Wang, D. et al. (2021). Spirulina-templated porous hollow carbon@magnetite core-shell microswimmers. *Applied Materials Today* 22: 100962. https://doi.org/10.1016/j.apmt.2021.100962

84 Ma, X., Jannasch, A., Albrecht, U.R., Hahn, K., Miguel-López, A., Schäffer, E. et al. (2015). Enzyme-powered hollow mesoporous janus nanomotors. *Nano Letters* 15 (10): 7043–7050.

85 Zhan, Z., Wei, F., Zheng, J., Yin, C., Yang, W., Yao, L. et al. (2020). Visible light driven recyclable micromotors for "on-the-fly" water remediation. *Materials Letters* 258: 126825. https://doi.org/10.1016/j.matlet.2019.126825

86 Wang, L., Borrelli, M., and Simmchen, J. (2021). Self-asymmetric Yolk–Shell photocatalytic ZnO micromotors. *ChemPhotoChem* 5 (10): 933–939.

87 He, X., Büchel, R., Figi, R., Zhang, Y., Bahk, Y., Ma, J. et al. (2019). High-performance carbon/MnO2 micromotors and their applications for pollutant removal. *Chemosphere* 219: 427–435.

88 Villa, K., Viktorova, J., Plutnar, J., Ruml, T., Hoang, L., and Pumera, M. (2020). Chemical microrobots as self-propelled microbrushes against dental biofilm. *Cell Reports Physical Science* 1 (9): 100181.

89 Zhang, J., Zheng, X., Cui, H., and Silber-Li, Z. (2017). The self-propulsion of the spherical Pt-SiO2 Janus micromotor. *Micromachines* 8 (4): 1–17.

90 Ying, Y., Plutnar, J., and Pumera, M. (2021). Six-degree-of-freedom steerable visible-light-driven microsubmarines using water as a fuel: application for explosives decontamination. *Small* 17 (23): 1–10.

91 Xu, H., Medina-Sánchez, M., and Schmidt, O.G. (2020). Magnetic micromotors for multiple motile sperm cells capture, transport, and enzymatic release. *Angewandte Chemie International Edition* 59 (35): 15029–15037.

92 Cao, H.X., Jung, D., Lee, H.S., Go, G., Nan, M., Choi, E. et al. (2021). Micromotor manipulation using ultrasonic active traveling waves. *Micromachines* 12: 192.

93 Mushtaq, F., Asani, A., Hoop, M., Chen, X.Z., Ahmed, D., Nelson, B.J. et al. (2016). Highly efficient coaxial TiO2-PtPd tubular nanomachines for photocatalytic water purification with multiple locomotion strategies. *Advanced Functional Materials* 26 (38): 6995–7002.

94 Pourrahimi, A.M., Villa, K., Ying, Y., Sofer, Z. and, Pumera M. (2018). ZnO/ZnO 2 /Pt janus micromotors propulsion mode changes with size and interface structure: enhanced nitroaromatic explosives degradation under visible light. *ACS Applied Materials & Interfaces* 10 (49): 42688–42697.

95 Villa, K., Manzanarcs Palcnzucla, C.L., Sofer, Z., Matějková, S., and Pumera, M. (2018). Metal-free visible-light photoactivated C 3 N 4 bubble-propelled tubular micromotors with inherent fluorescence and on/off capabilities. *ACS Nano* 12 (12): 12482–12491.

96 Ussia, M., Urso, M., Dolezelikova, K., Michalkova, H., Adam, V., and Pumera, M. (2021). Active light-powered antibiofilm ZnO micromotors with chemically programmable properties. *Advanced Functional Materials* 31 (27): 1–10.

97 Urso, M., Ussia, M., and Pumera, M. (2021). Breaking polymer chains with self-propelled light-controlled navigable hematite microrobots. *Advanced Functional Materials* 31 (28): 1–10.

98 Zhang, Q., Dong, R., Wu, Y., Gao, W., He, Z., and Ren, B. (2017). Light-driven Au-WO3@C janus micromotors for rapid photodegradation of dye pollutants. *ACS Applied Materials & Interfaces* 9 (5): 4674–4683.

99 Wu, Y., Dong, R., Zhang, Q., and Ren, B. (2017). Dye-enhanced self-electrophoretic propulsion of light-driven TiO2-Au janus micromotors. *Nano-Micro Letters* 9: 30.

100 Tong, J., Wang, D., Wang, D., Xu, F., Duan, R., Zhang, D. et al. (2020). Visible-light-driven water-fueled ecofriendly micromotors based on iron phthalocyanine for highly efficient organic pollutant degradation. *Langmuir* 36 (25): 6930–6937.

101 Shivalkar, S., Gautam, P.K., Verma, A., Maurya, K., Sk, M.P., Samanta, S.K. et al. (2021). Autonomous magnetic microbots for environmental remediation developed by organic waste derived carbon dots. *Journal of Environmental Management* 297 (May): 113322. https://doi.org/10.1016/j.jenvman.2021.113322

102 Serrà, A., Pip, P., Gómez, E., and Philippe, L. (2020). Efficient magnetic hybrid ZnO-based photocatalysts for visible-light-driven removal of toxic cyanobacteria blooms and cyanotoxins. *Applied Catalysis B: Environmental* 268 (July): 118745.

103 Sanchez, C., Arribart, H., and Guille, M.M.G. (2005). Biomimetism and bioinspiration as tools for the design of innovative materials and systems. *Nature Materials* 4 (4): 277–288.

104 Soudi, N., Nanayakkara, S., Jahed, N.M.S., and Naahidi, S. (2020). Rise of nature-inspired solar photovoltaic energy convertors. *Solar Energy* 208 (July): 31–45.

105 Serrà, A., Artal, R., García-Amorós, J., Gómez, E., and Philippe, L. (2020). Circular zero-residue process using microalgae for efficient water decontamination, biofuel production, and carbon dioxide fixation. *Chemical Engineering Journal* 388 (January): 124278.

106 Li, Y., Horsman, M., Wang, B., Wu, N., and Lan, C.Q. (2008). Effects of nitrogen sources on cell growth and lipid accumulation of green alga Neochloris oleoabundans. *Applied Microbiology and Biotechnology* 81 (4): 629–636.

107 Lupatini, A.L., Colla, L.M., Canan, C., and Colla, E. (2017). Potential application of microalga Spirulina platensis as a protein source. *Journal of the Science of Food and Agriculture* 97 (3): 724–732.

108 Artal, R., Philippe, L., Gómez, E., and Serrà, A. (2020). Recycled cyanobacteria ashes for sono-enhanced photo-Fenton wastewater decontamination. *Journal of Cleaner Production* 267 (September): 121881.

109 Tao, X., Wu, R., Xia, Y., Huang, H., Chai, W., Feng, T. et al. (2014). Biotemplated fabrication of Sn@C anode materials based on the unique metal biosorption behavior of microalgae. *ACS Applied Materials & Interfaces* 6 (5): 3696–3702.

110 Zhang, X., Yu, M., Liu, J., and Li, S. (2012). Bioinspired synthesis of a hollow metallic microspiral based on a Spirulina bioscaffold. *Langmuir* 28 (8): 3690–3694.

111 Kamata, K., Piao, Z., Suzuki, S., Fujimori, T., Tajiri, W., Nagai, K. et al. (2014). Spirulina-templated metal microcoils with controlled helical structures for THz electromagnetic responses. *Scientific Reports* 4: 1–7.

112 Xia, Y., Zhang, W., Huang, H., Gan, Y., Xiao, Z., Qian, L. et al. (2011). Biotemplating of phosphate hierarchical rechargeable LiFePO4/C spirulina microstructures. *Journal of Materials Chemistry* 21 (18): 6498–6501.

113 Liu, X., Cheng, H., Guo, Z., Zhan, Q., Qian, J., and Wang, X. (2018). Bifunctional, moth-eye-like nanostructured black titania nanocomposites for solar-driven clean water generation. *ACS Applied Materials & Interfaces* 10 (46): 39661–39669.

114 Guldin, S., Kohn, P., Ste, M., Song, J., Divitini, G., Ecarla, F. et al. (2013). Self-cleaning antire fl ective optical coatings.

115 Zeng, Y., Chen, X.F., Yi, Z., Yi, Y., and Xu, X. (2018). Fabrication of p-n heterostructure ZnO/Si moth-eye structures: Antireflection, enhanced charge separation and photocatalytic properties. *Applied Surface Science* 441: 40–48. https://doi.org/10.1016/j.apsusc.2018.02.002

116 Jin, X., Shi, G., Zhu, H., Ni, C., and Li, Y. (2019). Fabricating biomimetic antireflective coating based on TiO2 pyramids by soft lithography. *ChemistrySelect* 4 (45): 13392–13395.

117 Rodríguez, R.E., Agarwal, S.P., An, S., Kazyak, E., Das, D., Shang, W. et al. (2018). Biotemplated morpho butterfly wings for tunable structurally colored photocatalysts. *ACS Applied Materials & Interfaces* 10 (5): 4614–4621.

118 Yin, C., Zhu, S., Chen, Z., Zhang, W., Gu, J., and Zhang, D. (2013). One step fabrication of C-doped BiVO4 with hierarchical structures for a high-performance photocatalyst under visible light irradiation. *Journal of Materials Chemistry A* 1 (29): 8367–8378.

119 Deng, X., Chen, R., Zhao, Z., Cui, F., and Xu, X. (2021). Graphene oxide-supported graphitic carbon nitride microflowers decorated by sliver nanoparticles for enhanced photocatalytic degradation of dimethoate via addition of sulfite: Mechanism and toxicity evolution. *Chemical Engineering Journal* 425 (July): 131683. https://doi.org/10.1016/j.cej.2021.131683

120 Zhu, B., Song, D., Jia, T., Sun, W., Wang, D., Wang, L. et al. (2021). Effective visible light-driven photocatalytic degradation of ciprofloxacin over flower-like Fe3O4/Bi2WO6 composites.

121 Oxygen, R. and Ros, S. Insights into the photocatalytic bacterial inactivation by flower-like Bi 2 WO 6 under solar or visible light.

122 Li, X., Fang, S., Ge, L., Han, C., Qiu, P., and Liu, W. (2015). Environmental Synthesis of flower-like Ag / AgCl-Bi 2 MoO 6 plasmonic photocatalysts with enhanced visible-light photocatalytic performance. *Applied Catalysis B* 177: 62–69.

123 Sin, J., Quek, J., Lam, S., Zeng, H., and Lin, H. (2021). Journal of environmental chemical engineering punica granatum mediated green synthesis of cauliflower-like ZnO and decorated with bovine bone-derived hydroxyapatite for expeditious visible light photocatalytic antibacterial, antibiofilm and antioxidant a. *Journal of Environmental Chemical Engineering* 9 (4): 105736. https://doi.org/10.1016/j.jece.2021.105736

124 Wu, Q., Liu, X., Hou, S., Qiang, L., Zhang, K., and Yang, Z. (2021). Physicochemical and Engineering Aspects Biotemplate synthesis and photocatalysis performance of multilayer porous ZnWO 4 nano-photocatalyst with rose petals as template. *Colloids and Surfaces A* 629 (April): 127459.

125 Patil, S.A., Shrestha, N.K., Hussain, S., Jung, J., Lee, S., Bathula, C. et al. (2021). Catalytic decontamination of organic / inorganic pollutants in water and green H 2 generation using nanoporous SnS 2 microflower structured film. *Journal of Hazardous Materials* 417 (March): 126105. https://doi.org/10.1016/j.jhazmat.2021.126105

126 Ashfaq, M., Talreja, N., Chauhan, D., and Rodríguez, C.A. (2021). A novel bimetallic (Fe / Bi) -povidone-iodine microflowers composite for photocatalytic and antibacterial applications. *Journal of Photochemistry and Photobiology B: Biology* 219 (March): 112204. https://doi.org/10.1016/j.jphotobiol.2021.112204

127 Vasilaki, E., Katsarakis, N., Dokianakis, S., and Vamvakaki, M. (2021). rGO functionalized ZnO–TiO2 core-shell flower-like architectures for visible light photocatalysis.

128 Hao, L., Ju, P., Zhang, Y., Zhai, X., Sun, C., and Duan, J. (2021). Fabrication of hierarchical flower-like BiOI/MoS 2 heterostructures with highly enhanced visible-light photocatalytic activities. *Colloids and Surfaces A* 610 (September 2020): 125714. https://doi.org/10.1016/j.colsurfa.2020.125714

129 Liu, L., Ding, L., Liu, Y., An, W., Lin, S., Liang, Y. et al. (2016). Applied surface science enhanced visible light photocatalytic activity by Cu 2 O-coupled flower-like Bi 2 WO 6 structures. *Applied Surface Science* 364: 505–515. https://doi.org/10.1016/j.apsusc.2015.12.170

130 Wu, T., Li, X., Zhang, D., Dong, F., and Chen, S. (2016). Efficient visible light photocatalytic oxidation of NO with hierarchical nanostructured 3D flower-like BiOClxBr1-x solid solutions. *Journal of Alloys and Compounds* 671: 318–327.

131 Zhao, S., Zhang, Y., Zhou, Y., Qiu, K., Zhang, C., Fang, J. et al. (2018). Reactable polyelectrolyte-assisted preparation of flower-like Ag/AgCl/BiOCl composite with

enhanced photocatalytic activity. *Journal of Photochemistry and Photobiology A: Chemistry* 350: 94–102. https://doi.org/10.1016/j.jphotochem.2017.09.070
132 Zhang, B., Zhang, M., Zhang, L., Bingham, P.A., Li, W., and Kubuki, S. (2020). PVP surfactant-modified flower-like BiOBr with tunable bandgap structure for efficient photocatalytic decontamination of pollutants. *Applied Surface Science* 530 (July): 147223.
133 Lv, X., Yan, D.Y.S., Lam, F.L.Y., Ng, Y.H., Yin, S., and An, A.K. (2020). Solvothermal synthesis of copper-doped BiOBr microflowers with enhanced adsorption and visible-light driven photocatalytic degradation of norfloxacin. *Chemical Engineering Journal* 401 (June): 126012. https://doi.org/10.1016/j.cej.2020.126012
134 Di, J., Xia, J., Ji, M., Wang, B., Yin, S., Zhang, Q. et al. (2016). Environmental advanced photocatalytic performance of graphene-like BN modified BiOBr flower-like materials for the removal of pollutants and mechanism insight. *Applied Catalysis B* 183: 254–262.
135 Cui, W., An, W., Liu, L., Hu, J., and Liang, Y. Novel Cu 2 O quantum dots coupled flower-like BiOBr for enhanced photocatalytic degradation of organic contaminant. 2014;280:417–427.
136 Zhang, J. and Ma, Z. (2017). Flower-like Ag 3 VO 4/BiOBr n-p heterojunction photocatalysts with enhanced visible-light-driven catalytic activity. *Molecular Catalysis* 436: 190–198. https://doi.org/10.1016/j.mcat.2017.04.004
137 Liang, S., Wang, J., Wu, X., Zhu, S., and Wu, L. (2019). Phase transformation synthesis of a new Bi 2 SeO 5 flower-like microsphere for efficiently photocatalytic degradation of organic pollutants. *Catalysis Today* 327 (January 2018): 357–365. https://doi.org/10.1016/j.cattod.2018.03.021
138 Lam, S.M., Quek, J.A., and Sin, J.C. (2018). Mechanistic investigation of visible light responsive Ag/ZnO micro/nanoflowers for enhanced photocatalytic performance and antibacterial activity. *Journal of Photochemistry and Photobiology A: Chemistry* 353: 171–184. https://doi.org/10.1016/j.jphotochem.2017.11.021
139 Liu, X., Xing, Z., Zhang, Y., Li, Z., Wu, X., Tan, S. et al. (2017). Fabrication of 3D flower-like black N-TiO2-x@MoS2 for unprecedented-high visible-light-driven photocatalytic performance. *Applied Catalysis B: Environmental* 201: 119–127. https://doi.org/10.1016/j.apcatb.2016.08.031
140 Bariki, R., Majhi, D., Das, K., Behera, A., and Mishra, B.G. (2020). Facile synthesis and photocatalytic efficacy of UiO-66/CdIn2S4 nanocomposites with flowerlike 3D-microspheres toward aqueous phase decontamination of triclosan and H2 evolution. *Applied Catalysis B: Environmental* 270 (December 2019): 118882. https://doi.org/10.1016/j.apcatb.2020.118882
141 Chen, Y., Zhang, X., Wang, L., Cheng, X., and Shang, Q. (2020). Rapid removal of phenol/antibiotics in water by Fe-(8-hydroxyquinoline-7-carboxylic)/TiO2 flower composite: adsorption combined with photocatalysis. *Chemical Engineering Journal* 402 (December): 126260.
142 Qian, X., Zhao, Y., Alsaid, Y., Wang, X., Hua, M., Galy, T. et al. (2019). Artificial phototropism for omnidirectional tracking and harvesting of light. *Nature Nanotechnology* 14 (11): 1048–1055.
143 Serrà, A., Zhang, Y., Sepúlveda, B., Gómez, E., Nogués, J., Michler, J. et al. (2020). Highly reduced ecotoxicity of ZnO-based micro/nanostructures on aquatic biota: influence of

architecture, chemical composition, fixation, and photocatalytic efficiency. *Water Research* 169: 155210.
144 Huang, H.B., Wang, Y., Bin, J.W., Cai, F.Y., Shen, M., Zhou, S.G. et al. (2018). Lotus-leaf-derived activated-carbon-supported nano-CdS as energy-efficient photocatalysts under visible irradiation. *ACS Sustainable Chemistry & Engineering* 6 (6): 7871–7879.
145 Virkutyte, J., Jegatheesan, V., and Varma, R.S. (2012). Visible light activated TiO 2/ microcrystalline cellulose nanocatalyst to destroy organic contaminants in water. *Bioresource Technology* 113: 288–293.
146 Manikandan, V., Lee, J.H., Velmurugan, P., Jayanthi, P., Chang, W.S., Park, Y.J. et al. (2018). Fabrication and characterization of TiO2-loaded Moringa oleifera gum-activated carbon and the photo-catalytic degradation of phosphate in aqueous solutions. *Nanotechnology for Environmental Engineering* 3 (1): 1–13.
147 Bharathi, D., Ranjithkumar, R., Chandarshekar, B., and Bhuvaneshwari, V. (2019). Preparation of chitosan coated zinc oxide nanocomposite for enhanced antibacterial and photocatalytic activity: as a bionanocomposite. *International Journal of Biological Macromolecules* 129: 989–996. https://doi.org/10.1016/j.ijbiomac.2019.02.061
148 Sharma, G., Kumar, A., Sharma, S., Al-Muhtaseb, A.H., Naushad, M., Ghfar, A.A. et al. (2019). Fabrication and characterization of novel Fe0@Guar gum-crosslinked-soya lecithin nanocomposite hydrogel for photocatalytic degradation of methyl violet dye. *Separation and Purification Technology* 211 (October 2018): 895–908. https://doi.org/10.1016/j.seppur.2018.10.028
149 Dippong, T., Cadar, O., Levei, E.A., Deac, I.G., and Borodi, G. (2018). Formation of CoFe2O4/PVA-SiO2 nanocomposites: effect of diol chain length on the structure and magnetic properties. *Ceramics International* 44 (9): 10478–10485.
150 Calza, P., Avetta, P., Rubulotta, G., Sangermano, M., and Laurenti, E. (2014). TiO2-soybean peroxidase composite materials as a new photocatalytic system. *Chemical Engineering Journal* 239: 87–92. https://doi.org/10.1016/j.cej.2013.10.098
151 Munikrishnappa, C., Kumar, S., Shivakumara, S., Mohan Rao, G., and Munichandraiah, N. (2019). The TiO 2 -graphene oxide-Hemin ternary hybrid composite material as an efficient heterogeneous catalyst for the degradation of organic contaminants. *Journal of Science: Advanced Materials and Devices* 4 (1): 80–88. https://doi.org/10.1016/j.jsamd.2018.12.003
152 Chen, X., Lu, W., Xu, T., Li, N., Qin, D., Zhu, Z. et al. (2017). A bio-inspired strategy to enhance the photocatalytic performance of g-C3N4 under solar irradiation by axial coordination with hemin. *Applied Catalysis B: Environmental* 201: 518–526. https://doi.org/10.1016/j.apcatb.2016.08.020
153 Artal, R., Serr, A., Michler, J., Philippe, L., and Elvira, G. (2020). Electrodeposition of mesoporous Ni-Rich Ni-Pt films for highly E ffi cient methanol oxidation. 1–17.
154 Mohanta, D., Raha, S., Vikram Gupta, S., and Ahmaruzzaman, M. (2019). Bioinspired green synthesis of engineered CuSnO 3 quantum dots: an effective material for superior photocatalytic degradation of Rabeprazole. *Materials Letters* 240: 193–196. https://doi.org/10.1016/j.matlet.2018.12.104
155 Surendra, T.V. and Roopan, S.M. (2016). Photocatalytic and antibacterial properties of phytosynthesized CeO2 NPs using Moringa oleifera peel extract. *Journal of*

Photochemistry and Photobiology B: Biology 161: 122–128. https://doi.org/10.1016/j.jphotobiol.2016.05.019

156 Ishwarya, R., Vaseeharan, B., Kalyani, S., Banumathi, B., Govindarajan, M., Alharbi, N.S. et al. (2018). Facile green synthesis of zinc oxide nanoparticles using Ulva lactuca seaweed extract and evaluation of their photocatalytic, antibiofilm and insecticidal activity. *Journal of Photochemistry and Photobiology B: Biology* 178 (October 2017): 249–258. https://doi.org/10.1016/j.jphotobiol.2017.11.006

157 Gu, H., Chen, X., Chen, F., Zhou, X., and Parsaee, Z. (2018). Ultrasound-assisted biosynthesis of CuO-NPs using brown alga Cystoseira trinodis: characterization, photocatalytic AOP, DPPH scavenging and antibacterial investigations. *Ultrasonics Sonochemistry* 41 (September 2017): 109–119. https://doi.org/10.1016/j.ultsonch.2017.09.006

158 Begum, S. and Ahmaruzzaman, M. (2018). Green synthesis of SnO2 quantum dots using Parkia speciosa Hassk pods extract for the evaluation of anti-oxidant and photocatalytic properties. *Journal of Photochemistry and Photobiology B: Biology* 184 (May): 44–53.

159 Tripathi, K.M., Tran, T.S., Kim, Y.J., and Kim, T.Y. (2017). Green fluorescent onion-like carbon nanoparticles from flaxseed oil for visible light induced photocatalytic applications and label-free detection of Al(III) ions. *ACS Sustainable Chemistry & Engineering* 5 (5): 3982–3992.

160 Kumar, B., Smita, K., Debut, A., and Cumbal, L. (2018). Utilization of Persea americana (Avocado) oil for the synthesis of gold nanoparticles in sunlight and evaluation of antioxidant and photocatalytic activities. *Environmental Nanotechnology, Monitoring & Management* 10 (June): 231–237. https://doi.org/10.1016/j.enmm.2018.07.009

161 Goutam, S.P., Saxena, G., Singh, V., Yadav, A.K., Bharagava, R.N., and Thapa, K.B. (2018). Green synthesis of TiO2 nanoparticles using leaf extract of Jatropha curcas L. for photocatalytic degradation of tannery wastewater. *Chemical Engineering Journal* 336 (September 2017): 386–396. https://doi.org/10.1016/j.cej.2017.12.029

162 Li, K., Huang, Z., Zhu, S., Luo, S., Yan, L., Dai, Y. et al. (2019). Removal of Cr(VI) from water by a biochar-coupled g-C3N4 nanosheets composite and performance of a recycled photocatalyst in single and combined pollution systems. *Applied Catalysis B: Environmental* 243 (October 2018): 386–396.

163 Li, S., Wang, Z., Zhao, X., Yang, X., Liang, G., and Xie, X. (2019). Insight into enhanced carbamazepine photodegradation over biochar-based magnetic photocatalyst Fe3O4/BiOBr/BC under visible LED light irradiation. *Chemical Engineering Journal* 360 (November 2018): 600–611.

164 Lu, L., Shan, R., Shi, Y., Wang, S., and Yuan, H. (2019). A novel TiO2/biochar composite catalysts for photocatalytic degradation of methyl orange. *Chemosphere* 222: 391–398.

165 Zhang, Y., Fan, R., Zhang, Q., Chen, Y., Sharifi, O., Leszczynska, D. et al. (2019). Synthesis of CaWO4-biochar nanocomposites for organic dye removal. *Materials Research Bulletin* 110 (June 2018): 169–173.

166 Mian, M.M. and Liu, G. (2019). Sewage sludge-derived TiO2/Fe/Fe3C-biochar composite as an efficient heterogeneous catalyst for degradation of methylene blue. *Chemosphere* 215: 101–114.

167 Jia, P., Tan, H., Liu, K., and Gao, W. (2017). Enhanced photocatalytic performance of ZnO/bone char composites. *Materials Letters* 205: 233–235.

168 Abdurahman, M.H., Abdullah, A.Z., and Shoparwe, N.F. (2021). A comprehensive review on sonocatalytic, photocatalytic, and sonophotocatalytic processes for the degradation of antibiotics in water: synergistic mechanism and degradation pathway. *Chemical Engineering Journal* 413 (October 2020): 127412. https://doi.org/10.1016/j.cej.2020.127412

169 Bembibre, A., Benamara, M., Hjiri, M., Gómez, E., Alamri, H.R., Dhahri, R. et al. (2021). Visible-light driven sonophotocatalytic removal of tetracycline using Ca-doped ZnO nanoparticles. *Chemical Engineering Journal* 427 (June 2021): 132006.

170 Hapeshi, E., Fotiou, I., and Fatta-Kassinos, D. (2013). Sonophotocatalytic treatment of ofloxacin in secondary treated effluent and elucidation of its transformation products. *Chemical Engineering Journal* 224 (1): 96–105. https://doi.org/10.1016/j.cej.2012.11.048

171 Ahmad, M., Ahmed, E., Hong, Z.L., Ahmed, W., Elhissi, A., and Khalid, N.R. (2014). Photocatalytic, sonocatalytic and sonophotocatalytic degradation of Rhodamine B using ZnO/CNTs composites photocatalysts. *Ultrasonics Sonochemistry* 21 (2): 761–773. https://doi.org/10.1016/j.ultsonch.2013.08.014

172 Taghizadeh, M.T. and Abdollahi, R. (2011). Sonolytic, sonocatalytic and sonophotocatalytic degradation of chitosan in the presence of TiO2 nanoparticles. *Ultrasonics Sonochemistry* 18 (1): 149–157.

173 Selli, E. (2002). Synergistic effects of sonolysis combined with photocatalysis in the degradation of an azo dye. *Physical Chemistry Chemical Physics* 4 (24): 6123–6128.

174 Rad, T.S., Ansarian, Z., Soltani, R.D.C., Khataee, A., Orooji, Y., and Vafaei, F. (2020). Sonophotocatalytic activities of FeCuMg and CrCuMg LDHs: influencing factors, antibacterial effects, and intermediate determination. *Journal of Hazardous Materials* 399 (May): 123062. https://doi.org/10.1016/j.jhazmat.2020.123062

175 Neppolian, B., Ciceri, L., Bianchi, C.L., Grieser, F., and Ashokkumar, M. (2011). Sonophotocatalytic degradation of 4-chlorophenol using Bi2O3/TiZrO4 as a visible light responsive photocatalyst. *Ultrasonics Sonochemistry* 18 (1): 135–139.

176 Ma, C.Y., Xu, J.Y., and Liu, X.J. (2006). Decomposition of an azo dye in aqueous solution by combination of ultrasound and visible light. *Ultrasonics* 44 (SUPPL.): 375–378.

177 Darbandi, M., Eynollahi, M., Badri, N., Mohajer, M.F., and Li, Z.-A. (2022). NiO nanoparticles with superior sonophotocatalytic performance in organic pollutant degradation. *Journal of Alloys and Compounds* 889: 161706.

178 Sadeghi Rad, T., Khataee, A., Arefi-Oskoui, S., Sadeghi Rad, S., Orooji, Y., Gengec, E. et al. (2022). Graphene-based ZnCr layered double hydroxide nanocomposites as bactericidal agents with high sonophotocatalytic performances for degradation of rifampicin. *Chemosphere* 286 (P2): 131740. https://doi.org/10.1016/j.chemosphere.2021.131740

179 Liang, L., Tursun, Y., Nulahong, A., Dilinuer, T., Tunishaguli, A., Gao, G. et al. (2017). Preparation and sonophotocatalytic performance of hierarchical Bi2WO6 structures and effects of various factors on the rate of Rhodamine B degradation. *Ultrasonics Sonochemistry* 39 (February): 93–100. https://doi.org/10.1016/j.ultsonch.2017.03.054

180 Sun, M., Yao, Y., Ding, W., and Anandan, S. (2020). N/Ti3+ co-doping biphasic TiO2/Bi2WO6 heterojunctions: Hydrothermal fabrication and sonophotocatalytic degradation of organic pollutants. *Journal of Alloys and Compounds* 820: 153172. https://doi.org/10.1016/j.jallcom.2019.153172

181 Meroni, D., Jiménez-Salcedo, M., Falletta, E., Bresolin, B.M., Kait, C.F., Boffito, D.C. et al. (2020). Sonophotocatalytic degradation of sodium diclofenac using low power ultrasound and micro sized TiO2. *Ultrasonics Sonochemistry* 67 (April): 105123. https://doi.org/10.1016/j.ultsonch.2020.105123

182 Toe, C.Y., Scott, J., Amal, R., and Ng, Y.H. (2019). Recent advances in suppressing the photocorrosion of cuprous oxide for photocatalytic and photoelectrochemical energy conversion. *Journal of Photochemistry and Photobiology C: Photochemistry Reviews* 40: 191–211.

183 Li, X., Teng, W., Zhao, Q., and Wang, L. (2011). Efficient visible light-induced photoelectrocatalytic degradation of Rhodamine B by polyaniline-sensitized TiO 2 nanotube arrays. *Journal of Nanoparticle Research* 13 (12): 6813–6820.

184 Liu, C.F., Huang, C.P., Hu, C.C., Juang, Y., and Huang, C. (2016). Photoelectrochemical degradation of dye wastewater on TiO2-coated titanium electrode prepared by electrophoretic deposition. *Separation and Purification Technology* 165: 145–153. https://doi.org/10.1016/j.seppur.2016.03.045

185 Brillas, E. and Martínez-Huitle, C.A. (2015). Decontamination of wastewaters containing synthetic organic dyes by electrochemical methods. An updated review. *Applied Catalysis B: Environmental* 166–167: 603–643. https://doi.org/10.1016/j.apcatb.2014.11.016

186 Alulema-Pullupaxi, P., Espinoza-Montero, P.J., Sigcha-Pallo, C., Vargas, R., Fernández, L., Peralta-Hernández, J.M. et al. (2021). Fundamentals and applications of photoelectrocatalysis as an efficient process to remove pollutants from water: a review. *Chemosphere* 281 (December 2020).

187 Al-Jawad, S.M.H. (2017). Structural and optical properties of core–shell TiO_2/CdS prepared by chemical bath deposition. *Journal of Electronic Materials* 46 (10): 5837.

188 Nkwachukwu, O.V. and Arotiba, O.A. (2021). Perovskite oxide–based materials for photocatalytic and photoelectrocatalytic treatment of water. *Frontiers in Chemistry* 9 (April): 1–20.

189 Zhang, X., Ke, X., and Yao, J. (2018). Recent development of plasmon-mediated photocatalysts and their potential in selectivity regulation. *Journal of Materials Chemistry A* 6 (5): 1941–1966.

190 Nair, V., Muñoz-Batista, M.J., Fernández-García, M., Luque, R., and Colmenares, J.C. (2019). Thermo-photocatalysis: environmental and energy applications. *ChemSusChem* 12 (10): 2098–2116.

191 Tian, J., Sang, Y., Yu, G., Jiang, H., Mu, X., and Liu, H. (2013). A Bi2WO6-based hybrid photocatalyst with broad spectrum photocatalytic properties under UV, visible, and near-infrared irradiation. *Advanced Materials* 25 (36): 5075–5080.

192 Li, C., Wang, F., Zhu, J., and Yu, J.C. (2010). NaYF4:Yb,Tm/CdS composite as a novel near-infrared-driven photocatalyst. *Applied Catalysis B: Environmental* 100 (3–4): 433–439.

193 Wang, G., Huang, B., Ma, X., Wang, Z., Qin, X., Zhang, X. et al. (2013). Cu 2 (OH)PO 4, a near-infrared-activated photocatalyst. *Angewandte Chemie* 125 (18): 4910–4913.

194 Qin, W., Zhang, D., Zhao, D., Wang, L., and Zheng, K. (2010). Near-infrared photocatalysis based on YF3: Yb3+,Tm3+/TiO2 core/shell nanoparticles. *Chemical Communications* 46 (13): 2304–2306.

195 Song, H., Meng, X., Dao, T.D., Zhou, W., Liu, H., Shi, L. et al. (2018). Light-enhanced carbon dioxide activation and conversion by effective plasmonic coupling effect of Pt and Au nanoparticles. *ACS Applied Materials & Interfaces* 10 (1): 408–416.

196 Muñoz-Batista, M.J., Eslava-Castillo, A.M., Kubacka, A., and Fernández-García, M. (2018). Thermo-photo degradation of 2-propanol using a composite ceria-titania catalyst: physico-chemical interpretation from a kinetic model. *Applied Catalysis B: Environmental* 225 (December 2017): 298–306. https://doi.org/10.1016/j.apcatb.2017.11.073

197 Verma, R., Samdarshi, S.K., Bojja, S., Paul, S., and Choudhury, B. (2015). A novel thermophotocatalyst of mixed-phase cerium oxide (CeO2/Ce2O3) homocomposite nanostructure: Role of interface and oxygen vacancies. *Solar Energy Materials and Solar Cells* 141: 414–422.

198 Obregón, S. and Colón, G. (2014). A ternary Er3+-BiVO4/TiO2 complex heterostructure with excellent photocatalytic performance. *RSC Advances* 4 (14): 6920–6926.

199 Weng, B., Qi, M.Y., Han, C., Tang, Z.R., and Xu, Y.J. (2019). Photocorrosion inhibition of semiconductor-based photocatalysts: Basic principle, current development, and future perspective. *ACS Catalysis* 9 (5): 4642–4687.

200 Hossain, M.M., Ku, B.C., and Hahn, J.R. (2015). Synthesis of an efficient white-light photocatalyst composite of graphene and ZnO nanoparticles: application to methylene blue dye decomposition. *Applied Surface Science* 354: 55–65. https://doi.org/10.1016/j.apsusc.2015.01.191

201 Li, Q., Li, X., Wageh, S., Al-Ghamdi, A.A., and Yu, J. (2015). CdS/Graphene nanocomposite photocatalysts. *Advanced Energy Materials* 5 (14): 1–28.

202 Ma, F., Wu, Y., Shao, Y., Zhong, Y., Lv, J., and Hao, X. (2016). 0D/2D nanocomposite visible light photocatalyst for highly stable and efficient hydrogen generation via recrystallization of CdS on MoS2 nanosheets. *Nano Energy* 27: 466–474. https://doi.org/10.1016/j.nanoen.2016.07.014

203 Yu, H., Huang, X., Wang, P., and Yu, J. (2016). Enhanced photoinduced-stability and photocatalytic activity of CdS by dual amorphous cocatalysts: synergistic effect of Ti(IV)-hole cocatalyst and Ni(II)-electron cocatalyst. *The Journal of Physical Chemistry C* 120 (7): 3722–3730.

204 Wang, C., Wang, L., Jin, J., Liu, J., Li, Y., Wu, M. et al. (2016). Probing effective photocorrosion inhibition and highly improved photocatalytic hydrogen production on monodisperse PANI@CdS core-shell nanospheres. *Applied Catalysis B: Environmental* 188: 351–359.

205 Rudd, A.L. and Breslin, C.B. (2000). Photo-induced dissolution of zinc in alkaline solutions. *Electrochimica acta* 45 (10): 1571–1579.

206 Kwon, Y., Soon, A., Han, H., and Lee, H. (2015). Shape effects of cuprous oxide particles on stability in water and photocatalytic water splitting. *Journal of Materials Chemistry A* 3 (1): 156–162.

207 Bendavid, L.I. and Carter, E.A. (2013). First-principles predictions of the structure, stability, and photocatalytic potential of Cu2O surfaces. *The Journal of Physical Chemistry B* 117 (49): 15750–15760.

208 Higashi, M., Domen, K., and Abe, R. (2012). Highly stable water splitting on oxynitride TaON photoanode system under visible light irradiation. *Journal of the American Chemical Society* 134 (16): 6968–6971.

209 Jing, D. and Guo, L. (2006). A novel method for the preparation of a highly stable and active CdS photocatalyst with a special surface nanostructure. *The Journal of Physical Chemistry B* 110 (23): 11139–11145.

210 Grewe, T. and Tüysüz, H. (2015). Amorphous and crystalline sodium tantalate composites for photocatalytic water splitting. *ACS Applied Materials & Interfaces* 7 (41): 23153–23162.

211 Keijer, T., Bakker, V., and Slootweg, J.C. (2019). Circular chemistry to enable a circular economy. *Nature Chemistry* 11 (3): 190–195. https://doi.org/10.1038/s41557-019-0226-9

212 Salama, E.S., Kurade, M.B., Abou-Shanab, R.A.I., El-Dalatony, M.M., Yang, I.S., Min, B. et al. (2017). Recent progress in microalgal biomass production coupled with wastewater treatment for biofuel generation. *Renewable & Sustainable Energy Reviews* 79 (March): 1189–1211. https://doi.org/10.1016/j.rser.2017.05.091

213 Yaqoob, A.A., Noor, N.H., Binti, M., Serrà, A., and Ibrahim, M.N.M. (2020). Advances and challenges in developing efficient graphene oxide-based ZnO photocatalysts for dye photo-oxidation. *Nanomaterials* 10 (5): 932.

214 Daghrir, R., Drogui, P., and Robert, D. (2012). Photoelectrocatalytic technologies for environmental applications. *Journal of Photochemistry and Photobiology A: Chemistry* 238: 41–52. https://doi.org/10.1016/j.jphotochem.2012.04.009

6

Fundamentals and Functional Mechanisms of Electrocatalysis in Water Treatment

Ahmed M. Awad Abouelata

Chemical Engineering & Pilot Plant Department, Engineering Division, National Research Centre (NRC), 33 Bohooth St., 12622, Dokki, Giza, Egypt aa.awad@nrc.sci.eg

6.1 Introduction

Recently, there has been increasing interest in water pollution all over the world. The United Nations (UN) reported that 2 million tons per day of wastewater is flushed into the world's water systems. Wastewater with different contaminants produced annually is about 1500 km^3 (six times all the rivers of the world); these contaminants may be organic, biological, heavy metals, or radioactive materials. In any case, they are considered hazardous materials and a threat to human health [1]. Public health is threatened, and unsafe and inadequate water cause approximately 3.1% of all deaths worldwide according to the World Health Organization (WHO) [2]. Hazardous organic materials are toxic potential carcinogens, and even a trace amount leads to health risks [3, 4]. Conventional wastewater treatment methods, however, are often insufficient for the removal of recalcitrant organic materials and they may cause the formation of dangerous or even more toxic transformation products [5, 6].

In some regions, fresh and drinking water suffers from different sources of contamination such as heavy metals, organic compounds, carcinogenic materials, pathogens, and microorganisms, which threaten human health and often lead to chronic diseases.

The quality of drinking water attracts great public interest; the presence of heavy metals, e.g. iron, manganese, or arsenic compounds, in groundwater and eventually in drinking water is a complex environmental problem. When heavy metals are present in both surface and groundwater, it can cause various water quality problems and disposal difficulty.

When heavy metals react with oxygen from the air, they form other oxide particles commonly difficult to remove [7]. Groundwater contamination can occur in various paths, such as natural or geochemical contamination, by leaks in pipelines, from landfill leachates, etc. It is divided into three main contamination categories, by organic compounds,

Photocatalysts and Electrocatalysts in Water Remediation: From Fundamentals to Full Scale Applications, First Edition. Edited by Prasenjit Bhunia, Kingshuk Dutta, and S. Vadivel.

microorganisms, and inorganic pollutants. Groundwater containing metals of inorganic pollution comprises a danger to the environment due to the fact that metals are not biodegradable and can cause severe adverse effects on human health [8]. When Fe and Mn are present in both surface and groundwater, even at low concentrations, they can be linked to various water quality problems, and their removal is urgent.

Industrial drainage threatens the environment due to the presence of inorganic, organic, and biocontaminants originated from activities of different industries such as textile, pulp and paper, pharmaceutics, leather, petrochemical, cosmetics, etc. Unfortunately, the drains of these industries are discharged into natural surface water such as rivers, lakes, or seas. For instance, the cosmetic industry generates wastewater characterized by high levels of chemical oxygen demand (COD), suspended solids, fats and oils, and detergents [9, 10].

Municipal wastewaters include many organic materials with active ingredients, e.g. cosmetic and pharmaceutical products, which are used in large amounts throughout the world. Most of them come either from domestic sewage, industrial, or hospital discharges and enter municipal sewage treatment plants (STPs). New STPs can effectively remove carbon and nitrogen as well as perform microbial pollution control, but these plants have not been designed to remove these trace polluting materials [11]. Also, cooling towers and hot water systems are considered to be the source of the bacteria, which cause diseases. These bacteria grow in warm water, particularly in areas where water stagnates [12, 13].

Advanced oxidation processes (AOPs) such as chemical, photochemical, ozonation, sonication, and electrochemical oxidation have succeeded at partially destroying and removing contaminants from wastewater with either recalcitrant organic compounds or heavy metals. But occasionally, AOPs are not sufficient for complete mineralization of contaminants and hence leave behind harmful intermediate materials. The selection of an appropriate method for treatment of contaminated water is therefore vital. Consequently, disposal of contaminants can be achieved for retrieving fresh water and maintaining human health, agriculture purposes, and other natural activities.

Under the effect of electric potential applied in liquids like water, new active particles are formed, and the dissolved materials are dissociated into ions or partially polar components. Due to these facts, electrocatalysis is considered one of the vital processes for wastewater reclamation. Electrocatalysis comprises advanced direct oxidation, indirect oxidation, direct reduction, and indirect reduction.

6.2 Electrocatalysis Treatment

Electrocatalysis methods can be successfully used for wastewater treatment. External electric energy supplied as DC, AC, or electropulsation is passed into wastewater as electric current. The presence of organic, inorganic, heavy metals, or microorganisms under the force of an electric potential leads to the instability, degradation, and then complete destruction of these contaminants in wastewater. The passage of DC, AC, or pulse electric current motivates the energized ions to move through the wastewater solution creating polarization between both sides of the electrodes (anode and cathode). Under these conditions the refractory long chain organic contaminants are destroyed and decomposed to simple short chain organic materials, such as H_2O, CO_2, NO_2, and SO_2, by the breaking of chemical bonds.

Dissolved heavy metals in wastewater are oxidized to the nondissolved form of higher oxide number, which easily agglomerate, settle down and separate through filtration. Microorganisms in wastewater are destroyed by passage of electric current, where the cell walls of amoeba, algea, pathogens, and bacteria are broken and the internal contents are destroyed. Different electrocatalysis methods are summarized in Figure 6.1.

Different electrocatalysis techniques include the following methods:

1) Advanced electro-oxidation: has succeeded particularly for the treatment of organic contaminants in two categories.
   - Direct anodic oxidation
   - Indirect anodic oxidation via intermediates of oxygen evolution.
2) Electroreduction.
3) Electrocoagulation and electrocoagulation/flotation.
4) Electrodisinfection.
5) Electrodialysis.
6) Electro-Fenton.
7) Sonoelectrolysis.
8) Photoelectrochemical.

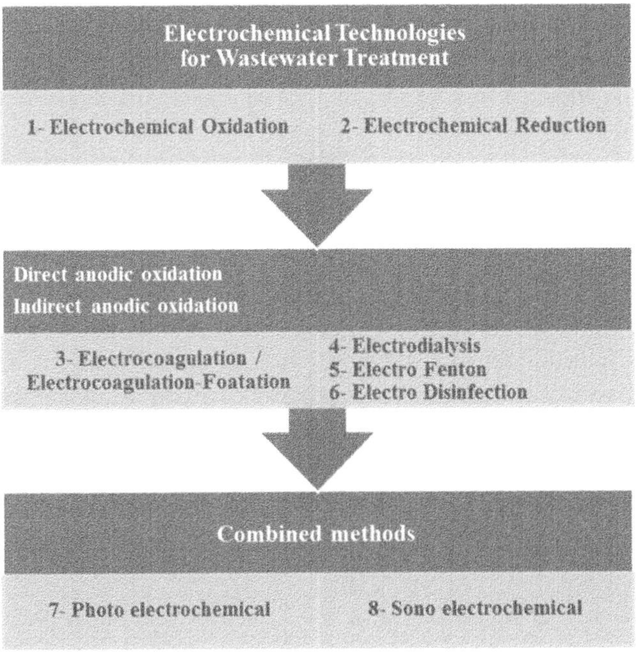

Figure 6.1 Diagram showing different electrochemical methods used for wastewater treatment.

6.3 Properties and Characteristics of Different Electrocatalysis Techniques

6.3.1 Electrochemical Oxidation

Several materials cause the contamination of water effluents, from sources such as drainage of industrial activities, dissolution of rocks, or decomposition of living organisms. These materials need external force for degradation and disposal of contaminants and selection of appropriate conditions for treatment. Electric potential could be applied by the passage of suitable electric current through wastewater, where refractory long chain and high molecular weight materials are dissociated into short chain and low molecular weight materials (incineration). Electrochemical methods are clean and effective techniques for anodic oxidation through direct or indirect paths. Highly oxidant $^{\bullet}OH$ free radicals are formed by water oxidation on high O_2-overvoltage anodes. Electrochemical oxidation methods have received wide recognition due to their effectiveness in the removal of toxic and biorefractory organic compounds [5, 14–16].

The degradation of refractory organic compounds could be achieved by anodic oxidation in an electrolytic cell by passage of an external current; however, recalcitrant organic bonds are broken, and long chain organic materials are incinerated to small safe materials, H_2O, and other gases [17]. Electrochemical advanced oxidation process (EAOP) methods are divided into two main types: direct and indirect anodic oxidation.

6.3.1.1 Direct Anodic Oxidation

Under the effect of high electric potential in wastewater and high electron transfer toward the anodic surface, pollutants of organic materials are adsorbed adjacent to the anode surface and directly destroyed until complete degradation.

In direct anodic oxidation (direct electron transfer to the anode), the contaminants are destroyed after adsorption on the anode surface without addition of any external substances. Such oxidation is theoretically possible at more negative potentials than those needed for water splitting and oxygen evolution. This process, however, usually results in electrode fouling due to the formation of polymeric layers on its surface and consequently leads to very poor chemical decontamination [18, 19]. Gattrell and Kirk investigated the electrooxidation of phenol with platinum and peroxidized platinum anodes using cyclic voltammetry and chronoamperometry. Their studies demonstrated that phenol can be irreversibly adsorbed on metallic platinum, quickly passivating the electrode [20].

In direct electrolysis, the contaminants are adsorbed and then oxidized on the anode surface without the involvement of any materials other than the electron (clean reagent):

$$R_{ads} - z\,e^{-} \rightarrow P_{ads} \tag{6.1}$$

Direct anodic oxidation can occur at low voltages, before O_2 evolution, but the reaction rate usually has low kinetics that rely on the electrocatalytic activity of the anode. High electrochemical rates have been noticed using noble metals such as Pd, Pt, and metal-oxide anodes such as IrO_2, Ru/TiO_2, and Ir/TiO_2.

The main problem of anodic oxidation at a fixed anodic potential before O_2 evolution is the low catalytic activity (the poisoning effect) due to the formation of a polymer layer on the anode surface. This deactivation, which depends on the adsorption properties of the anode surface and the concentration and nature of the organic compounds, is more accentuated in the presence of aromatic organic substrates such as phenol, chlorophenols, naphthol, etc. [21]. There is also a possibility of electrode fouling leading to the passivation of the electrode surface, which results in poor chemical decontamination [22–24]. The poisoning effect can be avoided by performing the oxidation in the presence of inorganic oxidation mediators or in the potential region of water discharge with simultaneous oxygen evolution.

The yield space–time (YST) of a reactor is defined as the mass of product produced by the reactor volume in a unit time:

$$\mathrm{YST} = \frac{iaM}{1000\,zF} CE \tag{6.2}$$

where i is the current density, a is the specific electrode surface area, M is the molecular weight, z is the number of electrons, F is Faraday's constant, and CE is the current efficiency. The YST gives an overall index of a reactor performance, especially the influence of the specific electrode area [25].

These techniques are used when conventional methods become insufficient because contaminants are refractory to oxidize or even partially oxidized, yielding reaction products with higher toxicity [26, 27]. On the other hand, hydroxyl radicals generated by EAO are able to nonselectively degrade several organometallic and organic contaminants in wastewater until their complete mineralization and conversion into inorganic ions, CO_2, and H_2O. The deficit of one electron in a hydroxyl radical valence orbit makes it an electrophile, so most reactions of hydroxyl radicals are electrophilic substitution in aromatic compounds and addition to alkenes (hydroxylation) [28].

6.3.1.2 Indirect Anodic Oxidation

Several studies that used different electrode materials such as boron-doped diamond (BDD) [29–32] and metal oxides such as PbO_2 [33, 34], SnO_2 [35], Sb-doped SnO_2 [36, 37], and Sb-doped $Sn_{0.8}W_{0.2}$-Ox [25] anodes for the electrochemical decontamination of wastewater via indirect anodic oxidation have been reported.

Recalcitrant organic compounds found as waste in the drainage of industries could be removed by indirect electrochemical advanced oxidation methods, using IrO2/Ti electrodes, where total removal of organic materials is accomplished by catalytic oxidation. The presence of ferrous ions in the solution of wastewater is very important for the oxidation and degradation of recalcitrant organic compounds, where Fe^{2+} can be oxidized to Fe^{3+} and free electrons are liberated to form a strong oxidant hydroxyl radical. In this process, the degradation and mineralization of COD in more than 7000 mg/L of organic wastewater was executed by using a high oxygen overpotential-modified IrO2/Ti electrode as anode. The IrO2/Ti electrode showed electrocatalytic oxidation behavior, where direct oxidation of organic materials occurred.

Also, indirect electrochemical advanced oxidation occurred by the electro-Fenton process in the presence of Fe^{2+} ions and Cl_2/OCl^- produced during electrolysis in the presence of NaCl.

Both direct and indirect electrochemical oxidations support and accelerate the degradation of highly harmful organic compounds in a short time. In the appropriate conditions, potent oxidant hydroxyl free radicals are formed, which are highly adsorbed onto the organic materials and lead to their decomposition and complete mineralization to less harmful compounds.

6.3.2 Electrochemical Reduction

The most attractive method to reduce vat and sulfur dyes is electrochemical reduction, because the addition of reducing agents is not required. This last method also avoids the generation of toxic products due to the reaction between the added reagents and the dye molecules. For all these reasons, electrochemical reduction processes are considered more suitable, where no reagent addition is required, no byproducts are formed, and no tertiary treatments are necessary to treat the final effluents. The electric energy supplied by an electric power device is the only requirement of electrochemical reduction [38]. The electrochemical reduction of textile dyes can be divided into direct and indirect reduction.

Both types of electrochemical reduction and operating conditions were studied and are summarized in Table 6.1.

These methods were deeply investigated for application at industrial scales, and several studies [48, 49] are available on functionalizing the surface of carbon electrodes to achieve higher reduction rates with the same color intensity and washing fastness as traditional methods [50]. The reuse of the reducing agent and also the dyeing bath in the indirect reduction electrochemical method, using a mediator, has been studied [45].

It can be concluded that the use of electrochemical techniques constitute a promising field for the different steps of textile processes, but their application to the treatment of vat and sulfur dyes is especially interesting for avoiding the use of reducing reagents.

6.3.3 Electrocoagulation/Flotation

The electrocoagulation (EC) process has been successfully used for the treatment of wastewater; it is attractive for its safety, selectivity, versatility, ease of control, amenability to automation, and environmental compatibility [51–53]. Electrocoagulation/flotation (ECF) includes the generation of coagulant in situ by the anodic dissolution of metal ions from an electrode with simultaneous formation of OH^- ions at the cathode [54, 55]. In ECF, contaminants are agglomerated toward the surface of the ECF cell by tiny bubbles of H_2 gas generated from the cathode [25]. Electrons, therefore, are the only employed agents in ECF, being responsible for facilitating wastewater treatment [56].

A simple ECF unit essentially involves an electrochemical cell, two electrodes as anode and cathode, and a DC power supply. Al and Fe are usually used as anode materials, the dissolution of which produces hydroxides, oxyhydroxides, and polymeric hydroxides

Table 6.1 Different direct electrochemical reduction methods and their conditions.

Treatment	Conditions
Direct electrochemical reduction	
Wastewater of sulfur dyes	High solubility [39]
Wastewater of sulfur dyes	Multicathode cell made of stainless steel electrodes [40] Formed via radical anion Effectiveness 70–90%
Wastewater of indigo dyes	Nickel cathode electrode [41, 42] Formed via radical anion Effectiveness 70–90%
Wastewater	Graphite electrodes [43] High surface area [42]
Wastewater of vat dyes	Graphite granules [42, 44]
Indirect electrochemical reduction	
Wastewater of dyes	Iron complexes act as mediators [45, 46]
Wastewater of vat dyes and indigo	Low solubility in water Have low contact to the surface of the cathode [39]
Wastewater of dyes	Electrolytic hydrogenation in situ by water electrolysis [47] H_2 reacts with dyes adsorbed at the electrode surface

[57–59]. In an ECF process, coagulants which are formed through electrolytic oxidation of the sacrificial electrode destabilize the contaminants followed by agglomeration of the destabilized phases to form flocs [56]. Then, they are floated by electrogenerated hydrogen or oxygen (typically hydrogen) bubbles toward the surface of the ECF cell. In ECF, numerous electrochemical reactions occur, summarized as follows [25, 60].

At the anode:

$$M_{(s)} \rightarrow M^{n+}{}_{(aq)} + ne^- \tag{6.3}$$

$$4H_2O_{(l)} \rightarrow 2H^+{}_{(aq)} + O_{2(g)} + 4e^- \tag{6.4}$$

At the cathode:

$$M^{n+}{}_{(aq)} + ne^- \rightarrow M_{(s)} \tag{6.5}$$

$$H_2O_{(l)} + 2e^- \rightarrow H_{2(g)} + 2OH^- \tag{6.6}$$

According to the reactions (6.2–6.5), the electro-generated metal ions (M^{n+}) immediately undergo further reactions producing hydroxides and polyhydroxides of strong affinity for dispersed particles and counter ions bringing about the coagulation [55, 56]. Besides, the gases evolved at the electrodes separate particles and coagulant aggregates by lifting them up through a flotation-like process while accelerating collisions between particles and coagulant by inducing more mixing [61]. Moreover, the pollutant's physicochemical characteristics influence its behavior within the system and eventual removal path. Accordingly, different results have been reported throughout the study, especially concerning application of different electrodes. In the scope of textile and organic wastewater treatment using EC, Phalakornkule et al. stated that Fe was superior to Al, while Ilhan et al. concluded that Al provided better removal in comparison with an Fe electrode [62, 63]. Furthermore, Linares-Hernandez et al. found that Fe was more effective in reducing COD, whilst Al was more effective in removing color [62–64].

6.3.4 Photoelectrochemical Combination

When Fe is present in both surface and groundwater, even at low concentrations, it can lead to various water quality problems, so its removal is important. Fe reacts with O_2 from the air to form reddish brown particles (rust) [7, 8].

Both photo and electrochemical oxidation technologies are lately considered more popular for water treatment. These processes are perceived as attractive options in solving the problem concerning iron removal of water especially if other pollutants such as ammonia, total dissolved solids, or natural organic matter (NOM) are found [65].

Free hydroxyl radicals ($OH^{\bullet}$) are a nonselective and very powerful oxidant agent able to oxidize organic and inorganic contaminants in water and are generated from chemical, electro, and photochemical processes (light irradiation).

A combined photoelectrochemical oxidation technique was studied considering the removal of Fe from water. The removal of Fe from a synthetic solution using a bench-scale photoelectrochemical oxidation system was investigated using different concentrations levels of iron at different current intensities. A combined UV-electrochemical oxidation process is a very effective and promising method for disposal of iron from groundwater. After a very short operating time $\leq$ 20 min, the removal of Fe reached 99% with low consumption of current intensity (0.25 A) without addition of an external oxidizing agent [66].

6.3.5 Electrodialysis

Electrodialysis (ED) is essential for transport of salt from first solution (diluate) to second solution (concentrate) by applying an electric current. This occurs in an electrodialysis cell, which provides all necessary parts for the process. Concentrate and diluate are separated by a membrane. An electric current is applied, moving the salt through the membrane.

The ED process is mainly a membrane separation process, and it could be used for the treatment of wastewater [67, 68]. In ED an electric potential is applied between anode and cathode as driving force for ion migration through the membrane. Due to passage of electric current, −ve and +ve ions are moved toward the respective electrodes based on their polarity. Cations are transferred though the compartment containing the membrane having −ve charge and vice versa for anions. The compartments are diluted and concentrated in alternative ways. The performance of the ED process depends on some parameters, e.g. current density, pH, flow rate, cell design, feed water ionic concentration, and properties of the ion exchange membrane.

Membrane fouling is an important factor that results in enhancement of energy consumption and decline of membrane flux. Selectivity of the membrane is adversely reduced. As most of coloids present in water are negatively charged, so most of the research is carried out on anion exchange membrane fouling [69, 70].

Direct current is applied between two electrodes. Consider a feed solution containing sodium chloride passed through an ED system and focus on one compartment. Cations pass through the cation exchange membrane and anions are restricted. Anions could only pass through an anion exchange membrane, which restricts passage of cations. The concentration of salts decreases in this compartment, but salt concentration keeps increasing in the adjacent compartment. Thus, salt concentration decreases in alternative compartments and enhances in the rest. Both desalinated water and concentrated brine leave from the adjacent compartments. An applied electric potential acts as driving force for the ED process [71].

ED is preferred when salt concentration is low, and it is regarded as an energy efficient process. The electric current requirement depends upon the number of ions transferred through the ion exchange membrane.

$$\text{Electric current required} = Z \times F \times Q \times \Delta C / \pounds \tag{6.7}$$

Where Z is the ion charge, F is Faraday's constant, Q is the feed solution flow rate, ΔC is the change in solute concentration, and $\pounds$ is the current utilization factor that accounts for energy efficiency. ED processes have been widely used at the commercial scale and for various applications [72].

The mechanism of the ED process can be demonstrated as follows. The ion exchange membranes are arranged between the anode and cathode in the ED system. Ionic compounds in the feed water start to move across the ion exchange membranes using an electrical driving force. The cations in the solution migrate toward the cathode and the anions migrate toward the anode by applying electrical current between the anode and cathode. Cations pass through the cation exchange membrane, but they are retained by the anion exchange membrane. Similarly, anions pass through the anion exchange membrane, but they are retained by the cation exchange membrane. The overall result of ED is, concentrated and diluted compartments occur in the system. The electrolyte solution is used to ensure the conductivity and remove gases produced by electrode reactions in the system [70, 73].

ED processes are used on a large scale commercially. ED has been widely used to demineralize, concentrate, and/or convert salt-containing solutions. Performance of the system depends on operating conditions and the ED cell structure [74, 75].

6.3.6 Electro-Fenton Process

The electro-Fenton (EF) processing technology is part of the widely known electrochemical advanced oxidation processes (EAOPs). In the EF method, the generated H_2O_2 at the cathode reacts with Fe^{2+} found in the medium, leading to the formation of hydroxyl radicals ($OH^{\bullet}$) by the Fenton reaction.

In the EF process, a sufficient amount of H_2O_2 is produced from the electrochemical reduction of O_2. These conditions are essential for the formation of Fenton's reagent reactions (indirect oxidation), where the highly oxidant OH• free radical can be formed according to the following steps:

1) Electrochemical reduction of dioxygen into $^{\bullet}O^{-2}$ and formation of H_2O_2 in acidic medium.
2) Reaction of H_2O_2 with Fe^{2+} ions yielding OH• radicals (electro-Fenton reaction).

In these conditions, Fe^{3+} ions are electrochemically reduced to Fe^{2+} ions (electrocatalytic system). The optimal pH range for the EF process was determined to be from pH 3 to 6, where high pH values will result in inhibition of the Fenton reaction, since the Fe^{2+} ions will be oxidized and form colloidal Fe^{3+} ions. Likewise, very low pH values < 1.5 would result in the decomposition of H_2O_2 into oxygen and water without forming OH• [76].

The EF process is deemed to be the most popular technique due to its simplicity and efficient performance in degradation of recalcitrant organic pollutants (ROPs) as well as water disinfection. Degradation of harmful organic pollutants is vital as they are considered detrimental, even when their presence is at the lowest concentration [77]. Through rigorous technical, environmental, and economic sustainability enhancements, the EF process could be implemented in a wide variety of industries as a highly effective and green water reclamation system. The EF process was originally derived from an advanced oxidation process (AOP) using Fenton's reagent, resulting in the reaction between H_2O_2 and Fe^{2+} to oxidize organic compounds [78, 79].

The EF process depends on the in situ generation of (•OH), a highly powerful oxidizing agent. These species are more effective oxidants ($E_0 = +2.8$ V) than the chemical reagents commonly adopted for this purpose, hypochlorous acid and permanganate ($E_0 \approx +1.5$ V), and H_2O_2 ($E_0 = +1.8$ V). The EF process is a promising technology, being more economical, efficient, and environmentally friendly for removing organic matter compared with conventional procedures [80–84].

The EF process has two different types. In the first type, Fenton's reagent has been added to the process from outside, and in the second H_2O_2 is added from the external process while Fe^{2+} is provided from the process of anode reduction results.

The EF process using the OH• derived from the reaction of H_2O_2 and Fe^{2+} by chemical reactions has the following equations:

$$Fe^{3+} + H_2O_2 \rightarrow Fe^{2+} + H^+ + HO\bullet^2 \tag{6.8}$$

$$Fe^{3+} + HO\bullet^2 \rightarrow Fe^{2+} + H^+ + O_2 \tag{6.9}$$

$$Fe^{3+} + R\bullet \rightarrow Fe^{2+} + R^+ \tag{6.10}$$

$$Fe^{2+} + HO\bullet^2 \rightarrow Fe^{3+} + HO^{2-} \tag{6.11}$$

The EF process is affected by different factors such as electrode type, pH, electrolyte, current density, temperature, ferrous ion concentration, hydrogen peroxide, O_2 flow rate, and electrode gap [85–87].

6.3.7 Sonoelectrochemical Combination

Sonication leads to effective oxidation and pyrolysis of a variety of hydrocarbons in aqueous solution. Ultrasound (US) has other benefits by causing extremely fast mass transport, formation of free OH• radicals, enhancing mixing and dissolution kinetics, and resisting the formation of solid products at the electrode surface.

US is a compression wave at frequencies that range from greater than about 15 kHz to 1 MHz. Acoustic cavitation (sonication) involves the application of sound waves being transmitted through wastewater as a wave of alternating cavitation cycles. An initial vapor phase in wastewater (microbubbles) develop when the pressure in wastewater drops below its vapor pressure. Bubble formation in liquids includes tiny gas pockets (i.e. flaws in the liquid matrix) that are not gas bubbles but can readily act as nuclei for cavitation bubbles at the relatively lower pressure in the compression cycle induced by sound waves. These nuclei activations result from the failure of the dynamic tensile strength of the liquid being sonicated due to gas or vapor imperfection of the predominately homogenous bulk matrix of the liquid [88].

The chemical effects of sonication are a result of acoustic cavitation and occur during rarefaction of molecules. They are torn apart, forming tiny (10–10 m diameter minimally) microbubbles, which grow to a critical size during the alternating cavitation cycles, and then implode releasing a large amount of energy in a short amount of time (<100 ns) [89]. The collapsing bubble interface results in the formation of hydroxyl (OH•) and hydrogen (H•) radicals [90]. The OH• and H• radicals formed during acoustic cavitation destroy hydrocarbon compounds very effectively [91].

$$H_2O \rightarrow H. + OH. \tag{6.12}$$

$$H. + OH. \rightarrow H_2O \tag{6.13}$$

$$2H. \rightarrow H_2 \tag{6.14}$$

$$2OH. \rightarrow H_2O_2 \tag{6.15}$$

$$2OH. \rightarrow O. + H_2O \tag{6.16}$$

$$2O. \rightarrow O_2 \tag{6.17}$$

$$O. + 2H. \rightarrow H_2O \tag{6.18}$$

$$O. + H_2O \rightarrow H_2O_2 \tag{6.19}$$

US is beneficial for electrochemical treatment due to:

i) Causing extremely fast mass transport.
ii) Enhancing the mixing and dissolution kinetics at low temperature.
iii) Affecting the formation of solid products at the electrode surface.

Marken et al. have investigated low temperature sonochemical processes involving the study of mass transport and the cavitation effects of US 20 kHz in liquid ammonia [92–94]. Also, Zhang and Coury [95–97] ascertained that the application of the sonochemical technique to electrochemical processes has provided many benefits and reported that irradiated glassy carbon electrodes exhibited enhancement of heterogeneous electron transfer rates for a variety of aqueous redox probes.

Advantages of the sonoelectrochemical method include [98, 99]:

1) Ultrasonic degassing at an electrode surface prevents gas bubble accumulation interfering with the passage of current.
2) Agitation via cavitation at the electrode surface assists ion transport across the electrode double layer throughout the electrochemical process and reduces ion depletion in the diffusion layer.
3) Cavitation at and near the electrode surface will result in continuous cleaning and activation of the electrode.
4) The ultrasonic energy may cause decomposition of the dye aggregates in a solution where organic compounds with relative high molecular mass have a tendency to form aggregates in the solution.
5) Very mild reaction conditions.
6) Excellent mass transfer of reaction solutions.
7) Keeping of the electrode activity during the process.
8) Safe and effective technology.
9) Able to treat toxic pollutants in a wide concentration range.

The combination of ultrasound and electrochemistry comprises a booming field due to the combined chemical (possible change in the mechanisms) and mechanical (the liquid/solid mass transfer and the activation of the electrodes surface) effects of ultrasound.

Sonoelectrochemical effects achieve many benefits including the surface effects, changed adsorption condition, improved mass transport, reduced electrode fouling, production of altered product distributions, increased current efficiencies and yields, requiring of less cell power, and increasing the limiting current. It offfers significant usefulness for a variety of different applications [100].

6.3.8 Electrodisinfection

Electrochemical disinfection of water is the eradication of microorganisms by using an electric current passed through water under treatment using suitable electrodes. At the phase boundary between the electrodes and the water, the electric current leads to the electrochemical production of disinfecting species from the water itself (e.g. O_3), or from species dissolved in the water (e.g. Cl^- is oxidized to free Cl_2) [101].

Sufficiently stable and efficient electrodes used for electrochemical disinfection of water have been developed and optimized. These are Ti electrodes with mixed oxide coatings based on Ir and/or Ru oxide [102–104], and doped-diamond electrodes [105].

This technique depends on current density, type of electrodes, and electrochemical production of free chlorine [106–110].

The advantage of the electrodisinfection process is the achievement of disinfection conditions in situ in the treatment device. This leads to avoiding the drawbacks normally associated with adding disinfectants, such as transport and storage of dangerous chemical compounds. Additional advantages over other conventional disinfection technologies include environmental compatibility, versatility, energy efficiency, safety, selectivity, amenability to automation, and cost effectiveness [111].

Electrodisinfection faces several challenges when being applied for large-scale treatment: electrode materials may be prone to erosion, complexation, oxidation, wearing, or inactivation, and the best electrode materials frequently involve precious metals of high costs [112]. Iron electrodes have been used, and seawater added as a chlorine source for disinfection [113].

Common electrode materials used in electrodisinfection include stainless steel, aluminum, graphite, carbon cloth, titanium, platinum, and diamond. Disinfection efficiency has been improved by coating the anode surface with ruthenium oxide (RuO2), mixed iridium oxide (MIO), and/or titanium oxide (TiO2), where the anode is turned highly resistant to corrosion and production of disinfecting species was improved [6].

HOCl is an efficient disinfectant species, and the treated water will have both residual Cl_2 and disinfection by-products. Cl_2 is chemically produced at the anode from Cl^- present in water [114]. The process has been summarized as follows [30, 115]:

$$2Cl^- \rightarrow Cl_2 + 2e^- \tag{6.20}$$

$$Cl_2 + H_2O \leftrightarrows HOCl + H^+ + Cl^- \tag{6.21}$$

$$HOCl \leftrightarrows ClO^- + H^+ \tag{6.22}$$

By using anodes with a high O_2 overvoltage, high current density, and low water temperature, ozone can be generated at the anode, and hydrogen peroxide may also be produced at the cathode:

$$3H_2O \rightarrow O_3 + 6e^- + 6H^+ \tag{6.23}$$

$$O_2 + 2H_2O + 2e^- \rightarrow H_2O_2 + 2OH^- \tag{6.24}$$

Research on synthetic ballast water electrodisinfection [116] demonstrated that killing of bacteria took place when using DC, even in the absence of chlorides.

Other researchers [117] were able to obtain efficient disinfection of zero-chloride water using AC. Barashkov et al. [118] claimed that disinfection was achieved due to the generation of hydroxyl radicals (•OH) and other intermediate highly active anion radicals ($\bullet O^{2-}$). Several investigators claim that free chlorine does not appear to be the main disinfecting agent in electrodisinfection [119]. Instead, these authors have found that the production of more powerful (than chlorine) microorganism killing agents, such as H2O2, [O], •OH, and $\bullet HO^2$, is what provides the high degree of disinfection observed in their experimental trials. Also, the bactericidal efficiency of the process generally increases when the retention time and current density are increased [120].

6.4 Case Studies and Successful Approaches

6.4.1 Reduction of Methylene Blue in Wastewater: Direct Anodic Oxidation [121]

Technique	Type of contaminant in wastewater	Main electrode	Type of electric power
Direct anodic oxidation	Methylene blue	Ti/Ni-Co oxide electrode	DC

The performance of $Ni_{0.6}Co_{0.4}$ oxide catalyst in the anodic oxidation of methylene blue dye (MB) was first studied by the electrochemical oxidation of 50 mg/L MB dye dissolved in 0.17 M Na_2SO_4 electrolyte (no chlorides were present).

Electrochemical oxidation of organic pollutants can be carried out through the direct method, where the contaminant is directly oxidized on the electrode surface. The mechanism of MB degradation on the $Ni_{0.6}Co_{0.4}$ oxide electrode in 0.17 M Na_2SO_4 aqueous solution was investigated by addition of methanol, which is a well-known scavenger of hydroxyl radicals [122, 123]. It was observed that the addition of 0.2 M methanol to the MB-containing solution did not have any effect on the apparent kinetics of degradation of MB. This confirms that the MB degradation on the $Ni_{0.6}Co_{0.4}$ oxide surface occurs mainly through direct oxidation of the molecule.

6.4.2 Disposal of Cosmetics Industry Drainage: Indirect Anodic Oxidation [124]

Technique	Type of contaminant in wastewater	Main electrode	Type of electric power
Indirect anodic oxidation	Cosmetics industry drainage	IrO_2/Ti electrode	DC

Most cosmetics and pharmaceuticals are hazardous materials, and the majorities are carcinogenic or contain impurities that cause cancer. These materials include coal tar colors, glycol ethers, parabens, phenylenediamine, diethanolamine (DEA), phthalates, paracetamol, and chlorophene. Also, natural and synthetic cosmetics contain wetting agents like DEA or triethanolamine, both of which are not carcinogenic, but if they exist along with some cosmetics products containing nitrites contaminants, they can cause a chemical reaction leading to the formation of highly carcinogenic nitrosamines [125]. The chemical compositions of some types of organic compounds contained in cosmetic wastewater are shown in Figure 6.2.

This study was focused on the preparation of a high surface area modified electrode IrO_2/Ti employed as anode in the electrolytic cell, where at high O_2 overpotential, strongly oxidant hydroxyl radical was formed by electro-Fenton reaction. The process can be successfully used to get rid of real cosmetics wastewater (COD = 7000 mg/l) by addition of ferrous ions and control of an acidic medium. Effects of Fe^{2+} concentration, NaCl concentration, and pH value on the removal efficiency were also studied by measuring COD value. The suggested cell design and electrochemical set up for wastewater treatment is shown in Figure 6.3.

Hydroxyl radicals are one of the most reactive chemical species, with relative oxidation power of 2.06 higher than both atomic oxygen 1.78 and hydrogen peroxide 1.31 [126]. Electrochemical methods are clean and effective techniques for direct and indirect oxidation [127]. In anodic oxidation these radicals are formed by water oxidation on high O_2 overpotential anodes [128, 129].

Ferrous ions play a big role in the degradation of refractory organic compounds, where the presence of Fe^{2+} ions and hydroxyl groups can support the formation of highly oxidant hydroxyl free radicals (•OH). They are strong oxidants in the solution, where highly dangerous long chain organic compounds are broken and turned to environmentally friendly small entities of organic compounds and gases of CO_2, NO_2, or SO_2. The addition of a small amount of Fe^{2+} ions (0.5 g/L) to the solution of cosmetics wastewater reduces the COD value from 7000 to less than 900 mg/L by electrochemical treatment, as shown in Figure 6.4.

The variation of initial pH value of wastewater influenced the removal efficiency. At low concentration of Fe^{2+} added, low removal efficiency was observed at high pH 9, and high

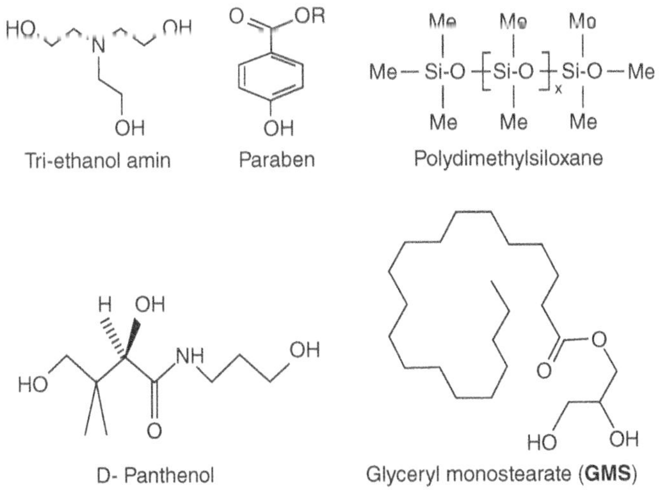

Figure 6.2 Chemical composition of some types of organic compounds contained in cosmetic wastewater. [124].

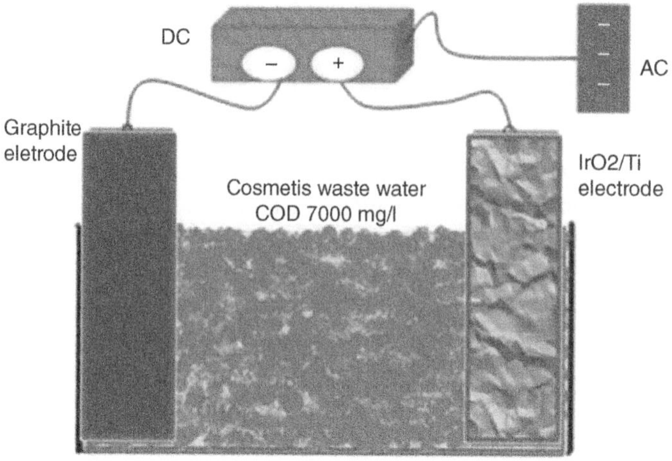

Figure 6.3 Electrochemical set up for wastewater treatment. [124].

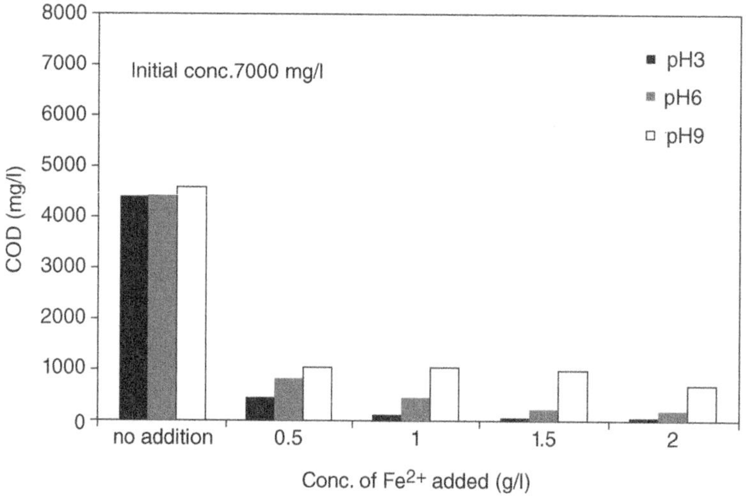

Figure 6.4 Effect of concentration of Fe^{2+} on the COD value of wastewater of cosmetics drainage at constant 20 V, 0.1 g/l NaCl, and 20 min. [124].

COD value 61 500 mg/l was also found. Higher removal efficiency and the lower COD value 6500 mg/l was achieved at pH 3. The presence of an acidic medium supports the formation of highly adsorptive Fe^{2+} ions for a long time in the solution without oxidation to the colloidal nonadsorptive Fe^{3+} ions. Also, during electrochemical oxidation in an acidic medium, hydrogen peroxide can be produced by two-electron reduction of oxygen at appropriate cathodic potential on a graphite electrode [130].

$$O_2 + 2H^+ + 2e^- \rightarrow 2H_2O_2 \tag{6.25}$$

Also, COD values were highly decreased by increasing the concentration of Fe^{2+} to 2 g/l, where COD was decreased to 800 and 100 mg/l at pH values 9 and 3, respectively. This indicates that the addition of 0.5 g/l of Fe^{2+} ions at pH 3 was the best economic condition for the removal of cosmetic organic materials through low-cost consumption of Fe^{2+} ions. The lowest removal efficiency was observed at very high pH 9 and was not suitable for oxidation of organic materials due to the presence of excess OH^- groups in the solution, which accelerate the oxidation of Fe^{2+} to Fe^{3+} without complete oxidation of organic materials. In the case of pH 3, highly oxidant ▪OH free radicals can be formed due to the presence of Fe^{2+} ions (Figure 6.5).

At a constant electric potential of 20V, electric current was continuously increased by addition of Fe^{2+} ions at constant concentration of NaCl 0.1 g/l and at different pH values. It was observed that without addition of Fe^{2+} ions, a low value of current 0.16A was obtained at different pH values, as shown in Figure 6.5, while it was improved to 0.30A by addition of 0.5 g/l of Fe^{2+} at different pH values. By continuous addition of Fe^{2+} ions, the concentration increased to 1 g/l, and the current was increased and varied according to pH value. In an acidic medium of pH 3, a high electric current was obtained due to the presence of a high concentration of H^+ ions. After the increase of Fe^{2+} ion concentration, electric current was continuously increased, and the ionic mobility in the solution was increased to overcome the solution resistance. Adequate time is required for total accomplishment of oxidation reactions, where most harmful organic materials are completely destroyed and/or converted into small safe materials. These reactions can be accelerated by passage of a high current 0.8A to obtain the lowest COD value, while enough time is required to obtain the lowest COD value with passage of low current 0.2A.

As shown in Figure 6.6, the increase of duration time of the electrochemical treatment of cosmetics wastewater led to the success of waste organic material removal. The time consumed for removal of these materials is highly influenced by the electric current; at high

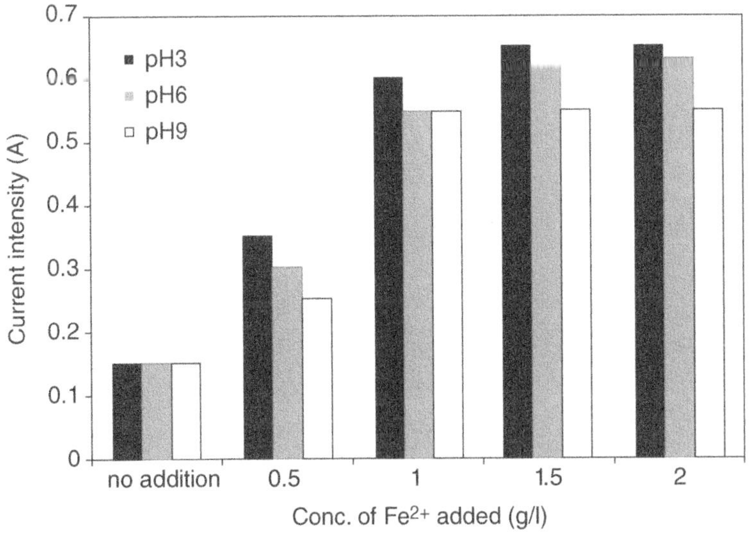

Figure 6.5 Effect of concentration of Fe^{2+} on the current intensity during electrochemical treatment at constant concentration of NaCl 0.1 g/l and time 20 min. [124].

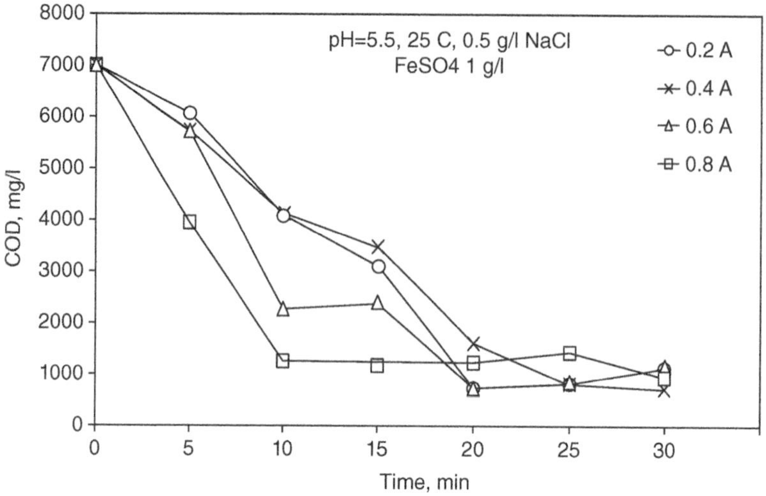

Figure 6.6 Effect of time on the removal efficiency during electrochemical treatment at constant concentration of Fe^{2+} 1 g/l and NaCl 0.1 g/l. [124].

applied current 0.8 A, the lowest COD 1000 mg/l was obtained after 10 min, and at low applied current 0.2 A, the lowest COD 950 mg/l was obtained after 20 min. At constant electric potential, the increase of Fe^{2+} ions enhanced the removal efficiency in the solution of cosmetic wastewater. Figure 6.7 depicts that the removal efficiency was improved by continuous addition of Fe^{2+}, where the COD value was reduced to 1400 mg/l by addition of a low concentration of Fe^{2+} ions 0.5 g/l at constant potential 20 V, NaCl 0.1 g/l, and pH 6. At the same conditions, COD was continuously reduced to 960 mg/l by increasing Fe^{2+} concentration to 2 g/l. This indicates the necessity of adding Fe^{2+} to produce a greater amount of free radicals that reinforce the oxidation of organic compounds, as per the following reactions:

$$\mathbf{Fe(OH)_2 \rightarrow Fe(OH)_3 + e^-} \quad (6.26)$$

$$\mathbf{OH^- + e^- \rightarrow \bullet OH} \quad (6.27)$$

Hydroxyl radicals react with organic compounds by three types of reactions [131]:

1) Hydrogen atom abstraction

$$\mathbf{RH + \bullet OH \rightarrow H_2O + \bullet R} \quad (6.28)$$

2) Electrophilic addition on double bond

$$\mathbf{PhX + \bullet OH \rightarrow PhXOH \bullet} \quad (6.29)$$

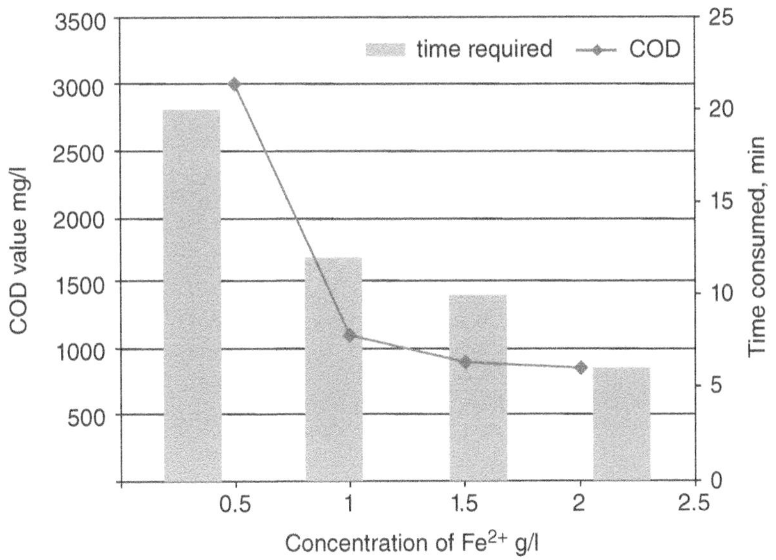

Figure 6.7 Effect of concentration of Fe^{2+} on COD value and time consumption at 0.1 g/l NaCl, 20V, and pH 3. [124].

3) Electron transfer.

$$RX + OH\bullet \rightarrow RX\bullet^{+} + OH \quad (6.30)$$

On the other hand, as the concentration of Fe^{2+} ions increased the electric current was increased. Accordingly, the time required for destruction of organic materials was increased at constant potential 20 V. Without addition of Fe^{2+} ions, the time consumption was higher than 20 min due to low electric current at constant NaCl concentration 0.1 g/l. While time was decreased gradually from 20 to 12 min by continuous addition of Fe^{2+} to 1 g/l. The excess addition of Fe^{2+} to more than 1 g/l revealed a continuous decrease of time due to continuous increase of current by addition of Fe^{2+} to a concentration of more than 1 g/l.

The addition of salts such as NaCl played an important role in the increase of ionic strength and electrical conductivity through the solution. Hence, electric current was increased leading to the enhancement of removal efficiency. At constant Fe^{2+} concentration of 0.5 g/l and time 20 min, low removal efficiency was observed without addition of NaCl, where COD of 5490 mg/l and current 0.09A were measured, as shown in Figure 6.8. The increase of NaCl concentration led to the improvement of the removal efficiency, where the measured COD value was reduced from 7000 to 2120 mg/l as the current increased from 0.09 to 0.18 A. Moreover, the electric dissociation of NaCl gives the advantage to oxidants formation such as Cl_2, OCl^-, and ClO_3, which support the oxidation of organic materials. Salt addition to the electrolytic solution can enhance the ionic transfer and overcome the resistance against electric current; this indicates the advantage of salt addition for the increase of $Fe(OH)_2$ formed in an acidic medium, which plays an important role for organic pollutant degradation by copolymerization and coagulation. The

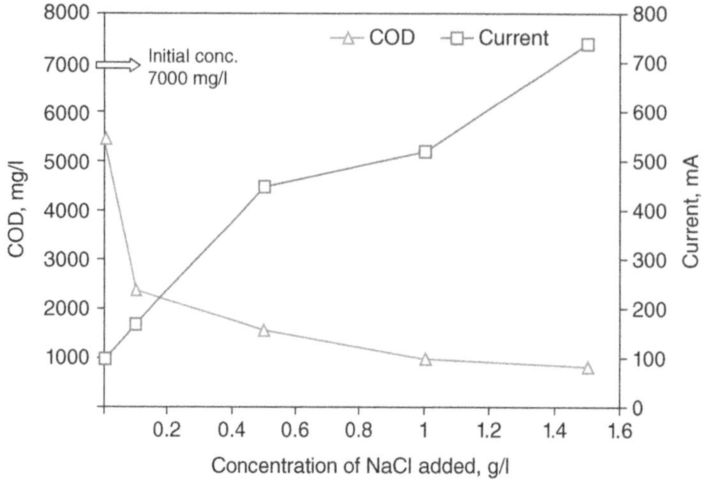

Figure 6.8 Effect of NaCl added on the current intensity during electrochemical treatment at constant Fe^{2+} concentration 0.5 g/l, 20 V, pH 3, and time 20 min. [124].

removal efficiency of cosmetics organic compounds from wastewater was also improved by indirect oxidation of organic materials; this takes place through chlorine generation and hydrolysis according to the following reactions:

$$2Cl^- \rightarrow Cl_2 + 2e^- \tag{6.31}$$

$$Cl_2 + H_2 \rightarrow HOCl + Cl^- + H^+ \tag{6.32}$$

$$HOCl \rightarrow OCl^- + H^+ \tag{6.33}$$

$$2HOCl + Cl^- \rightarrow ClO_3^- + 2HCl \tag{6.34}$$

The electrolytic solution is composed of chlorine, hypochlorous acid, and hypochlorite, as well as a small amount of O_2 evolved at high current density [132, 133].

Accordingly, the advantage of NaCl addition was also extended to improve the removal efficiency in a low time and consequently, the economic cost and energy consumption were reduced. Also, the auto generation of Cl_2 gas in the solution of wastewater can be exploited for disinfection purposes, if the wastewater contains bacteria and biocontamination cannot be tolerated. The optimum conditions and results are summarized in Table 6.2.

Table 6.2 Characteristics of wastewater before and after electrochemical treatment [124].

Variable	Before EAO	After EAO
COD	7000 mg/l	150 mg/l
pH	4	7.2
Color	Yellowish brown	clear
Turbidity	500 NTU	75 NTU

6.4.3 Decomposition of Azo Dyes: Electrochemical Reduction

Technique	Type of contaminant in wastewater	Main electrode	Type of electric power
Electrochemical reduction	Azo dyes	Platinated titanium electrode (Pt/Ti)	DC

The electrochemical reduction method has been discussed in a restricted number of papers, because its efficiency in pollutants degradation is weak in comparison to direct and indirect electrooxidation methods [134]. Bechtold et al. deem this method particularly appropriate for the treatment of highly colored wastewaters such as the residual pad-batch dyeing bath with reactive dyes [135]. The reduction of dyes takes place by producing hydrazine (in the partial reduction), and its total reduction generates amino compounds, Eq. (6.35). They recommend the importance of a separated cell in the case of dye baths containing chlorides. This separation is important to avoid the formation of chlorine and chlorinated products.

$$\underset{\text{Hydrolyzed Azo Dye}}{\mathbf{R_1\text{-}N{=}N\text{-}R_2}} \rightarrow \underset{\text{Hydrazine}}{\mathbf{R_1\text{-}NH\text{-}NH\text{-}R_2}} \rightarrow \underset{\text{Amines}}{\mathbf{R_1\text{-}NH_2 + R_2\text{-}NH_2}} \quad (6.35)$$

In the same way, Vanerkova et al. propose a reduction mechanism for azo dye degradation with platinated titanium electrodes (Pt/Ti) in the presence of NaCl [136]. In this study, the action of hypochlorite generated by oxidation of chloride is also discussed. Zanoni et al. [137] studied the hydrolysis under reduction process of two anthraquinone reactive dyes. They demonstrate that the acidic medium provides the best conditions, and that the presence of borate in the solution modifies the reduction process.

6.4.4 Removal of Ammonia of Onshore and Offshore Formation Water: Electrocoagulation/Flotation

Technique	Type of contaminant in wastewater	Main electrode	Type of electric power
Electrocoagulation/ Flotation	Ammonia of onshore and offshore formation water from petroleum production fields	Fe electrodes	DC

Ammonia is one harmful pollutant discharged into seawater as a result of oil industry activity from both onshore and offshore sites [138–140].

High levels of ammonia leads to toxicity for living organisms in seawater. It is also highly toxic in seawater at high pH in the presence of microorganisms. Toxicity due to ammonia may ensue from both ionized ammonia ($^+NH_4$) and un-ionized ammonia (NH_3). Researchers have shown that un-ionized ammonia is the most harmful species to seawater and threatens fish [141]. Ammonia limits are narrow for living organisms in seawater according to reports of the European Inland Fisheries Advisory Commission [142]. Fish and other living sea organisms are influenced, and a large variety of physiological disturbances were observed due to excess ammonia [143].

The most common method for the treatment of ammonia from wastewater is biological nitrification, but this is not suitable for ammonia of high concentration [144]. Electrochemical methods have been recommended as having potential applications in wastewater treatment [145]. The electrochemical method has some important advantages such as high efficiency and relatively low investment costs [146]. A study aimed to compare between both types of real formation of water samples (onshore and offshore) produced during oil production from an oil field located in the Suez Gulf, Egypt [147].

Most offshore platforms discharge formation water directly into the ocean or sea surface. It is subject to EPA regulations on entrained or dissolved oil and other chemicals included in the formation water.

The electrochemical method is essential for removal of excess ammonia that threatens fish, crustaceans, and other living organisms in seas or oceans, where excess ammonia in the formation water of both onshore and offshore sites is recommended to be removed in situ. This study was concerned with the formation water produced from the petroleum stations in the Suez Gulf, Gabal El Zeit region in the Red Sea, where the presence of high concentrations of ammonia is considered to be a big predicament that must be removed before disposal.

Before discharging into seawater, electrochemical (EC) treatment was used for remediation of onshore and offshore formation water containing high concentrations of ammonia, 5.54 and 110 mg /L, respectively (Figure 6.9). After EC treatment, ammonia was successfully removed by using three types of electrodes, as shown in Figure 6.10, where the measurements of ammonia concentration revealed no detection of ammonia after EC treatment of either onshore or offshore sites, as shown in Figure 6.10. This indicates the complete removal of ammonia from the formation water.

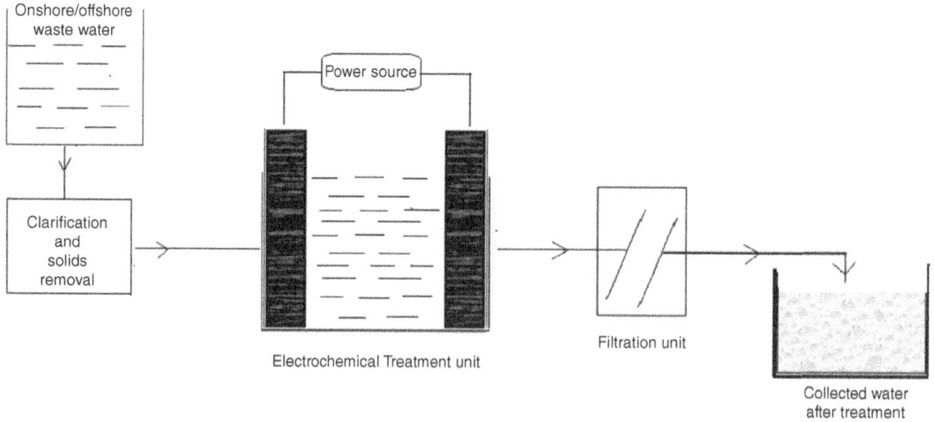

Figure 6.9 Flow chart of electrochemical treatment of formation water. [147] / With permission of Elsevier.

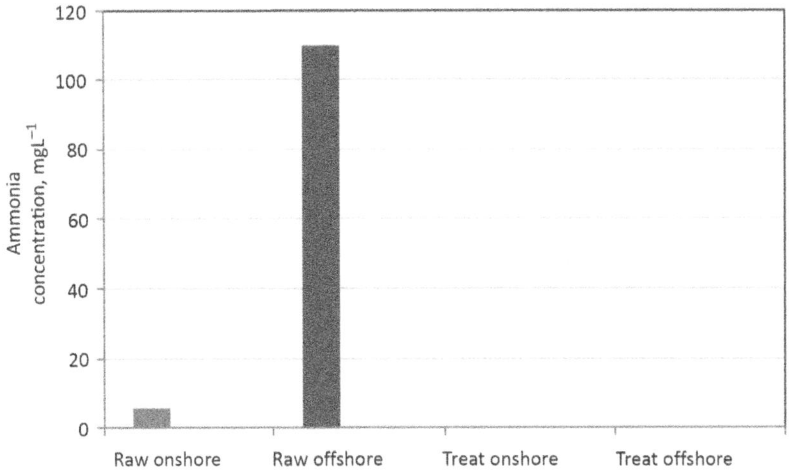

Figure 6.10 The total removal of ammonia of both onshore and offshore formation water, using the electrocoagulation method. [147] / With permission of Elsevier.

6.4.5 Alleviation of Iron and Manganese from Ground Water Using the Combined Photoelectrochemical Method

Technique	Type of contaminant in wastewater	Main electrode	Type of electric power
Combined photoelectro-chemical (CPE)	Iron and manganese from ground water	Al electrodes and UV lamp	DC

Aesthetic and operational problems like bad color and taste, staining, scale formation in the water distribution system, and high turbidity may occur due to the presence of Fe in a supply of ground water. Mn is a very common element that can be found everywhere on earth and is one of the most abundant metals in soils, where it occurs as oxides and hydroxides, and cycles through its various oxidation states. Mn is one out of three toxic essential trace elements, which means that it is not only necessary for humans to survive, but it is also toxic when too high doses are present in a human body [148].

Mn is difficult to remove from ground water by using normal methods, since it requires a high potential to overcome its high activation energy of manganese oxide formation, where MnO_2 is formed by highly oxidizing and high pH conditions [65, 149].

A study was carried out in a cylindrical quartz photo reactor, with a coaxial and immersed medium pressure UV mercury lamp used as the UV emitter and light source (Heraeus TQ150, input energy of 150 W) emitting a polychromatic radiation in the range from 100 to 280 nm wavelength. The UV lamp was equipped with a cooling water jacket to maintain the temperature of the reaction of wastewater treatment at room temperature. The reaction vessel was filled with synthetic solution containing both Fe and Mn. The electrochemical characterization of the solution was carried out by using a DC power supply GW-3030 and two electrodes, graphite cathode and aluminum anode. The measurements were performed at room temperature, and the mixing was accomplished by using a continuous magnetic stirrer [150]. A schematic is shown in Figure 6.11.

By using the CPE method and a mixture solution of 5 ppm $FeSO_4.7H_2O$ and 5 ppm $MnSO_4.H_2O$, a high removal efficiency was achieved after a short time; the concentration of Fe was decreased from 5 ppm to less than 0.1 ppm (R = 98%) for 5 min duration time, while the removal of Mn was decreased from 5 ppm to 1.7 ppm (R = 66%) for 5 min duration time, and was highly decreased to 0.2 ppm (R = 96%) after a duration time of 20 min (Figure 6.12). It is obvious that the emoval efficiency of Mn was less than Fe after the same time; since the removal of Mn required an oxidation potential higher than iron, the removal efficiency of Mn can be improved by using a higher potential and adequate time. This indicates that the electrochemical potential supported by UV irradiation (CPE) enhanced the oxidation of both soluble Fe^{2+} and Mn^{2+} to insoluble Fe^{3+} and Mn^{4+} ions in a short time, which can be adsorbed on the surface of formed $Al(OH)_3$ floc. The lowest concentration achieved of Mn after CPE treatment was 0.2 ppm (R = 96%), but this is still higher than the recommended standard values according to the reports of the World Health Organization (WHO). A second treatment may thus be required for the removal of a residue of Mn to attain the standard concentration of 0.05 ppm (R = 99.6%), so another method of electrochemical oxidation was employed, where the solution was totally exposed to an electric field between anode and cathode. Some reactions took place at the surface of the electrodes and in the bulk of the water solution, where Al^{3+} ions were generated by anodic oxidation, with reduction of hydrogen ions at the surface of cathode, and water molecules electrolyzed to OH^- and H^+.

$$Al \rightarrow Al^{3+} + 3e^- \tag{6.36}$$

$$H_2O \rightarrow OH^- + H^+ \tag{6.37}$$

$$Al^{3+} + OH^- \rightarrow Al(OH)_3 \tag{6.38}$$

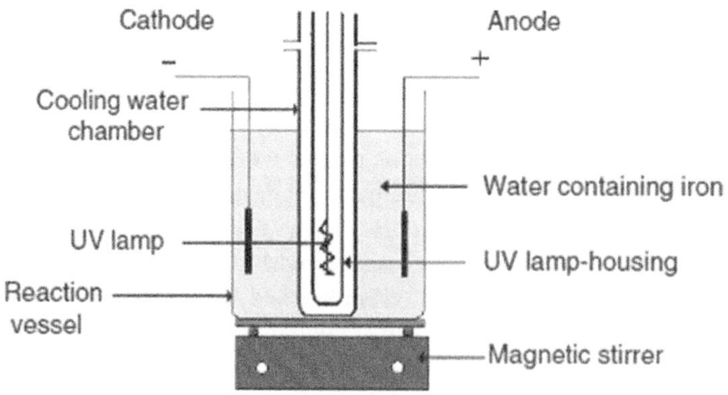

Figure 6.11 The schematic diagram of the experimental set-up for the CPE method. [150].

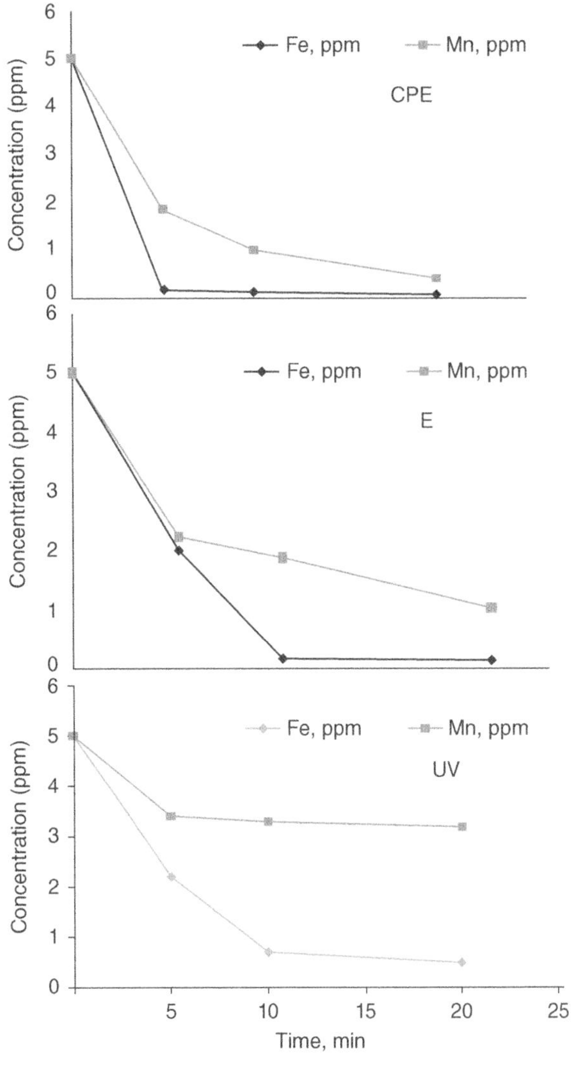

Figure 6.12 Comparison of removal efficiency for Fe and Mn by CPE, electrochemical and photochemical methods from a synthetic solution of Fe and Mn (initial concentration 5 ppm, 150W, 0.25 A). [150].

The presence of manganese in the solution of iron may behave as a catalyst for the oxidation of Fe^{2+} to Fe^{3+}, and Mn^{2+} is oxidized to Mn^{4+} and coprecipitated together.

It can thus be considered that the presence of two or several types of dissolved metals is more applicable in the electrochemical or CPE methods, where the dissolved ions enhance the conductivity through water and hence reduce the potential required. They represent catalyst and mediators for oxidation according to their redox potential.

The CPE method was used for oxidation of soluble forms of Fe^{2+} and Mn^{2+} to the insoluble forms Fe^{3+} and Mn^{4+}. The CPE method revealed more efficiency than sole EC and UV methods. The presence of both dissolved Fe and Mn has the advantage of less resistivity of waste water. Low concentration of NaCl (15–45 ppm) was added to increase the conductivity and electric current along with its bactericide effect after electrolysis to chlorine. The effect of Fe^{2+} and Mn^{2+} concentrations revealed that the higher the concentration of dissolved iron and manganese ions, the higher the removal efficiency obtained. The study showed the rapid oxidation of Fe^{2+} more than Mn^{2+} due to the lower oxidation potential of Fe versus Mn, and the catalytic oxidation behavior of Mn may accelerate the oxidation of Fe.

6.4.6 Disposal of Phenol from Industrial Wastewater: Electro-Fenton Method

Technique	Type of contaminant in wastewater	Main electrode	Type of electric power
Electro-Fenton	Phenol	Stainless and nickel electrodes	DC

Among different types of contaminants, phenolic compounds are toxic to the flora and fauna that live in bodies of water [151]. Phenolic compounds are released into water by wastewater from industrial activities, such as oil refineries, coke industries, pharmaceutical industry, pulp and paper industries, and food processing industries and various chemical factories [152].

A total of 250 mg/L of phenol compounds in a solution were processed by the EF method; within only 5 minutes, the total organic allowance reached 52.2%, but after adding the allowable activated carbon, 75%. The electrodes used were iron electrodes with optimal stoichiometry pH 5.2, $H_2O_2 = 37.2$ mM, conductivity 125 μS/cm, stirring speed 100 rpm, electrolyte NaCl, current 0.8 mA/cm^2, and distance between electrodes 4 cm [153]. Stainless steel and nickel were used as electrodes to set aside phenol compounds with the EF method with degradation results of 95.2% using stainless steel for 90 minutes and 72% using nickel anodes for 120 minutes; current used was 900 mA, with addition of Fe^{2+} as much as 5 mg/L [154].

6.4.7 Remediation of Wastewater Contaminated with *E. coli* Using Electrodisinfection

Technique	Type of contaminant in wastewater	Main electrode	Type of electric power
Electrodisinfection	Marrero Wastewater Treatment Plant, in Marrero, LA contaminated with *E. coli* in the samples	Iridium oxide-coated titanium and stainless electrodes	DC or AC

Modern ballast water electrodisinfection units, for example, take advantage of the high chloride ion and salt concentrations present in seawater to develop free chlorine species to promote disinfection [30]. In this case, hypochlorous acid species are the leading disinfectant species, and the treated water will have both residual chlorine and disinfection by-products. Chlorine is chemically produced at the anode from chlorides present in water [114]. The process is summarized as follows [6]:

$$2Cl^- \rightarrow Cl^2 + 2e^- \quad (6.39)$$

$$Cl_2 + H_2O \leftrightarrows HOCl + H^+ + Cl^-, \quad (6.40)$$

$$HOCl \leftrightarrows ClO^- + H^+ \quad (6.41)$$

By using anodes with a high oxygen overvoltage, high current density, and low water temperature, ozone can be generated at the anode, and hydrogen peroxide may also be produced at the cathode:

$$3H_2O \rightarrow O_3 + 6e^- + 6H^+ \quad (6.42)$$

$$O_2 + 2H_2O + 2e^- \rightarrow H_2O_2 + 2OH^- \quad (6.43)$$

Some technologies have been developed specifically to avoid hydrogen production and focus on hydroxyl ions generation as described in (6.6):

$$O_2 + 2H_2O + 4e^- \rightarrow 4OH^- \quad (6.44)$$

Another important characteristic of electrochemical disinfection reactions is that the generation of disinfectant byproducts is less than in conventional chlorination [155].

Research performed at the University of New Orleans on synthetic ballast water electrodisinfection [116] demonstrated that killing of bacteria took place when using DC, even in the absence of chlorides.

Wastewater collected from the effluent channel of one of the secondary clarifiers at the Marrero Wastewater Treatment Plant, in Marrero, LA, was used as the experimental fluid [156]. Determination of *E. coli* in the samples was done according to EPA Method 1603.

Electrodisinfection experiments were carried out using both DC and AC. The most important difference between DC and AC arises from the disinfection efficiency provided by similar concentrations of free and total chlorine. Using DC, the coliform removal was close to 100% with free and total chlorine concentrations of 0.06 mg/L Cl_2 and 0.142 mg/L Cl_2, respectively.

Based on the results of the study, it is the opinion of the authors that DC electrodisinfection is not only more efficient than its AC counterpart, but also a valid alternative to conventional chlorination. Iridium oxide-coated titanium is a very stable electrode material that can be used for electrodisinfection of secondary wastewater effluents. In an electrodisinfection reactor equipped with this type of electrode, water disinfection can be achieved even in the presence of low chloride concentrations, as is the case in the present research (average Cl^- = 150 mg/L).

6.4.8 Disposal of 2,4-dichlorophenoxyacetic Acid (2,4-D) Using Electrosonication

Technique	Type of contaminant in wastewater	Main electrode	Type of electric power
Electrosonication	2,4-dichlorophenoxyacetic acid (2,4-D) and 2,4-dichlorophenol (2,4-DCP)	Nickel foil electrodes	DC

Two energies are useful when used simultaneously in order to destroy and remove hazardous materials in wastewater: electric potential and ultrasonic waves. One of these hazardous materials is 2,4-dichlorophenoxyacetic acid (2,4-D), which is the most widely used herbicide worldwide [157]. In spite of its mild toxicity (LD_{50} 370–666 mg/kg), it can be converted into highly toxic chloro-organic products [158]. 2,4-D is a poorly biodegradable pollutant, and a maximum concentration level of 70 ppb is recommended in drinking water by WHO [159].

Regarding the decomposition of 2,4-D and 2,4-dichlorophenol (2,4-DCP), recent reports indicate that the intermediates might be even more harmful than the parent compounds themselves.

Sonoelectrochemistry was also successful in wastewater detoxification [160]; however, as far as we know, the combination of ultrasonic waves, electrical field, and Fenton's process, has never been examined for water treatment. This approach seems promising, due to the power of combining its individual components, which may enhance the overall performance.

One study has evaluated the performance of the SEF process for decomposition of 2,4-D and 2,4-DCP in water for full degradation means (full removal) [161].

The faster degradation of 2,4-D and 2,4-DCP within SF and SEF schemes is believed to be due to the high efficiency for the production of $^{\bullet}OH$ radicals as well as to the ultrasonic cleaning of the electrode's active surfaces during these processes.

$$\text{Fenton Reaction:} \quad H_2O_2 + Fe^{2+} \rightarrow OH^- + HO^{\bullet} + Fe^{3+} \tag{6.45}$$

$$\text{Ultrasonic reaction:} \quad H_2O \rightarrow H + HO^{\bullet} \tag{6.46}$$

Among very reactive species suitable for SE oxidation treatments (such as $O_2^{\bullet}$, $OH^{\bullet}$, $HO_2^{\bullet}$, $ROO^{\bullet}$) the hydroxyl radical is certainly the most reactive. It is a very strong one-electron oxidizing agent, which seldom reacts as an electron transfer reagent; it is very reactive in hydrogen atom abstraction and in electrophilic addition processes [162].

The oxidative and reductive electrochemical processes and the reactions involving HO are significantly more effective under the SEF process.

The SEF scheme appears to be more efficient either in decomposing the parent compounds or their respective intermediates, and the time required for full degradation was considerably shorter for the SEF scheme.

6.5 Conclusion

- Electrochemical catalysis is considered one of the essential and successful processes used for wastewater treatment and providing clean water for drinking, agriculture, or industrial uses.
- Different techniques of electrochemical catalysis could be applied, such as advanced electrochemical oxidation, electrochemical reduction, electrocoagulation, electroflotation, electrodialysis, the electro-Fenton process, and electrodisinfection, according to the type, available possibilities and heaviest of contaminants.
- Electrochemical processes combined with other processes and energy sources, e.g. UV radiation (electrophotochemical) and ultrasonic waves (electrosonication) are effective methods for treatment of wastewater and disposal of refractory materials.
- Advanced electrochemical oxidation was successfully used for the removal of harmful organic materials from real cosmetics industry wastewater. High removal efficiency was achieved by using high O_2 overpotential IrO_2/Ti electrodes as anode. COD was measured and indicated the successful degradation of persistent compounds, where the refractory and organic pollutants were oxidized to less harmful, environmentally friendly materials. The applied technique is economically promising, since removal efficiency of 85% can be achieved by electric energy consumption less than 0.3 KWh/l.
- It has been demonstrated that a combined electrochemical oxidation process and UV radiation is a very effective and promising method for removal of heavy metals from groundwater e.g. Fe. Within a very short reaction time not exceeding 20 min, the removal of iron reached 99% with less consumption of current intensity (0.25A) without the use of external oxidizing agent. Both free hydroxyl radicals and hydrogen peroxide oxidizing agents, which are responsible for removing Fe from ground water, are generated through this process. All these characteristics make this approach an appropriate solution for removal of Fe and other heavy metals from groundwater.
- High concentration of ammonia in wastewater of petroleum stations was totally removed using electrochemical catalysis. Ammonia of onshore and offshore wastewater was ceased after electrochemical treatment using three types of electrodes Al, Fe, and Ti/IrO_2. The process succeeded at removal of ammonia and remediation of other pollutants in one step. Both ionized ammonia (NH_4^+) and un-ionized ammonia (NH_3) were not detected after treatment. TSS was also successfully removed by complete incineration of suspended solids by electric current. TDS, COD, and BOD were reduced, and high removal efficiency was achieved in the case of using Ti/IrO_2 electrodes.
- Electrochemical processes (EC) and the electro-Fenton method (EC/Fenton) were studied for the treatment of real textile industry wastewater. All the processes were able to decolorize textile wastewater up to over 85%. Although EC/Fenton generates sludge, it has the most efficiency in terms of COD and color removal. EC processes and EC/Fenton processes were very dependent on pH because both processes have a single mechanism. The coagulation mechanism in EC and EC/Fenton processes indicated more capability in organic matter removal from wastewater.
- A combined photoelectrochemical method (EC/UV) was used for oxidation of the soluble forms Fe^{2+} and Mn^{2+} to the insoluble forms Fe^{3+} and Mn^{4+}. The combined method

(CPE) revealed more efficiency than sole EC and UV methods. The presence of both dissolved iron and manganese has the advantage of less resistivity of the solution of wastewater. The effect of Fe^{2+} and Mn^{2+} concentrations revealed that the higher the concentration of dissolved iron and manganese ions, the higher the removal efficiency obtained.

References

1 UN WWAP. (2003). United Nations World Water Assessment Programme. The World Water Development Report 1: Water for People, Water for Life. Paris, France: UNESCO.

2 World Health Organization (WHO). *World Health Report: Reducing Risks, Promoting Healthy Life*. France. http://www.who.int/whr/2002/en/whr02_en.pdf. Retrieved 14 July 2009.

3 Mathur, N., Bhatnagar, P., and Bakre, P. (2006). Assessing mutagenicity of textile dyes from Pali(Rajasthan) using Ames bioassay. *Applied Ecology and Environmental Research* 4: 111–118.

4 Indu, M., Gupta, A., and Sahoo, C. (2014). Electrochemical oxidation of methylene blue using lead acid battery anode. *APCBEE Procedia* 9: 70–74.

5 Panizza, M., Barbucci, A., Ricotti, R., and Cerisola, G. (2007). Electrochemical degradation of methylene blue. *Separation and Purification Technology* 54: 382–387.

6 Kraft, A. (2008). Electrochemical water disinfection: a short review. *Platinum Metals Review* 52 (3): 177–185.

7 Spellman, F.R. (2001). *Handbook for Waterworks Operator Certification*, 2, 6–11, 81–83. Lancaster, USA: Technomic Publishing Company Inc.

8 Oram, B. (1995). *Iron and Manganese in Drinking Water*. Wilkes University.

9 Apfel, G. (1972). Pollution control and abatement in the cosmetic industry. *CTFA Cosmetic Journal* 4: 28–32.

10 Ritter, U. (1989). Environmental pollution aspects at plants for cosmetics production. *Wachse* 115: 383–386.

11 Reif, R., Suárez, S., Omil, F., and Lema, J.M. (2008). Fate of pharmaceuticals and cosmetic ingredients during the operation of a MBR treating sewage. *Desalination* 221 (1–3): 511–517.

12 Kim, B.R., Anderson, J.E., Mueller, S.A., Gaines, W.A., and Kendall, A.M. (2002). Literature review-efficacy of various disinfectants against Legionella in water systems. *Water Research* 36: 4433–4444.

13 Kool, J.L., Carpenter, J.C., and Fields, B.S. (1999). Effect of monochloramine disinfection of municipal drinking water on risk of nosocomial Legionnaires' disease. *The Lancet* 353 (23): 272–277.

14 Anglada, A., Urtiaga, A., and Ortiz, I. (2009). Contributions of electrochemical oxidation to waste-water treatment: Fundamentals and review of applications. *Journal of Chemical Technology & Biotechnology* 84: 1747–1755.

15 Ghasemian, S. and Omanovic, S. (2017). Fabrication and characterization of photoelectrochemically-active Sb-doped Snx-W(100-x)%-oxide anodes: toward the removal of organic pollutants from wastewater. *Applied Surface Science* 416: 318–328.

16 Zhao, W., Xing, J., Chen, D., Jin, D., and Shen, J. (2016). Electrochemical degradation of Musk ketone in aqueous solutions using a novel porous Ti/SnO2-Sb2O3/PbO2 electrodes. *Journal of Electroanalytical Chemistry* 775: 179–188.
17 Martinez-Huitle, C.A. and Ferro, S. (2006). Electrochemical oxidation of organic pollutants for the wastewater treatment: direct and indirect processes. *Chemical Society Reviews* 35: 1324.
18 Rodrigo, M.A., Michaud, P.A., Duo, I., Panizza, M., Cerisola, G., and Comninellis, C. (2001). Oxidation of 4-Chlorophenol at Boron-Doped Diamond Electrode for Wastewater Treatment. *Journal of the Electrochemical Society* 148: 60.
19 Chatzisymeon, E., Dimou, A., Mantzavinos, D., and Katsaounis, A. (2009). Electrochemical oxidation of model compounds and olive mill wastewater over DSA electrodes: 1. The case of Ti/IrO2 anode. *Journal of Hazardous Materials* 167: 268.
20 Gattrell, M. and Kirk, D.J. (1993). A Study of the Oxidation of Phenol at Platinum and Preoxidized Platinum Surfaces. *Journal of the Electrochemical Society* 140: 1534.
21 Rodrigo, M.A., Michaud, P.A., Duo, I., Panizza, M., Cerisola, G., and Comninellis, C. (2001). Oxidation of 4-chlorophenol at boron-doped diamond electrodes for wastewater treatment. *Journal of the Electrochemical Society* 148: D60–D64.
22 Feng, Y., Yang, L., Liu, J., and Logan, B.E. (2016). Electrochemical technologies for wastewater treatment and resource reclamation. *Environmental Science: Water Research & Technology* 2: 800–831.
23 Gattrell, M. and Kirk, D. (1993). A study of the oxidation of phenol at platinum and preoxidized platinum surfaces. *Journal of the Electrochemical Society* 140: 1534.
24 Chatzisymeon, E., Dimou, A., Mantzavinos, D., and Katsaounis, A. (2009). Electrochemical oxidation of model compounds and olive mill wastewater over DSA electrodes: 1. The case of Ti/IrO2 anode. *Journal of Hazardous Materials* 167: 268–274.
25 Chen, G. (2004). Electrochemical technologies in wastewater treatment. *Separation and Purification Technology* 38: 11–41.
26 Arapoglou, D., Vlyssides, A., Israilides, C., Zorpas, A., and Karlis, P. (2003). Detoxification of Methyl-Parathion Pesticide in Aqueous Solutions by Electrochemical Oxidation. *Journal of Hazardous Materials* 98: 191.
27 Harrington, T. and Pletcher, D. (1999). The Removal of Low Levels of Organics from Aqueous Solutions Using Fe(II) and Hydrogen Peroxide Formed In Situ at Gas Diffusion Electrodes. *Journal of the Electrochemical Society* 146: 2983.
28 Casado, J., Fornaguera, J., and Galan, M.I. (2005). Mineralization of Aromatics in Water by Sunlight-Assisted Electro-Fenton Technology in a Pilot Reactor. *Environmental Science & Technology* 39 (6): 1843–1847.
29 Panizza, M. and Cerisola, G. (2005). Application of diamond electrodes to electrochemical processes. *Electrochimica Acta* 51: 191–199.
30 Lacasa, E., Tsolaki, E., Sbokou, Z., Rodrigo, M.A., Mantzavinos, D., and Diamadopoulos, E. (2013). Electrochemical disinfection of simulated ballast water on conductive diamond electrodes. *Chemical Engineering Journal* 223: 516–523.
31 Katsoni, A., Mantzavinos, D., and Diamadopoulos, E. (2014). Sequential treatment of diluted olive pomace leachate by digestion in a pilot scale UASB reactor and BDD electrochemical oxidation. *Water Research* 57: 76–86.

32 Solano, A.M.S., de Araújo, C.K.C., de Melo, J.V., Peralta-Hernandez, J.M., da Silva, D.R., and Martínez-Huitle, C.A. (2013). Decontamination of real textile industrial effluent by strong oxidant species electrogenerated on diamond electrode: viability and disadvantages of this electrochemical technology. *Applied Catalysis B: Environmental* 130: 112–120.

33 Zhou, M., Dai, Q., Lei, L., Ma, C.A., and Wang, D. (2005). Long life modified lead dioxide anode for organic wastewater treatment: electrochemical characteristics and degradation mechanism. *Environmental Science & Technology* 39: 363–370.

34 Liu, Y., Sun, T., Su, Q., Tang, Y., Xu, X., Akram, M., and Jiang, B. (2020). Highly efficient and mild electrochemical degradation of bentazon by nano-diamond doped PbO_2 anode with reduced Ti nanotube as the interlayer. *Journal of Colloid and Interface Science* 575: 254–264.

35 Polcaro, A., Palmas, S., Renoldi, F., and Mascia, M. (1999). On the performance of Ti/SnO2 and Ti/PbO2 anodesin electrochemical degradation of 2-chlorophenolfor wastewater treatment. *Journal of Applied Electrochemistry* 29: 147–151.

36 Wu, T., Zhao, G., Lei, Y., and Li, P. (2011). Distinctive tin dioxide anode fabricated by pulse electrodeposition: high oxygen evolution potential and efficient electrochemical degradation of fluorobenzene. *The Journal of Physical Chemistry C* 115: 3888–3898.

37 Zhang, J., Wei, X., Miao, J., Zhang, R., Zhang, J., Zhou, M., and Lu, W. (2020). Enhanced performance of an Al-doped SnO_2 anode for the electrocatalytic oxidation of organic pollutants in water. *Materials Today Communications* 24: 101164.

38 Mireia, S. and Gutiérrez-Bouzán Carmen, M. (2012). Electrochemical techniques in textile processes and wastewater treatment. *International Journal of Photoenergy* 629103: 12 p. doi:10.1155/2012/629103.

39 Schrott, W. (2004). Electrochemical dyeing. *Textile Asia* 35 (2): 45–47.

40 Bechtold, T., Burtscher, E., and Turcanu, A. (1998). Direct cathodic reduction of Leuco Sulfur Black 1 and Sulfur Black 1. *Journal of Applied Electrochemistry* 28 (11): 1243–1250.

41 Roessler, A., Crettenand, D., Dossenbach, O., Marte, W., and Rys, P. (2002). Direct electrochemical reduction of indigo. *Electrochimica Acta* 47 (12): 1989–1995.

42 Roessler, A. and Jin, X. (2003). State of the art technologies and new electrochemical methods for the reduction of vat dyes. *Dyes and Pigments* 59 (3): 223–235.

43 Roessler, A., Crettenand, D., Dossenbach, O., and Rys, P. (2003). Electrochemical reduction of indigo in fixed and fluidized beds of graphite granules. *Journal of Applied Electrochemistry* 33 (10): 901–908.

44 Roessler, A. and Crettenand, D. (2004). Direct electrochemical reduction of vat dyes in a fixed bed of graphite granules. *Dyes and Pigments* 63 (1): 29–37.

45 Thomas, B. and Aurora, T. (2006). Iron-complexes of bis(2-hydroxyethyl)-amino-compounds as mediators for the indirect reduction of dispersed vat dyes—cyclic voltammetry and spectroelectrochemical experiments. *Journal of Electroanalytical Chemistry* 591 (1): 118–126.

46 Kulandainathan, M.A., Muthukumaran, A., Patil, K., and Chavan, R.B. (2007). Potentiostatic studies on indirect electrochemical reduction of vat dyes. *Dyes and Pigments* 73 (1): 47–54.

47 Roessler, A., Dossenbach, O., Marte, W., and Rys, P. (2002). Electrocatalytic hydrogenation of vat dyes. *Dyes and Pigments* 54 (2): 141–146.

48 McCreery, R.L. (2008). Advanced carbon electrode materials for molecular electrochemistry. *Chemical Reviews* 108 (7): 2646–2687.

49 Engstrom, R.C. (1982). Electrochemical pretreatment of glassy carbon electrodes. *Analytical Chemistry* 54 (13): 2310–2314.

50 Kulandainathan, M.A., Kiruthika, K., Christopher, G., Babu, K.F., Muthukumaran, A., and Noel, M. (2008). Preparation of iron-deposited graphite surface for application as cathode material during electrochemical vat-dyeing process. *Materials Chemistry and Physics* 112 (2): 478–484.

51 Thakur, C., Srivastava, V.C., and Mall, I.D. (2009). Electrochemical treatment of a distillery wastewater: Parametric and residue disposal study. *Chemical Engineering Journal* 148 (2–3): 496–505.

52 Modirshahla, N., Behnajady, M.A., and Mohammadi-Aghdam, S. (2008). Investigation of the effect of different electrodes and their connections on the removal efficiency of 4-nitrophenol from aqueous solution by electrocoagulation. *Journal of Hazardous Materials* 154 (1–3): 778–786.

53 Rajeshwar, K., Ibanez, J.G., and Swain, G.M. (1994). Electrochemistry and the environment. *Journal of Applied Electrochemistry* 24 (11): 1077–1091.

54 Bayramoglu, M., Kobya, M., Can, O.T., and Sozbir, M. (2004). Operating cost analysis of electrocoagulation of textile dye wastewater. *Separation and Purification Technology* 37 (2): 117–125.

55 Emamjomeh, M.M. and Sivakumar, M. (2009). Review of pollutants removed by electrocoagulation and electrocoagulation/flotation processes. *Journal of Environmental Management* 90 (5): 1663–1679.

56 Mollah, M.Y.A., Morkovsky, P., Gomes, J.A.G., Kesmez, M., Parga, J., and Cocke, D.L. (2004). Fundamentals, present and future perspectives of electrocoagulation. *Journal of Hazardous Materials* 114 (1–3): 199–210.

57 Kobya, M., Demirbas, E., Dedeli, A., and Sensoy, M.T. (2010). Treatment of rinse water from zinc phosphate coating by batch and continuous electrocoagulation processes. *Journal of Hazardous Materials* 173 (1–3): 326–334.

58 Holt, P.K., Barton, G.W., and Mitchell, C.A. (2005). The future for electrocoagulation as a localised water treatment technology. *Chemosphere* 59 (3): 355–367.

59 Ghanbari, F., Moradi, M., Mohseni-Bandpei, A., Gohari, F., Mirtaleb Abkenar, T., and Aghayani, E. (2014). Simultaneous application of iron and aluminum anodes for nitrate removal: a comprehensive parametric study. *International Journal of Environmental Science and Technology* 11: 1653–1660. 10.1007/s13762-014-0587-y

60 Mouedhen, G., Feki, M., Wery, M.D.P., and Ayedi, H.F. (2008). Behavior of aluminum electrodes in electrocoagulation process. *Journal of Hazardous Materials* 150 (1): 124–135.

61 Merzouk, B., Yakoubi, M., Zongo, I., Leclerc, J.P., Paternotte, G., Pontvianne, S., and Lapicque, F. (2011). Effect of modification of textile wastewater composition on electrocoagulation efficiency. *Desalination* 275 (1–3): 181–186.

62 Phalakornkule, C., Polgumhang, S., Tongdaung, W., Karakat, B., and Nuyut, T. (2010). Electrocoagulation of blue reactive, red disperse and mixed dyes, and application in treating textile effluent. *Journal of Environmental Management* 91 (4): 918–926.

63 Ilhan, F., Kurt, U., Apaydin, O., and Gonullu, M.T. (2008). Treatment of leachate by electrocoagulation using aluminum and iron electrodes. *Journal of Hazardous Materials* 154 (1–3): 381–389.

64 Linares-Hernández, I., Barrera-Díaz, C., Roa-Morales, G., Bilyeu, B., and Ureña-Núñez, F. (2009). Influence of the anodic material on electrocoagulation performance. *Chemical Engineering Journal* 148 (1): 97–105.

65 Zaw, M. and Chiswell, B. (1999). Iron and manganese dynamics in lake water. *Water Research* 33: 1900.

66 Brillas, E., Sauleda, R., and Casado, J. (1998). Degradation of 4-chlorophenol by anodic oxidation, electro-Fenton, photoelectro-Fenton, and peroxi-coagulation processes. *Journal of the Electrochemical Society* 145: 759.

67 Akhter, M., Habib, G., and Qamar, S.U. (2018). Application of electrodialysis in waste water treatment and impact of fouling on process performance. *Journal of Membrane Science & Technology* 8: 2. 10.4172/2155-9589.1000182

68 Ashrafi, O., Yerushalmi, L., and Haghighat, F. (2015). Wastewater treatment in the pulp-and-paper industry: A review of treatment processes and the associated greenhouse gas emission. *Journal of Environmental Management* 158: 146–157.

69 American Water Works Association (Ed). (1995). Electrodialysis and electrodialysis reversal: M38. *Journal of the American Water Works Association* 38.

70 Caprarescu, S., Purcar, V., and Vaireanu, D.I. (2012). Separation of copper ions from synthetically prepared electroplating wastewater at different operating conditions using electrodialysis. *Separation Science and Technology* 47 (16): 2273–2280.

71 Khan, M.I., Luque, R., Akhtar, S., Shaheen, A., Mehmood, A. et al. (2016). Design of anion exchange membranes and electrodialysis studies for water desalination. *Materials* 9: 365.

72 Strathmann, H. (2010). Electrodialysis, a mature technology with a multitude of new applications. *Desalination* 264: 268–288.

73 Mohammadi, T., Razmi, A., and Sadrzadeh, M. (2004). Effect of operating parameters on Pb2+ separation from wastewater using electrodialysis. *Desalination* 167: 379–385.

74 Chang, D.I., Choo, K.H., Jung, J.H., Jiang, L., Ahn, J.H., Nam, M.Y., Kim, E.S., and Jeong, S.H. (2009). Foulant identification and fouling control with iron oxide adsorption in electrodialysis for the desalination of secondary effluent. *Desalination* 236 (1–3): 152–159.

75 Ruiz, B., Sistat, P., Huguet, P., Pourcelly, G., Araya-Farias, M., and Bazinet, L. (2007). Application of relaxation periods during electrodialysis of a casein solution: impact on anion-exchange membrane fouling. *Journal of Membrane Science* 287 (1): 41–50.

76 Neyens, E. and Baeyens, J. (2003). A review of classic Fenton's peroxidation as an advanced oxidation technique. *Journal of Hazardous Materials* B 98: 33–50.

77 Casado, J. (2019). Toward industrial implementation of Electro-Fenton and derived technologies for wastewater treatment: a review. *Journal of Environmental Chemical Engineering* 7. 10.1016/j.jece.2018.102823

78 Babuponnusami, A. and Muthukumar, K. (2014). A review on Fenton and improvements to the Fenton process for wastewater treatment. *Journal of Environmental Chemical Engineering* 2: 557–572. 10.1016/j.jece.2013.10.011

79 Fenton, H.J.H. (1894). Oxidation of tartatic acid in presence of Iron. *Journal of the Chemical Society, Perkin Transactions I* 65: 899–910.

80 Brillas, E., Sir′es, I., and Oturan, M.A. (2009). Electro-fenton process and related electrochemical technologies based on fenton's reaction chemistry. *Chemical Reviews* 109 (12): 6570–6631.

81 Dirany, A., EfremovaAaron, S., Oturan, N., Sir′es, I., Oturan, M.A., and Aaron, J.J. (2011). Study of the toxicity of sulfamethoxazole and its degradation products in water by a bioluminescence method during application of the electro-Fenton treatment. *Analytical and Bioanalytical Chemistry* 400 (2): 353–360.

82 Garcia-Segura, S., Centellas, F., Arias, C. et al. (2011). Comparative decolorization of monoazo, diazo and triazo dyes by electro-Fenton process. *Electrochimica Acta* 58 (1): 303–311.

83 Isarain-ch′avez, E., Garrido, J.A., Rodr′ıguez, R.M. et al. (2011). Mineralization of metoprolol by electro-fenton and photoelectrofenton processes. *Journal of Physical Chemistry A* 115 (7): 1234–1242.

84 M′endez-mart′ınez, A.J., D′avila-jim′enez, M.M., Ornelas-d′avila, O. et al. (2012). Electrochemical reduction and oxidation pathways for Reactive Black 5 dye using nickel electrodes in divided and undivided cells. *Electrochimica Acta* 59 (138): 140–149.

85 Panizza, M. and Oturan, M.A. (2011). Degradation of Alizarin Red by electro-Fenton process using a graphite-felt cathode. *Electrochimica Acta* 56 (20): 7084–7087.

86 Ruiz, E.J., Arias, C., Brillas, E., Hern′andez-ram′ırez, A., and Peralta-hern′andez, J.M. (2011). Mineralization of Acid Yellow 36azo dye by electro-Fenton and solar photoelectro-Fenton processeswith a boron-doped diamond anode. *Chemosphere* 82 (4): 495–501.

87 Ozcan, A.S., Erdem, B., and Ozcan, A. (2005). Adsorption of Acid Blue 193 from aqueous solutions onto BTMA-bentonite. *Colloids and Surfaces A* 266 (1–3): 73–81.

88 Atchley, A. and Crum, L. (1988). Acoustic cavitation and bubble dynamics. In: *Ultrasound: Its Chemical, Physical and Biological Effects* (ed. K. Suslick), 1–64. New York, NY: VCH Publishers.

89 Suri, R. and Kamrajapuram, A. (2003). Heterogeneous ultrasonic destruction of aqueous organic contaminants. *Water Science and Tecnology* 47 (9): 137–142.

90 Petrier, C. and Casadonte, D. (2001). The sonochemical degradation of aromatic and chloroaromatic contaminants. In: *Advances in Sonochemistry* (ed. T.J. Mason), 91–109. Stamford, CT: JAI Press Inc.

91 Mohammad, J. and Peters, R. (2003). Combined sonication + vapor stripping for treatment of petroleum hydrocarbon-contaminated groundwater In: *Annual American Institute of Chemical Engineers Meeting*. San Francisco, CA.

92 Del Campo, F.J., Neudeck, A., Compton, R.G., and Marken, F. (1999). Low-temperature sonoelectrochemical processes: Part 1. Mass transport and cavitation effects of 20 kHz ultrasound in liquid ammonia. *Journal of Electroanalytical Chemistry* 477: 77–78.

93 Del Campo, F.J., Neudeck, A., Compton, R.G., Marken, F., Bull, S.D., and Davis, S.G. (2001). Low-temperature sonoelectrochemical processes Part 3.Electrodimerisation of 2-nitrobenzyl chloride in liquid ammonia. *Journal of Electroanalytical Chemistry* 507: 144–151.

94 Del Campo, F.J., Neudeck, A., Compton, R.G., Marken, F., and Aldaz, A. (2001). Low-temperature sonoelectrochemical processes: Part 3. Electrodimerisation of 2-nitrobenzylchloride in liquid ammonia. *Journal of Electroanalytical Chemistry* 506: 170–177.

95 Banks, C.E. and Compton, R.G. (2003). Ultrasonically enhanced voltammetric analysis and applications: an overview. *Electroanalysis* 15: 329.

96 Banks, C.E. and Compton, R.G. (2003). Voltammetric exploration and applications of ultrasonic cavitation. *ChemPhysChem* 4 (2): 169–178. 10.1002/cphc.200390027

97 Zhang, H.H. and Coury, L.A. (1993). Effects of high-intensity ultrasound on glassy-carbon electrodes. *Analytical Chemistry* 65: 1552–1558. 10.1021/ac00059a012

98 Gonzalez-Garcia, J., Esclapez, M.D., Bonete, P., Hernandez, Y.V., Garreton, L.G., and Saez, V. (2010). Current topics on sonoelectrochemistry. *Ultrasonics* 50: 318–322.

99 Pollet, B.G., Lorimer, J.P., Hihn, J.Y., Phull, S.S., Mason, T.J., and Walton, D.J. (2002). The effect of ultrasound upon the oxidation of thiosulphate on stainless steel and platinum electrodes. *Ultrasonics Sonochemistry* 9: 267–274.

100 Mason, T.J., Lorimer, J.P., and Walton, D.J. (1990). Sonoelectrochemistry. *Ultrasonics* 28: 333–337.

101 Kraft, A. (2008). Electrochemical water disinfection: a short review electrodes using platinum group metal oxides. *Platinum Metals Review* 52 (3).

102 Hayfield, P.C.S. (1998). Development of the Noble Metal/Oxide Coated Titanium Electrode Part I: The Beginning of the Story. *Platinum Metals Review* 42 (1): 27.

103 Hayfield, P.C.S. (1998). Development of the Noble Metal/Oxide Coated Titanium Electrode Part II: The Move from Platinum/Iridium to Ruthemum Oxide Electrocatalysts. *Platinum Metals Review* 42 (2): 46.

104 Hayfield, P.C.S. (1998). Development of the Noble Metal/Oxide Coated Titanium Electrode Part III: Coated Titanium Anodes in Widely Ranging Oxygen Evolving Situations. *Platinum Metals Review* 42 (3): 116.

105 Kraft, A. (2007). Doped Diamond: A Compact Review on a New, Versatile Electrode Material. *International Journal of Electrochemical Science* 2 (5): 355.

106 Kraft, A., Stadelmann, M., Blaschke, M., Kreysig, D., Sandt, B., Schröder, F., and Rennau, J. (1999). Electrochemical water disinfection Part I: Hypochlorite production from very dilute chloride solutions. *Journal of Applied Electrochemistry* 29 (7): 859.

107 Kraft, A., Blaschke, M., Kreysig, D., Sandt, B., Schröder, F., and Rennau, J. (1999). Electrochemical water disinfection. Part II: Hypochlorite production from potable water, chlorine consumption and the problem of calcareous deposits. *Journal of Applied Electrochemistry* 29 (8): 895.

108 Kraft, A., Wünsche, M., Stadelmann, M., and Blaschke, M. (2003). Electrochemical water disinfection, *Recent Research & Development Electrochemical* 6: 27.

109 Nakajima, N., Nakano, T., Harada, F., Taniguchi, H., Yokoyama, I., Hirose, J., Daikoku, E., and Sano, K. (2004), Evaluation of disinfective potential of reactive free chlorine in pool tap water by electrolysis. *Journal of Microbiological Methods* 57 (2): 163.

110 Bergmann, M.E.H. and Koparal, A.S. (2005), Studies on Electrochemical Disinfectant Production Using Anodes Containing RuO_2. *Journal of Applied Electrochemistry* 35 (12): 1321.

111 Rajeshwar, K. and Ibáñez, J. (1997). *Environmental Electrochemistry: Fundamentals and Applications in Pollution Abatement*. San Diego, CA: Academic Press.

112 Ibáñez, J. (2004). Electrochemistry encyclopedia. Department of Chemistry and Chemical Engineering, Mexican Microscale Chemistry Center, Iberoamericana University, Mexico D.F., Mexico. http://electrochem.cwru.edu/ed/encycl

113 Vik, E.A., Carlson, D.A., Eikum, A.S., and Gjessing, E.T. (1984). Electrocoagulation of potable water. *Water Research.* 18 (11): 1355–1360.

114 Huang, X., Qu, Y., Cid, C.A., Finke, C., Hoffmann, M.R., Lim, K., and Jiang, S.C. (2016). Electrochemical disinfection of toilet wastewater using wastewater electrolysis cell. *Water Research* 92: 164–172.

115 Snoeyink, V.L. and Jenkins, D. (1980). *Aquatic Chemistry*. New York, NY: Wiley.

116 McCraven, E.K. (2009). Electro-disinfection of ballast water. MS Thesis. New Orleans, LA, USA: University of New Orleans

117 Johnstone, P.T. and Bodger, P.S. (2000). Disinfection of deionised water using AC high voltage. *IEE Proceedings-Science, Measurement and Technology* 147 (3): 141–144.

118 Barashkov, N.N., Eisenberg, D., Eisenberg, S., Shegebaeva, G.S., Irgibaeva, I.S., and Barashkova, I.I. (2010). Electrochemical chlorine-free AC disinfection of water contaminated with salmonella typhimurium bacteria. *Russian Journal of Electrochemistry* 46 (3): 320–325.

119 Reimanis, M., Mezule, L., Ozolins, J., Malers, J., and Juhna, T. (2012). Drinking water disinfection with electrolysis. *Latvian Journal of Chemistry* 4: 296–304. 10.2478/v10161-012-0016-9

120 Pulido, M. (2005). Evaluation of an electro-disinfection technology as an alternative to chlorination of municipal wastewater effluents. MS Thesis. New Orleans, LA, USA: University of New Orleans.

121 Nwanebu, E.O., Liu, X., Pajootan, E., Yargeau, V., and Omanovic, S. (2021). Electrochemical degradation of methylene blue using a Ni-Co oxide anode. *Catalysts* 11: 793. 10.3390/catal11070793

122 Ghasemian, S., Asadishad, B., Omanovic, S., and Tufenkji, N. (2017). Electrochemical disinfection of bacteria-laden water using antimonydoped tin-tungsten-oxide electrodes. *Water Research* 126: 299–307.

123 Chung, S.-K. and Toshihiko, O. (1998). Hydroxyl radical scavengers from white mustard (Sinapis alba). *Food Sci. Biotechnol.* 7: 209–213.

124 Awad Abouelata, A.M. and Abdel Ghany, N.A. (2015). Electrochemical advanced oxidation of cosmetics waste water using IrO2/Ti-modified electrode. *Desalination and Water Treatment* 53 (3): 681–688. 10.1080/19443994.2013.848671

125 Wang, R., Moody, R.P., Koniecki, D., and Zhu, J. (2009). Low molecular weight cyclic volatile methylsiloxanes in cosmetic products sold in Canada: Implication for dermal exposure. *Environment International* 35 (6): 900–904.

126 Barbusinski, K. (2005). The modified Fenton process for decolorization of dye wastewater. *Polish Journal of Environmental Studies* 14 (3): 281–285.

127 Comninellis, C. (1994). Electrocatalysis in the electrochemical conversion/combustion of organic pollutants for waste water treatment. *Electrochimica Acta* 39: 1857–1862.

128 Belhadj Tahar, N. and Savall, A. (1998). Mechanistic aspects of phenol electrochemical degradation by oxidation on a Ta/PbO2 anode. *Journal of the Electrochemical Society* 145 (10): 3427–3434.

129 Brillas, E., Sire´s, I., Arias, C., Cabot, P.L., Centellas, F., Rodrı'guez, R.M., and Garrido, J.A. (2005). Mineralization of paracetamol in aqueous medium by anodic oxidation with a boron-doped diamond electrode. *Chemosphere* 58: 399–406.

130 Busca, G., Berardinelli, S., Resini, C., and Arrighi, L. (2008). Technologies for the removal of phenol from fluid streams: A short review of recent developments. *Journal of Hazardous Materials* 160: 265–288.

131 Oturan, M.A. and Brillas, E. (2007). Electrochemical advanced oxidation processes (EAOPs) for environmental applications. *Portugaliae Electrochimica Acta* 25: 1–18.

132 Errami, M., El Mouden, O.I.D., Salghi, R., Zougagh, M., Zarrouk, A., Hammouti, B., Chakir, A., Al-Deyab, S.S., and Bouri, M. (2012). Detoxification of bupirimate pesticide in aqueous solutions by electrochemical oxidation. *Der Pharma Chemica* 4 (1): 297–310.

133 Talaat, H.A., Ghaly, M.Y., Kamel, E.M., Awad Abouelata, A.M., and Ahmed, E.M. (2011). Combined electro-photochemical oxidation for iron removal from ground water. *Desalination and Water Treatment* 28: 265–269.

134 Martínez-Huitle, C.A. and Brillas, E. (2009). Decontamination of wastewaters containing synthetic organic dyes by electrochemical methods: a general review. *Applied Catalysis B* 87 (3–4): 105–145.

135 Bechtold, T., Mader, C., and Mader, J. (2002). Cathodic decolourization of textile dyebaths: tests with full scale plant. *Journal of Applied Electrochemistry* 32 (9): 943–950.

136 Vaněrková, D., Sakalis, A., Holčapek, M., Jandera, P., and Voulgaropoulos, A. (2006). Analysis of electrochemical degradation products of sulphonated azo dyes using high-performance liquid chromatography/tandem mass spectrometry. *Rapid Communications in Mass Spectrometry* 20 (19): 2807–2815.

137 Zanoni, M.V.B., Fogg, A.G., Barek, J., and Zima, J. (1997). Electrochemical investigations of reactive dyes; cathodic stripping voltammetric determination of anthraquinone-based chlorotriazine dyes at a hanging mercury drop electrode. *Analytica Chimica Acta* 349 (1–3): 101–109.

138 Admiraal, W. (1977). Tolerance of estuarine benthic diatoms to high concentrations of ammonia, nitrite ion, nitrate ion and orthophosphate. *Marine Biology* 43: 307–315.

139 Helmy, Q. and Kardena, E. (2015). Petroleum oil and gas industry waste treatment; common practice in Indonesia. *Journal of Petroleum & Environmental Biotechnology* 6: 241. 10.4172/2157-7463.1000241

140 Pearce, K. and Whyte, D. (2005). Environmentek/csir,"WATER and wastewater management in the oil refining and RE-refining industry". Final Report to the Water Research Commission, WRC Report TT 180/05.

141 Durborow Robert, M., Crosby David, M., and Brunson Martin, W. (1992). Ammonia in fish ponds. Southern Regional Aquaculture Center. SRAC Publication No. 463.

142 Poxton, M.G. and Allouse, S.B. (1982). Water quality criteria for marine fisheries. *Aquacultural Engineering* 1 (3): 153–191.

143 Jensen, F.B. (1995). Uptake and effects of nitrite and nitrate in animals. In: *Nitrogen Metabolism and Excretion* (ed. P.J. Walsh and P. Wright), 289–303. Boca Raton: CRC Press.

144 Jiang, X., Cheng, Z., Gao, Z., Ma, W., Ma, X., and Wang, R. (2015). Removal of ammonia from wastewater by natural freezing method In: *International Conference on Chemical, Material and Food Engineering (CMFE).*

145 Saracco, G. (2002). Indo-Italian workshop on emerging technologies for industrial wastewater and environment (September). Jüttner K, Galla U, Schmieder H. (2000). Electrochemical approaches to environmental problems in the process industry. *Electrochimica Acta* 45: 2575–2594.

146 Wang, Y., Xu, G., Jinglu, L., Yang, Y., Lei, Z., and Zhang, Z. (2012). Efficient electrochemical removal of ammonia with various cathodes and Ti/RuO2-Pt anode. *Journal of Applied Sciences* 2: 241–247.

147 Awad Abouelata Ahmed, M., Elhadad Adel, M.A., and Samir, H. (2018). In situ, one step removal of ammonia from onshore and offshore formation water of petroleum production fields. *Chemosphere* 205: 203–208.

148 World Health Organization. (1993). Guidelines for drinking-water quality: volume 1: recommendations, 2nd ed. ISBN 9241544600, https://apps.who.int/iris/handle/10665/259956

149 Xuejun, C., Zhemin, S., Xiaolong, Z., Yaobo, F., and Wenhua, W. (2005). Advanced treatment of textile wastewater for reuse using electrochemical oxidation and membrane fi ltration. *Water SA* 31: 127–132.

150 Talaat Hala, A., Ghaly Montaser, Y., Kamel Eman, M., Ahmed Enas, M., and Awad Abouelata Ahmed, M. (2010). Simultaneous removal of iron and manganese from ground water by combined photo-electrochemical method. *Journal of American Science* 6 (12).

151 Akyol, A., Can, O.T., Demirbas, E., and Kobya, M. (2013). A comparative study of electrocoagulation and electro-Fenton for treatment of wastewater from liquid organic fertilizer plant, *Separation and Purification Technology* 112: 11.

152 Nidheesh, P.V. and Gandhimathi, R. (2012). Trends in electro-Fenton process for water and wastewater treatment: An overview, *Desalination* 299: 1.

153 Khatri, I., Singh, S., and Garg, A. (2018). Performance of electro-Fenton process for phenol removal using Iron electrodes and activated carbon, *Journal of Environmental Chemical Engineering* 6: 7368.

154 Radwan, M., Gar Alalm, M., and Eletriby, H. (2018). Optimization and modeling of electro-Fenton process for treatment of phenolic wastewater using nickel and sacrificial stainless steel anodes, *Journal of Water Process Engineering* 22: 155.

155 Saha, J. and Gupta, S.K. (2015). Electrochlorinator - An advanced disinfection system for drinking water supply. *Discovery* 41 (188): 86–92.

156 La Motta, E.J., Rincón, G.J., Acosta, J., and Chávez, X. (2017). Electro-disinfection of municipal wastewater: Laboratory scale comparison between direct current and alternating current. *International Journal of Engineering Research and Application* 7 (1, Part-5) 6–12, ISSN: 2248-9622.

157 Industry Task Force II on 2,4-D research Data Web Site (1999). http://www4.coastalent.com/24d.

158 Kaiumova, D., Kaiumov, F., Opelz, G., and Susal, C. (2001). Toxic effects of the herbicide 2,4-dichlorophenoxyacetic acid on lymphoid organs of the rat, *Chemosphere* 43: 801.

159 Sun, Y. and Pignatello, J.J. (1993). Organic intermediates in the degradation of 2,4-dichlorophenoxyacetic acid by Fe^{3+}/H_2O_2 and Fe^{3+}/H_2O_2/UV, *Journal of Agricultural and Food Chemistry* 41: 308.

160 Trabelsi, F., A€ıt-lyazidi, H., Ratsimba, B., Wilhelm, A., Delmas, H., Fabre, P.L., and Berlan, J. (1996). Oxidation of phenol in wastewater by sonoelectrochemistry, *Chemical Engineering Science* 51: 1857.

161 Yasman, Y., Bulatov, V., Gridin, V.V., Agur, S., Galil, N., Armon, R., and Schechter, I. (2004). A new sono-electrochemical method for enhanced detoxification of hydrophilic chloroorganic pollutants in water, *Ultrasonics Sonochemistry* 11: 365–372.

162 Oturan, M. and Pinson, J.J. (1995). Hydroxylation by Electrochemically Generated OH.bul. Radicals. Mono- and Polyhydroxylation of Benzoic Acid: Products and Isomer Distribution, *The Journal of Physical Chemistry A* 99: 13948, and references cited there.

7

Different Synthetic Routes of Electrocatalysts and Fabrication of Electrodes

Sethumathavan Vadivel[1], Harshvardhan Mohan[2], Saravana Kumar Rajendran[3], and Pavithra Muthu Kumar Sathya[4]

[1] *Department of Chemistry, Saveetha School of Engineering, Saveetha Institute of Medical and Technical Sciences, Chennai, India*
[2] *Department of Chemistry, Research Institute of Physics and Chemistry, Jeonbuk National University, Jeonju, Republic of Korea*
[3] *Departamento de Ingeniería Mecánica, Facultad de Ingeniería, Universidad de Tarapacá, Avda. General Velásquez, Arica, Chile*
[4] *Department of Microbiology, PSG College of Arts & Science, Tamilnadu, India*

7.1 Introduction

Renewable sources of energy have been gaining increased attention in recent times owing to the increased depletion of fossil fuels. The discontinuity in supply of renewable sources remains a threat, however, demanding the need for introduction of novel techniques. Fuel cells have been gaining increased attention in this regard for their heightened energy exchange efficiency (40 to 60%) and the potential toward direct conversion of chemical energy to electrical energy [1–5]. Adding to this is the fact that water happens to be the only by-product released from fuel cells, making them more environmentally friendly. The reactions in fuel cells involve methanol oxidation [6, 7], ethanol oxidation [8, 9], formic acid oxidation [10, 11], and oxygen reduction reaction (ORR) [12, 13]. Of all the reactions, cathodic reactions occupy the maximum kinetics loss and incur more cost due to the complication in mechanisms involved owing to the deployment of multielectrodes. Platinum-based electrodes find extensive application in commercial fuel cells owing to higher ORR reaction kinetics and increased catalytic efficiency; however, scarce prevalence, decreased durability, and increased cost act as a pitfall [14] for large scale application.

The aforesaid problems shall be overcome by following these ways:

i) Nanosizing the Pt electrocatalysts. This shall not only reduce the cost by 50% but also positively impact the catalytic activity by exposing more active sites for the catalytic reaction.
ii) Forming core shell structures by alloying Pt with low-cost metals like Fe, Co, and Ni. This in addition to cost cutting would favor the catalytic reaction by regulating the electronic structures between components.
iii) Deploying transition metal oxides [15], sulphides [16], hydroxides [17], and nitrides [18] toward the synthesis of nonprecious metal catalysts that shall replace Pt-based catalysts.
iv) Creating avenues for the deployment of metal-free catalysts.

Photocatalysts and Electrocatalysts in Water Remediation: From Fundamentals to Full Scale Applications, First Edition. Edited by Prasenjit Bhunia, Kingshuk Dutta, and S. Vadivel.

The best of all the aforesaid techniques happens to be the generation of metal-free catalysts [19]. Several researchers have emphasized the synthesis of metal-free catalysts like heteroatom doped carbon-based nanomaterials [20], especially over the span of the last ten years. Web of Science data with respect to similar studies indicates the attractiveness that the field of study has necessitated in the recent past. Unlike precious metal-based catalysts, carbon generally demonstrates inertness toward ORR. The introduction of heteroatoms onto the carbon skeleton has been found to favor the creation of active reactive sites through perturbation of the spin density of carbon atoms [21]. Alongside these advantages and benefits are present several areas demanding attention as well:

i) Methods facilitating efficient and controlled doping of heteroatoms onto carbon skeletons.
ii) Inadequacy of ways and means to precisely define the configuration of the doped structure [22].
iii) Addressing the variations in ORR active sites with respect to the different configurations of doped carbon materials.
iv) Lack of information on ORR mechanisms for doped carbons [23].

In general, fuel cells involve the conversion of chemical energy into electricity deploying hydrogen-rich fuels, without eliminating any greenhouse gas. Close to 30% of costs incurred toward fuel cells goes for acquisition of Pt, which happens to be the key component of electrocatalysts [1]. Substituting for Pt shall be of great help in reducing the sum of money involved. The following are identified as the key features to be present in electrocatalysts [2]:

i) Increased stability in the assigned pH range
ii) Greater photon and electronic conductivity
iii) Higher resistance against corrosion
iv) Heightened specific surface area
v) Appropriate porosity in the way of minimizing water flooding in electrodes.

As of now Pt/C catalysts have been identified to best serve all the aforesaid features. A major setback to this heterostructure happens to be the proneness to electrocorrosion. This shall be attributed to the insufficient fuel gas purging resulting in the ageing of both Pt electroparticles and the carbon support [24]. This shall end up either in Pt dissolution or support carbon oxidation leading to Ostwald ripening and Pt agglomeration [25]. Adding to this is the sensitivity toward CO poisoning the heterostructure possesses [26], necessitating the need for an alternate material with greater stability and heightened tolerance to CO.

7.2 Fundamental Principles of Alkaline Water Oxidation

Two half-reactions constitute conventional alkaline water electrolysis:

i) Hydrogen evolution reaction (HER)
ii) Oxygen evolution reaction (OER).

Similar to ORR, OER has been identified as a sluggish process.

7.3 Electrochemical Evaluating Parameters of Electrocatalysts for OER Performance

A three-electrode cell with galvanostatic control, rotating disc electrode (RDE), support stainless steel (SS), and substrates like nickel foam (NF) is deployed for the purpose of electrochemical OER measurements in the laboratory scale [27]. A catalytic ink, a mixture of catalyst powder, binder, and solvent, is coated onto the glassy carbon on the RDE surface. The set-up involves a counter electrode (carbon rod), reference electrodes (Ag/AgCl), saturated calomel electrode (Hg/HgO), and reversible hydrogen electrode (RHE) [78–81]. Electrocatalysts are coated on a conductive substrate such as NF, SS [28], carbon cloth (CC)[86], fluorine-doped tin oxide (FTO) [29], or carbon fibre paper (CFP) [88], which serves as a working electrode in OER reactions. Studies have revealed that conductive substrates have better catalytic performance than RDEs owing to their large surface area, excellent electrical conductivity, and good corrosion resistance at high pH values [30]. The deployment of binder materials has been discovered to hinder the electrode conductivity in RDEs. Adding to this is the possibility of particle agglomeration over long term use of RDEs [31]. The type of substrate employed to coat the catalyst powder has a direct influence over the reaction efficiency. NF and SS have been identified as strong OER contenders, while glassy carbon has been identified as demonstrating lesser OER response. Hu et al. [32] attempted to evaluate the catalytic activity of NF and SS for OER and identified that both the substrates demonstrated remarkable catalytic activity [33]. The forthcoming sections attempt to discuss the factors facilitating the assessment of catalytic activity like overpotential, Tafel slope, turnover frequency (TOF), electrochemical impedance spectrum (EIS), stability, and electrochemically active surface area (ECSA).

The structure and morphology of the synthesized heterostructure, a function of synthesis technique, is a predominant factor to be analyzed. It is the synthesis technique that defines that cation oxidation state of the material [34] that happens to be a strong determinant of catalytic potential. The different synthesis techniques deployed toward NiCo and CoFe electrocatalysts have been demonstrated. The synthesis techniques fall under two domains: physical and chemical. Chemical techniques follow a bottom-up pattern – smaller units combined to form composites – while physical techniques deploy a top-down pattern – bulkier units broken down into smaller. Chemical techniques involve hydrothermal, solvothermal, sol-gel, coprecipitation, chemical vapor deposition (CVD), etc.

The solvothermal method necessitates the provocation of chemical reactions by increasing temperature and pressure in a closed system. Several remarkable advancements have been made in the recent past, and this method has been applauded for its versatility toward the obtainment of the exact anticipated structure and morphology. Temperature, time, ligand effect, reductant agents, etc. play a significant role in determining the process efficiency [35]. The solvothermal method finds predominant application in nonaqueous media; when the medium is aqueous the method is named hydrothermal. Both the methods involve dissolving precursors in a solvent of choice and heating in an autoclave at the solvent's boiling point. The process is facile enough to be accomplished in merely a step, exhibiting greater control over the structural and morphological aspects of the synthesized material by tuning the parameters as desired. Crystallization in solvothermal processes is restricted by temperature and pressure. The method makes feasible the homogenous dispersion of nanoparticles with greater degrees of purity. The method has been proven to be of great advantage toward large scale applications owing to the reduced cost and ease of operation involved [36, 37].

The sol-gel method is a wet technique deployed for nanoparticle synthesis. This involves several steps like hydrolysis of alkoxides, condensation, drying, and ageing of the mixture. Calcination is generally deployed toward the closure of the process [38].

Coprecipitation involves heating the precursor in a solvent alongside a precipitant, which is removed post-treatment deploying centrifugation and calcination. This technique poses the risk of particle agglomeration, thus hindering catalyst performance [39].

CVD is a process put to use for gaseous and vaporous materials through a gas–solid interface to form solid products with few structural defects.

In addition to the chemical methods are physical techniques like electrospinning, laser ablation in liquid (LAL), and electrodeposition.

Electrospinning finds application in the synthesis of one-dimensional nanofibers like nanocables, porous tubes, and nanorods. The method has immense potential application toward the large-scale production of electrocatalysts. The technique involves the electrospinning of a polymer solution mixed with metal precursors as desired onto a given substrate. The method has been proven efficient owing to low cost and minimal waste involved [40].

LAL is a rapid technique making possible the efficient control of different material properties like composition, size, crystallinity, etc. [41]

Electrodeposition is a relatively simpler technique whereby layered double hydroxides (LDHs) are prepared in a short time span on the electrode surface, such as NF, SS mesh (SSM), and CFP [42]. Despite the rapidity of the process, the decreased ability to control the structural parameters is a perplexity. Of all the substrates used, NF has been identified as the ideal substrate owing to its robustness in structure, greater surface area, and appreciable conductivity [43].

A substantial number of researchers have employed NF to grow materials for efficient electrocatalysis [44]. Additionaly, SSM and CFP were reported to have a remarkable potential for growing catalytic materials on the surface and showed a magnificent catalytic activity for OER. An electrocatalyst that displays low overpotential and Tafel slope and high stability/durability is required for stunning water oxidation. To obtain such appealing performance, it is necessary to produce a material with plenty of active catalytic sites, excellent electrical conductivity, and a robust structure with abundant defects. Having discussed the major physical and chemical methods, the predominant success of any technology happens to be the efficiency in drawing high quality water at an affordable price from unused water. Considering the prevailing scenario, biological treatments cannot be solely sufficient owing to the very many disadvantages like prolonged duration, inadequacy in getting rid of sludge, requirement of extensive land area for treatment modalities, etc. Electrochemical remediation systems rightly come in as an alternative to surpass these disadvantages in degradation of toxic contaminants from water.

The following sections attempt to discuss the several electrochemical treatment techniques like electrocoagulation, electroflotation, and electrocoagulation/flotation with regards to the parameters influencing them. Of the three, electrocoagulation has been identified to be the best choice not just for the satisfactory removal rates but also for the cost efficiency and simplicity.

7.4 Electrocoagulation

Electrocoagulation involves the electrochemical synthesis of destabilization agents like Al and Fe so as to neutralize the electric charge and hence facilitate the removal of pollutants. On attainment of respective charges, the particles agglomerate into one. The process bears the advantage of less sludge production and minimal or no use of chemicals throughout the process. Al plates can be used as 3+ ions by connecting the plates to a low power supply (PS), producing ions which attract all the negatively charged particles, especially the bacteria, causing their coagulation and sedimentation. During EC, the coagulant is generated in situ by electrolytic oxidation of an anode of appropriate material.

7.5 Electroflotation

Electroflotation involves the electrical generation of tiny bubbles of hydrogen and oxygen which act as a medium of interaction with pollutants enabling them to coagulate and float on the water surface.

7.6 Electrocoagulation/flotation

This is a combination of both the discussed methods and has a broader application especially toward the removal of inorganic pollutants. The method finds application in both surface and ground water treatment.

Electrolysis as a treatment method finds lesser application in comparison to chemical methods despite the convenience and greater efficiency. Electrodes with aluminium (Al), iron (Fe), steel (St), and graphite are generally the best suited to electrochemical water treatment.

7.7 Electro-oxidation in Wastewater Treatment

Electrochemical oxidation of wastewater is an age-old technique, the recent research advancements in it being attempts to improve stability, oxidation efficiency, and reaction kinetics. Electro-oxidation was introduced in the 1960s and 1970s and continues to be a widely explored area for its high surface area, high catalytic activity, high stability to anodic corrosion, lower energy consumption, and excellent mechanical and chemical resistance [41]. The conductive layer encompasses a mixture of both inert and active metal oxides. The active oxides act as electrocatalyst while inert oxides act as modulators of electrochemical properties so as to improve catalytic activity [42]. Oxidation here can be both direct or in situ.

Fabrication of materials goes by the following steps:

i) Conventional microfabrication to prepare metallic electrodes
ii) Electrodeposition of metal oxide on the existing surface.

The common electrodes used in electrochemical degradation are IrO_2, RuO_2 while SnO_2, and TiO_2. SnO_2 is identified as a strong contender for its increased stability, high potential for OER, and slow resistivity. Researchers report longer lifetime and greater electrochemical area than other traditional methods [45]. SnO_2 electrodes have been identified as facilitating complete oxidation of organic compounds. Antimony-doped SnO_2 electrodes behave as a metal-like material making them promising contenders for electro-oxidation of industrial wastewater. The relatively high chemical and electrochemical stability, high conductivity, and high oxygen evolution overpotential support their electrolytic behavior [43].

Iridium has a great scope for its resistance to corrosion and oxidation even at high temperatures. Iridium oxide possesses excellent catalytic properties and demonstrates a greater degree of stability even in strong acidic solutions [44].

7.7.1 Electrode Characterization

In characterization, properties such as surface morphology, composition, and microcrystal structure of the coatings, plus roughness, and relative average deposited films' growth rates are measured. This involves the utility of several techniques like weight measurement, scanning electron microscopy (SEM), X-ray diffraction (XRD), profilometry, etc. The electrode performances are generally characterized using cyclic voltammetry (CV).

7.8 Doped Diamond Electrodes

Diamond, owing to its larger gap (5 eV) becomes an unsuitable and insulating electrode material. Diamond can be made conducting by doping with elements like boron, phosphorus, or nitrogen. Hydroxyl radicals are formed on boron-doped diamond (BDD) electrodes during electrolysis by means of spin trapping. These electrogenerated hydroxyl radicals are very reactive, and in the absence of organic compounds react to form hydrogen peroxide, which is further oxidized to O_2.

7.8.1 Chemical Vapor Deposition

Doped-diamond films are deposited onto conducting substrates through plasma assisted CVD, deploying hydrogen as carrier, methane as carbon source, and other gases as dopant sources. The substrate surface is activated using diamond nanoparticles, which serve as nucleation sites for growth of the diamond film before the coating process is carried out. Surface activation is performed either through polishing with diamond powder or immersion and ultrasonication in diamond nanoparticle suspension [46]. The substrate temperature is maintained at about 750 to 825°C, with growth rates between 0.2 and 3 mm/h, and the thickness of diamond film between 1 and 10 mm. The films in general are roughly surfaced nanocrystalline structures.

7.8.2 Properties of Doped-diamond Electrodes

Doping levels, nondiamond carbon content, and surface termination act as strong determinants of the electrochemical properties of diamond electrodes. On the other hand, boron, owing to a low charge carrier activation energy of 0.37 eV, happens to be the most widely used diamond dopant. Diborane or trimethyl borane is made use of to facilitate boron doping at the diamond polishing stage [47]. A close linkage has been identified between conductivity and doping levels. Typical boron doping concentration for diamond is between 500 ppm and about 10 000 ppm or 1019 – 1021 atoms cm^{-3}. The superconducting properties of BDD has been established even at very low temperatures [48].
BDD finds extensive application in wastewater treatment for the following reasons:

- i) High hydrogen and oxygen evolution potential
- ii) Low capacitance
- iii) Void of surface oxidation and reduction reactions
- iv) Greater chemical and electrochemical stability
- v) Resistance to corrosion
- vi) Efficiency toward anodic production of hydroxyl radicals without electrode modification
- vii) Near complete mineralization of organic contaminants
- viii) Lower operation costs.

7.8.3 Limitations of BDD

Mass transport limitations for miniscule pollutant concentration reduces the efficiency of BDD, confirming the inefficiency of it toward diluted wastes. The higher price of conductive diamond anodes is also a setback [20, 49].

7.9 Conclusion

The type and characteristic of the anode material used has a strong impact on electrochemical oxidation of industrial wastes. Conductive diamond has been identified as a potential contender toward decontamination of organic wastes for several reasons such as high overpotential for oxygen and hydrogen evolution, high electrochemical stability, resistance to anodic corrosion, and high current efficiencies. Titanium based anodes and BDD are found to be promising electrocatalytic materials with enhanced mineralization of refractory organics and are suitable candidates for commercial applications.

References

1 Najam, T., Cai, X., Aslam, M.K., Tufail, M.K., and Shah, S.S.A. (2020). Nano-engineered directed growth of Mn_3O_4 quasi-nanocubes on N-doped polyhedrons: efficient electrocatalyst for oxygen reduction reaction. *International Journal of Hydrogen Energy* 45 (23): 12903–12910.

2 Makimura, Y. and Ohzuku, T. (2003). Lithium insertion material of LiNi1/2Mn1/$2O_2$ for advanced lithium-ion batteries. *Journal of Power Sources* 119–121: 156–160.

3 Thackeray, M.M. (1997). Manganese oxides for lithium batteries. *Progress in Solid State Chemistry* 25 (1–2): 1–71.

4 Guo, D., Dou, S., Li, X., Xu, J., Wang, S., Lai, L. et al. (2016). Hierarchical MnO_2/rGO hybrid nanosheets as an efficient electrocatalyst for the oxygen reduction reaction. *International Journal of Hydrogen Energy* 41 (10): 5260–5268.

5 Cheng, F., Su, Y., Liang, J., Tao, Z., and Chen, J. (2010). MnO_2-based nanostructures as catalysts for electrochemical oxygen reduction in alkaline media. *Chemistry of Materials* 22 (22): 898–905.

6 Zheng, H., Modibedi, M., Mathe, M., and Ozoemena, K. (2017). The thermal effect on the catalytic activity of MnO2(a,b, andc) for oxygen reduction reaction. *Materials Today: Proceedings* 4 (11): 11624–11629.

7 Ominde, N., Bartlett, N., Yang, X.Q., and Qu, D. (2010). Investigation of the oxygen reduction reaction on the carbon electrodes loaded with MnO_2 catalyst. *Journal of Power Sources* 195 (13): 3984–3989.

8 Brenet, J.P. (1979). Electrochemical behaviour of metallic oxides. *Journal of Power Sources* 4 (3): 183–190.

9 Post, J.E. (1999). Manganese oxide minerals: crystal structures and economic andenvironmental significance. *PNAS* 96 (7): 3447–3454.

10 Brock, S.L., Duan, N., Tian, Z.R., Giraldo, O., Zhou, H., and Suib, S.L. (1998). A review of porous manganese oxide materials. *Chemistry of Materials* 10 (10): 2619–2628.

11 Wang, X. and Li, Y. (2002). Selected-control hydrothermal synthesis °fa- and b-MnO2 single crystal nanowires. *Journal of the American Chemical Society* 124 (12): 2880–2881.

12 Huang, X., Lv, D., Yue, H., Attia, A., and Yang, Y. (2008). Controllable synthesis alpha and. Beta- MnO_2: cationic effect on hydrothermal crystallization. *Nanotechnology* 19 (22): 225606.

13 Shi, X., Ahmad, S., Pérez-Salcedo, K., Escobar, B., Zheng, H., and Kannan, A.M. (2019). Maximization of quadruple phase boundary for alkaline membrane fuel cell using non-stoichiometrica-MnO2 as cathode catalyst. *International Journal of Hydrogen Energy* 44 (2): 1166–1173.

14 Cao, Y.L., Yang, H.X., Ai, X.P., and Xiao, L.F. (2003). The mechanism of oxygen reduction onMnO2-catalyzed air cathode in alkaline solution. *Journal of Electroanalytical Chemistry* 557 (15): 127–134.

15 Wei, C., Sun, S., Mandler, D., Wang, X., Qiao, S.Z., and Xu, Z.J. (2019). Approaches for measuring the surface areas of metal oxide electrocatalysts for determining their intrinsic electrocatalytic activity. *Chemical Society Reviews* 48 (9): 2518–2534.

16 Yang, Y., Luo, L.M., Zhang, R.H., Du, J.J., Shen, P.C., Dai, Z.X. et al. (2016). Free-standing ternary PtPdRu nanocatalysts with enhanced activity and durability form ethanol electrooxidation. *Electrochimica Acta* 222 (20): 1094–1102.

17 Vidal-Iglesias, F.J., Arán-Ais, R.M., Solla-Gullón, J., Herrero, E., and Feliu, J.M. (2012). Electrochemical characterization of shape-controlled Pt nanoparticles indifferent supporting electrolytes. *ACS Catalysis* 2 (5): 901–910.

18 Chinnadurai, D., Nallal, M., Kim, H.J., Li, O.L., Park, K.H., and Prabakar, K. (2020). Mn3+activesurface site enriched manganese phosphate nano-polyhedrons for enhanced bifunctional oxygen electrocatalyst. *ChemCatChem* 12 (8): 2348–2355.

19 Zhao, R., Liu, Z., Gong, M. et al. (2016). Ethylenediamine tetramethylene phosphonic acid assisted synthesis of palladium nanocubes and their electrocatalysis of formic acid oxidation. *Journal of Solid State Chemistry* 21: 1297–1303.

20 Vandamme, D., Foubert, I., and Muylaert, K. (2013). Flocculation as a low-cost method for harvesting microalgae for bulk biomass production. *Trends in Biotechnology* 31: 233–239.

21 Zhao, R., Fu, G., Zhou, T. et al. (2014). Multi-generation overgrowth induced synthesis of three-dimensional highly branched palladium tetrapods and their electrocatalytic activity for formic acid oxidation. *Nanoscale* 6: 2776–2781.

22 Chong, L., Wen, J., Kubal, J. et al. (2018). Ultralow-loading platinum-cobalt fuel cell catalysts derived from imidazolate frameworks. *Science* 362: 1276–1281.

23 Li, J., Yin, H.M., Li, X.B. et al. (2017). Surface evolution of a Pt-Pd-Au electrocatalyst for stable oxygen reduction. *Nature Energy* 2: 17111–17119.

24 Escudero-Escribano, M., Malacrida, P., Hansen, M.H. et al. (2016). Tuning the activity of Pt alloy electrocatalysts by means of the lanthanide contraction. *Science* 352: 73–76.

25 Huang, X., Zhao, Z., Cao, L. et al. (2015). High-performance transition metal-doped Pt_3Ni octahedra for oxygen reduction reaction. *Science* 348: 1230–1234.

26 Xu, X., Pan, Y., Zhong, Y. et al. (2020). Ruddlesden-popper perovskites in electrocatalysis. *Materials Horizons* 7: 2519–2565.

27 Pan, Y., Xu, X., Zhong, Y. etal. (2020). Directevidenceofboostedoxygenevolutionoverperovskite by enhanced lattice oxygen participation. *Nature Communications* 11: 1–10.

28 Yu, H., Davydova, E.S., Ash, U. et al. (2019). Palladium-ceria nanocatalyst for hydrogen oxidation in alkaline media: optimization of the Pd-CeO_2 interface. *Nano Energy* 57: 820–826.

29 Miller, H.A., Vizza, F., Marelli, M. et al. (2017). Highly active nanostructured palladium-ceria electrocatalysts for the hydrogen oxidation reaction in alkaline medium. *Nano Energy* 33: 293–305.

30 Zhang, W., Yang, Y., Huang, B. et al. (2019). Ultrathin PtNiM (M= Rh, Os, and Ir) nanowires as efficient fuel oxidation electrocatalytic materials. *Advanced Materials* 31: 1805833–1805841.

31 Huang, L., Zheng, C.Y., Shen, B. et al. (2020). High-index-facet metal-alloy nanoparticles as fuel cell electrocatalysts. *Advanced Materials* 32: 2002849–2002854.

32 Badalyan, A. and Stahl, S.S. (2016). Cooperative electrocatalytic alcohol oxidation with electron-proton-transfer mediators. *Nature* 535: 406–410.

33 Wang, K., Du, H., Sriphathoorat, R. et al. (2018). Vertex-type engineering of Pt-Cu-Rh het- erogeneous nanocages for highly efficient ethanol electrooxidation. *Advanced Materials* 30: 1804074–1804080.

34 Luo, S., Chen, W., Cheng, Y. et al. (2019). Trimetallic synergy in intermetallic PtSnBi nanoplates boosts formic acid oxidation. *Advanced Materials* 31: 1903683–1903689.

35 Yang, N., Zhang, Z., Chen, B. et al. (2017). Synthesis of ultrathin PdCu alloy nanosheets used as a highly efficient electrocatalyst for formic acid oxidation. *Advanced Materials* 29: 1700769–1700774.

36 Xi, Z., Li, J., Su, D. et al. (2017). StabilizingCuPdnanoparticlesviaCuPdcouplingto$Wo_{2.72}$ nanorods in electrochemical oxidation of formic acid. *Journal of the American Chemical Society* 139: 15191–15196.

37 Liu, Z., Zhao, Z., Peng, B. et al. (2020). Beyond extended surfaces: understanding the oxygen reduction reaction on nanocatalysts. *Journal of the American Chemical Society* 142: 17812–17827.

38 He, Y., Guo, H., Hwang, S. et al. (2020). Singlecobaltsitesdispersedinhierarchicallyporous nanofiber networks for durable and high-power PGM-free cathodes in fuel cells. *Advanced Materials*: 2003577–2003590.

39 Liu, G., Li, X., Ganesan, P. et al. (2010). Studies of oxygen reduction reaction active sites and stability of nitrogen-modified carbon composite catalysts for PEM fuel cells. *Electrochimica Acta* 55: 2853–2858.

40 Zhou, X., Yang, Z., Nie, H. et al. (2011). Catalyst-free growth of large scale nitrogen- Doped carbon spheres as efficient electrocatalysts for oxygen reduction in alkaline medium. *Journal of Power Sources* 196: 9970–9974.

41 Andersen, R.A. and Kawachi, M. (2005). Traditional microalgae isolation techniques. In : *Algal Culturing Techniques* (ed. R.A. Andersen), 83–100. Elsevier.

42 Borowitzka, M.A. and Moheimani, N.R. (eds.) (2013). *Algae for Biofuels and Energy*. Netherlands, Dordrecht: Springer.

43 Apel, A.C., Pfaffinger, C.E., Basedahl, N., Mittwollen, N., Gö̈bel, J., Sauter, J., Brück, T., and Weuster-Botz, D. (2017). Open thin-layer cascade reactors for saline microalgae production evaluated in a physically simulated Mediterranean summer climate. *Algal Research* 25: 381–390.

44 Doucha, J. and Lívanský, K. (2009). Outdoor open thin-layer microalgal photobioreactor: potential productivity. *Journal of Applied Phycology* 21: 111–117.

45 Rizwan, M., Mujtaba, G., Memon, S.A., Lee, K., and Rashid, N. (2018). Exploringthepotentialof microalgae for new biotechnology applications and beyond: a review. *Renewable & Sustainable Energy Reviews* 92: 394–404.

46 Bumbak, F., Cook, S., Zachleder, V., Hauser, S., and Kovar, K. (2011). Best practices in heterotrophic high-cell-density microalgal processes: achievements, potential and possible limitations. *Applied Microbiology and Biotechnology* 91: 31–46.

47 Uduman, N., Qi, Y., Danquah, M.K., Forde, G.M., and Hoadley, A. (2010). Dewatering of microalgal cultures: a major bottleneck to algae-based fuels. *Journal of Renewable and Sustainable Energy* 2: 012701.

48 Matos, C.T., Santos, M., Nobre, B.P., and Gouveia, L. (2013). Nannochloropsis sp. biomass recovery by Electro-Coagulation for biodiesel and pigment production. *Bioresource Technology* 134: 219–226.

49 Granados, M.R., Aci′en Fernandez, F.G., Gomez, C., Ferna′ndez-Sevilla, J.M., and Molina Grima, E. (2012). Evaluation of flocculants for the recovery of freshwater microalgae. *Bioresource Technology* 118: 102–110.

50 Chung, H.T., Cullen, D.A., Higgins, D. et al. (2017). Direct atomic-level insight into the active sites of a high-performance PGM-free ORR catalyst. *Science* 357: 479–483.

8

Electrocatalytic Degradation of Organic Pollutants from Water

Anamaria Baciu[1], Sorina Negrea[2,3], and Florica Manea[1,]*

1 Politehnica University Timisoara, Faculty of Industrial Chemistry and Environmental Engineering, P-ta Victoriei No. 2, Timisoara, Romania
[2] National Institute of Research and Development for Industrial Ecology (INCD ECOIND), Timisoara Branch, Romania
[3] "Gheorghe Asachi" Technical University of Iasi, Department of Environmental Engineering and Management, Iasi, Romania
** Corresponding author*

8.1 Introduction

Continuous increase of water polluted with organic compounds is due mainly to the development of anthropogenic activities and represents one of the most important worldwide concerns. A large part of the organic pollutants in water are persistent organic pollutants considered as priority hazardous or emerging pollutants, characterized by a toxic, mutagenic, and carcinogenic character. They are refractory to the biological treatment and to other conventional water treatment processes (coagulation, sorption, and chemical oxidation) [1]. Also, natural organic matter (i.e. humic and fulvic acids) should be considered, especially for surface waters used as a main drinking water source, because these are precursors of toxic byproducts generated in the disinfection stage of drinking water treatment technology [2].

In recent years, several advanced water treatment processes have been developed to address the removal of persistent organic pollutants, of which electrochemical oxidation represents one of the most important within the advanced oxidation processes (AOPs). The main characteristic of AOPs is in situ production of hydroxyl radicals (·OH), which are highly reactive, and strongly and nonselectively attack the majority of organic pollutants including highly recalcitrant ones, being able to assure the advanced degradation until their complete mineralization [3]. Electro(catalytic) oxidation is interesting for mineralization of wastewater containing low organic load expressed by the chemical oxygen demand (COD) parameter ($< 5\ g{\bullet}L^{-1}$). It represents a very promising alternative to the other strong oxidation processes due to its advantages related to environmental compatibility, versatility, ease of automation, and relative low cost [4, pp.1–23, 5]. In this context, the electrochemical oxidation process is considered for the finishing step in water/wastewater treatment technology, but also electrochemical conversion of recalcitrant organic pollutants to biodegradable ones must be considered for biological-based wastewater treatment technology. To conduct the electrochemical oxidation process in water/wastewater treatment for the removal of organic pollutants, the oxidation principles, the anode and electrochemical reactor configurations, and the possibility to couple/combine with other unitary processes should be considered [6–8].

Photocatalysts and Electrocatalysts in Water Remediation: From Fundamentals to Full Scale Applications, First Edition. Edited by Prasenjit Bhunia, Kingshuk Dutta, and S. Vadivel.

8.2 Principles and Fundamental Aspects of Electrooxidation

The electrooxidation process in water treatment technology aims to degrade/oxidize the pollutants from water/wastewater within the electrolysis cell.

Several aspects related to the electrooxidation mechanism and reaction types are considered for a synthetic image of the complex electrochemical process and its constitutive stages (see Figure 8.1).

Thus, two approaches have to be considered, related to the type of the pollutants' oxidation, which allows a classification of the electrooxidation process, proposed by Comninellis [9]:

- Partial oxidation of the organic pollutants generating oxidation subproducts, the so-called electrochemical conversion method, which can be useful in wastewater treatment if the oxidation subproducts are more biodegradable and can be further degraded by the biological oxidation.
- Complete oxidation/degradation of organic pollutants assuring mineralization, called electrochemical incineration and combustion. The main principle of this process considers the generation of hydroxyl radicals (•OH), which are very powerful oxidants, and includes electrochemical oxidation to the AOP.

Another classification of the electrooxidation process takes into account the oxidation mechanism: direct electrooxidation that occurred via electron transfer at the electrode surface, and mediated or indirect electrooxidation via the oxidant species generated electrochemically. Sometimes, based on the electrode composition and the operation conditions during the electrooxidation of wastewaters, both oxidation mechanisms may coexist at different proportions [10, pp. 71–73].

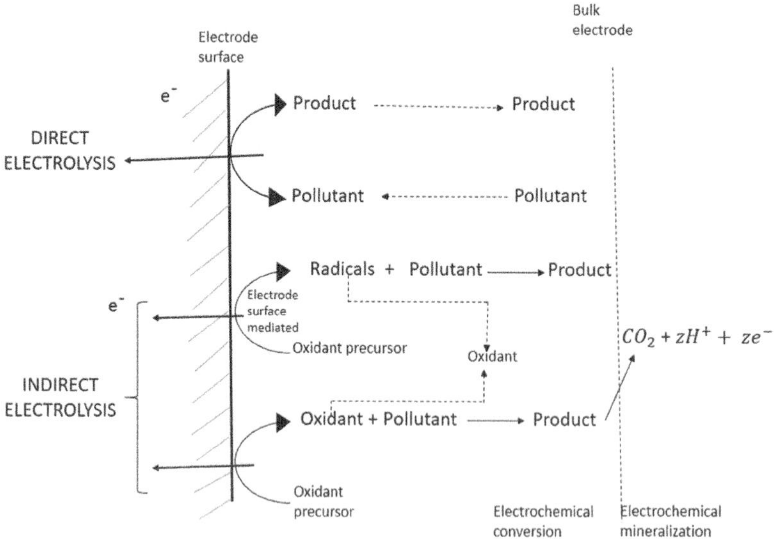

Figure 8.1 A synthetic image of the electrooxidation process.

Moreover, taking into consideration this last approach addressing the oxidation mechanism, it is obvious that the electrooxidation process is initiated on the electrode surface and can be continued both on the electrode surface and in the supporting electrolyte bulk. This suggests another classification of the electrooxidation processes [11, 12]:

- Heterogeneous oxidation of the pollutants on the anode surface, which is a complex process based on several transport and electron transfer stages. The stages of the transport consist of: the transport of the pollutants from the bulk to the electrode surface followed by the adsorption of the pollutant onto the electrode surface, the electrochemical degradation reaction of the pollutants by electron transfer directly or mediated onto the electrode surface, and finally, the desorption of the electrooxidation products and their transport to the bulk solution (water).
- Homogeneous oxidation of pollutants in the bulk solution via oxidants produced on the anode surface from components of the electrolyte, as indirect electrooxidation process. The oxidants are produced by the heterogeneous anodic oxidation of water or ions from water composition or a supplementary dose to generate the oxidants, which act in the bulk of the electrochemical cell.

The most important oxidant is the hydroxyl radical, which occurs in accordance with Eq. (8.1) related to the water pH. The hydroxyl radical generation makes the electrooxidation process belong to the category of AOPs.

$$H_2O \rightarrow \cdot OH_{ads} + H^+ + e^-; \text{ or } HO^- \rightarrow \cdot OH_{ads} + e^- \tag{8.1}$$

Considering water composition, several anodic oxidation reactions with oxidant generation take place according to Eqs. (8.2)–(8.13). The presence of chloride leads to the generation of chlorine and its derivates (hypochlorite, chlorite, and chlorate) according to Eqs. (8.5)–(8.10). Also, strong persulfate, perphosphate, and percarbonate oxidants are generated in the presence of sulfate, phosphate, and carbonates anions in accordance with Eqs. (8.11)–(8.13).

$$3H_2O \rightarrow O_3 + 6H^+ + 6e^- \tag{8.2}$$

$$2Cl^- \rightarrow Cl_2 + 2e^- \tag{8.3}$$

$$Cl_2 + H_2O \leftrightarrow HClO + Cl^- + H^+ \tag{8.4}$$

$$HClO \leftrightarrow ClO^- + H^+ \tag{8.5}$$

$$6HClO + 3H_2O \rightarrow 2ClO_3^- + 4Cl^- + 12H^+ + 1.5O_2 + 6e^- \tag{8.6}$$

$$Cl^- + \cdot OH \rightarrow ClO^- + H^+ + e^- \tag{8.7}$$

$$ClO^- + \cdot OH \rightarrow ClO_2^- + H^+ + e^- \tag{8.8}$$

$$ClO_2^- + \cdot OH \rightarrow ClO_3^- + H^+ + e^- \tag{8.9}$$

$$ClO_3^- + \cdot OH \rightarrow ClO_4^- + H^+ + e^- \tag{8.10}$$

$$2SO_4^{-} \rightarrow S_2O_8^{2-} + 2e^{-} \tag{8.11}$$

$$2PO_4^{3-} \rightarrow P_2O_8^{4-} + 2e^{-} \tag{8.12}$$

$$2CO_3^{2-} \rightarrow C_2O_6^{4-} + 2e^{-} \tag{8.13}$$

8.3 Electrode Materials and Cell Configuration

The electrochemical advanced oxidation process produces in situ hydroxyl radicals, which are capable of degrading and mineralizing the organic pollutants, by using adequate electrode materials with respect to the cell configuration and flow type. The divided and undivided cell configurations have to be considered in the electrochemical reactor design under flow plug and perfect mixing. In divided cells, a porous diaphragm or an ion conducting membrane is used for the separation of the anolyte and catholyte, and its choice is as important as the appropriate choice of electrode materials for optimized functioning of the electrochemical process. A classification of the electrode types linked to the structural conformation of the electrode is presented in Figure 8.2. Two-dimensional (2D) and three-dimensional (3D) electrodes can be designed with respect to the rector configuration and operation as static or moving electrodes with different geometries and arrangement. Plates, cylinders, and discs are considered for designing the static electrodes, which can be arranged in parallel or rotated for moving electrodes. Porous electrodes and packed or fluidized bed electrodes are considered for 3D electrodes.

Both the effectiveness and selectivity of the electrochemical oxidation process thus depend strongly on the nature of the anode materials. For example, some anode materials are fitting for electrochemical conversion by partial and selected oxidation of organic pollutants, while other anodes are suitable for electrochemical combustion producing the complete mineralization of the compounds to CO_2 [3, 13].

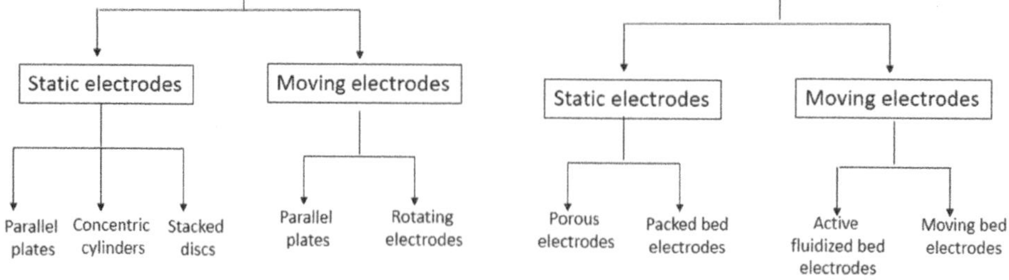

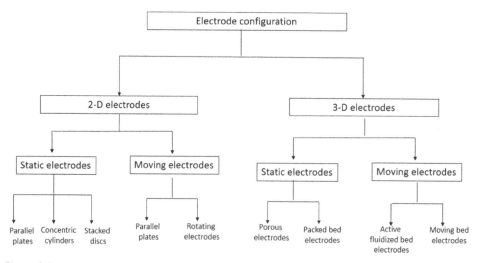

Figure 8.2 Electrode configuration.

For water treatment using the electrooxidation process, two main categories of electrodes should be considered: carbon-based electrodes and dimensionally stable anodes (DSA).

For environmental application, one of the criteria for efficiency of the electrode material is the manufacturing strategy. The most studied techniques are thermal decomposition, electrodeposition, and their modifications. These innovative strategies, which involve new heating methods, different solvents, and modification in their composition, show significant time and energy savings, are introduced as competitive methodologies to produce an upgrade in electrode materials, which fits with the trend for environmental protection applications [14].

In addition, diverse characteristics related to the stability, lifetime, and service life of the electrodes are considered in the organic pollutant oxidation, in order to improve the treatment efficiency. The life time of the electrode is essential as it must be highly efficient for a long period of time, which depends on the synthesis method and the electrical charge applied. Stability and the service life of the electrode depend on its structural properties and the composition; for example, the coating of the electrode with noble metals improves the electrode stability. Porous electrodes have a higher life time than a smoother surface, and also have a larger specific active area and a higher oxygen evolution overpotential, which improves the rate of the electrochemical activity. This aspect is taken into account in the production of a stable electrode. When the electrode stability is affected, the amount of oxygen evolution increases, which is a very important aspect monitored during electrode operation within the electrooxidation process. It is obvious that if the stability and the oxidation strength are high, the service life value is shorter [15–19].

Considering the selection of the electrode material for the design of the electrooxidation process, the electrochemical behavior can be easily checked in relation with the oxygen evolution overpotential and the specific electroactive area through the cyclic voltammetry (CV) technique, which is the first electrochemical experiment performed for the further development of the electrochemical process. A comparison of CV results is shown as an example in Figure 8.3, and a major difference between carbon nanotube-modified graphite (CNT/graphite electrode) and SnO_2/Ti dimensionally stable anodes can be easily noticed. SnO_2/Ti anodes exhibit higher overpotential value for the oxygen evolution in comparison with CNT/graphite (about +2.20 V versus +1.00 V vs Ag /AgCl). Also, more porosity of the CNT/graphite, as based on SEM images, is reflected in the background current through the capacitive component given by the morphostructural characteristics of the electrode surface (Insets a and b from Figure 8.3).

Regardless of the anode material, the first step in the electrochemical oxidation mechanism is anodic water discharge to form hydroxyl radicals, which are adsorbed onto the electrode surface (M), without concern of solution pH [20, 21], (Eq. 8.14):

$$M + H_2O \rightarrow M(\cdot OH) + H^+ + e^- \quad (8.14)$$

The next step is then determined by the nature of the electrode material, which is classified as active or nonactive as a function of its catalytic activity.

At active electrodes, thus, the hydroxyl radicals are adsorbed onto the electrodes to form a higher metal oxide, signifying that the electrochemical conversion takes place where the organic pollutants are selectively converted to more oxidized intermediates, while the

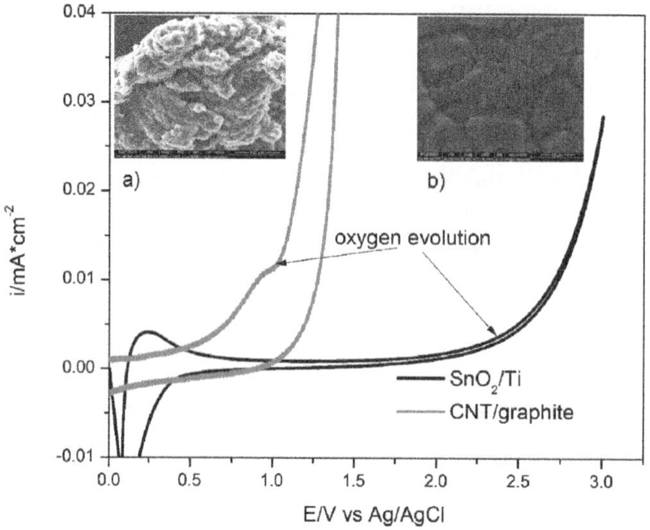

Figure 8.3 Cyclic voltammograms recorded in 0.1 M Na_2SO_4 supporting electrolyte with scan rate of 0.05 $V \cdot s^{-1}$ at CNT/graphite and SnO_2/Ti electrodes. (a) SEM image of CNT/graphite. (b) SEM image of SnO_2/Ti.

higher metal oxide is decomposed to release oxygen that oxidizes the organic pollutant (R) in accordance with Eqs. (8.15) and (8.16):

$$M(\cdot OH) \rightarrow MO + H^+ + e^- \quad (8.15)$$

$$MO + R \rightarrow M + RO \quad (8.16)$$

Examples of active electrodes include DSA (RuO_2, IrO_2), Pt, and doped Ti/TiO_2 electrode.

At nonactive electrodes, total oxidation may become the dominant reaction pathway with simultaneous oxidation of radicals to yield oxygen, assuring the electrochemical combustion through complete oxidation of organic pollutants to CO_2 (Eq. 8.17):

$$M(\cdot OH) + R \rightarrow M + CO_2 + H_2O + H^+ + e^- \quad (8.17)$$

Boron-doped diamond, tin, and lead oxides are typical examples of nonactive electrodes [22].

8.3.1 Carbon-based Electrodes

The carbon atoms, possessing a valence of four, have the ability to form single, double, and triple covalent bonds among themselves or with other elements. The carbon atoms can undergo sp, sp^2, or sp^3 hybridizations with a narrow band gap between their 2s and 2p electronic shells [23]. Graphite with sp^2 hybridization and diamond with sp^3 hybridization are the two most widely known allotropic forms of carbon.

Since the 1840s, graphite has been known to exist. It is an ideal electrode substrate in many aspects because of its wide anodic potential range, efficiency in wastewater/water treatment, low residual current, chemical inertness, and relatively low cost compared with other materials. It is intensively used on a laboratory scale for investigation of new processes, such as graphene synthesis, mainly as a precursor for other C forms. Furthermore, graphite electrodes exhibit fast response and can be easily fabricated in different configuration sizes. [24–26].

Graphite consists of stacks of graphene layers that are weakly coupled by van der Waals forces. The amorphous form is an assortment of carbon atoms in a noncrystalline, irregular manner, which is essentially graphite but not held in a crystalline macrostructure. Depending on the atomic arrangements of 3D carbon atoms (bulk), it forms mechanically weak, optically opaque, highly electrically conductive graphite and/or mechanically strong, optically transparent, insulating diamond. Carbon nanofoams, glassy carbons, and linear acetylenic carbons are the other forms of known 3D carbon allotropes [27–29].

Graphite electrodes, however, exhibit several weaknesses, e.g. low oxidation activity of organic pollutants, high average corrosion rate, and low mineralization efficiency. It must be underlined that the efficiency of the electrooxidation process is highly dependent on operating conditions related to concentration and type of electrolyte solution, organic pollutant concentration, and the duration of the electrooxidation process [30–32].

Since the 1980s, the scientific and engineering community have been using the diamond form of carbon in electrochemistry in various applications including the electrooxidation process for water treatment. Taking into account the carbon-based electrode in electrochemical water treatment, the **boron-doped diamond (BDD) electrode** is consider an excellent anode material because of its high chemical inertness with low adsorption properties and strong tendency to be stable in very aggressive media, high mechanical strength, long-term response stability under anodic polarization in aqueous electrolytes, and high overpotential for the oxygen evolution reaction (OER). In the synthesis of BDD electrodes, the sp^3/sp^2 carbon hybridization ratio has been found to be an important parameter, linked to promotion of the active or nonactive electrode behavior and to the boron inclusion into the diamond lattice, resulting in different electrochemical activity [33–36]. In addition, the quantity of boron doping controls the diamond behavior; thus, a low doping level makes it exhibit semiconductor properties, whereas at high doping level diamond exhibits semimetallic conductivity [22].

Considering the major advantages of nanoparticle properties related to electrochemical aspects such as enlarging the specific surface area of the electrode and enhancing the electron transfer rate in the overall electrode process, **nanostructured carbon materials** have been considered for integration within the electrode composition for application in water treatment.

Nanostructured carbon materials are promising building blocks with unique structure and tunable physicochemical properties suitable for innovative solutions to environmental challenges. One of the hot topics for research, nanoscience/nanotechnology is considered to have a big potential for solving the technical challenges associated with environmental application especially in electrochemical water treatment. The nanostructuration thus allows changing the traditional concept of a carbon-based electrode material, by their combination, which brings new properties of well-defined nanostructured carbon electrode composition. This aspect is reflected in a new electrocatalytic treatment that exceeds the performance of conventional carbon electrodes, new additional functions that can enhance the heterogeneous catalysis mechanism at electrode surfaces, or increase the limiting step rate of the electrochemical reaction. Nanotechnology applied to the environmental field has charted a path for innovation in water treatment, making the advanced electrochemical process a greener process than conventional ones. The proactive perspective of this greener process application of carbon-based materials applies it to the prevention of environmental degradation, improving public health, optimizing energy efficiency, and from a retroactive point of view, to environmental remediation, water reuse, and pollutant transformation/mineralization [37, 38].

The geometrical structure of the particles in carbon nanomaterials is the basic criterion for their classification. The particles can have shapes of tubes, horns, spheres, or ellipsoids. The tube-shaped carbon nanoparticles are known as carbon nanotubes, while the spherical or ellipsoidal nanoparticles are present in fullerenes [39, 40]. Besides these, carbon nanofibres, graphene, carbon nanodiamonds (CNDs), and carbon dots (CDs) are the most significant allotropic modifications of nanocarbon. The 0D nanodiamonds, 1D nanotubes, and 2D graphene nanosheets can act as a prototype for nanocomposites [41], which should be considered more valuable for the design of the nanstructured carbon-based electrode materials characterized by better stability due to the ligand/resin used for composite formation.

Carbon nanotubes (CNTs) are part of the fullerenes family, having a nanostructure derived from that of graphite (a type of hybridization of sp^2 carbon atoms) in which a simple curvature is introduced by several topological defects, giving it one-dimensional character and a molecular dimension. The characteristic diameter of nanotubes is from 1–50 nm, and their length can be up to 10 μm; they have variable number of layers and chirality vectors [42]. CNTs present the natural choice for the development of advanced nanostructured electrodes, considering several aspects, e.g. excellent properties like rigidity, strength, elasticity, high thermal and electrical conductivities, manufacturing costs, in particular for multiwalled carbon nanotubes, and the possibility to improve properties during the synthesis process (oxide doping), post-treatment, and functionalization [43].

Based on their structure, they can be divided into single-walled carbon nanotubes (SWCNTs) and multiwalled carbon nanotubes (MWCNTs). SWCNTs are cylindrical assemblies composed only of carbon atoms with dimensions from 0.4–2 nm. The electrochemical properties of SWCNTs depend upon the value of roll-up vectors (n, m), and they can be metallic, semiconductive armchair, or zigzag. Also, the SWCNTs can exhibit electrical conductivity or semiconductive properties that depend upon the diameter of the tubes [44].

MWCNTs can be considered a collection of SWCNTs with different diameters ranged from 2 to 100 nm that are concentrically dispersed. The length and diameter of these structures differ in direction, and their properties are very different. The distances between two successive walls are close to the value of the distance between two graphene strips in a graphite crystal. Each MWCNT can consist of 2 to 50 SWCNTs. They have inner diameters between 1.5 and 15 nm, and outer diameters between 2.5 and 30 nm [45]. Based on the rolling of graphene sheets, the CNTs are classified as zigzag, armchair, or chiral, depending on the number of unit vectors in the crystal lattice of graphene along two directions in a honeycomb structure. The chirality has a significant effect on the properties of CNTs, determining whether a particular CNT has metallic or semiconducting behavior [46].

CNT-based materials operated in an electrochemical electrode configuration are promising materials for improving the energy consumption of advanced electrooxidation because of the high electroactive surface area and suitable properties for optimization of mass transport of organic pollutants at the electrode/solution interface. CNT-based electrodes present selective oxidation features. While fast degradation of target pollutants is usually reported, only partial mineralization is achieved because of the electrochemical recalcitrance of some degradation by-products. One of the most explored ways to improve the effectiveness of CNTs is to synthesize modified CNTs in order to increase the overvoltage of OER or to promote interaction between the electrode surface and target pollutants. The main drawback of these electrodes is their low overvoltage for OER that reduces their capacity for generation of nonselective oxidant species and assures the selective oxidation of organic pollutants and low mineralization rate [33].

Carbon nonofibers (CNFs), based on sp^2 hybridized orbitals, differ mainly from CNTs by the orientation of graphene planes [47] and can be viewed as overlapping (truncated) or conical graphite disks. CNFs can also occur with hollow cylindrical structures with a diameter of about 100 nm and a length of up to several tens of microns (length / diameter > 100). Their crowded layered conical structure is arranged in different ways such as ribbon-like, platelet, or herringbone fiber. Their mechanical strength and electric properties are just like those of CNTs [138], but due to their unique structure, carbon nanofibers have the behavior of semiconductors, having chemically reactive end planes. As a consequence of stacking of graphene sheets with different shapes in different arrangements, CNFs have more edge sites on their outer walls in comparison to CNTs. The presence of edge sites makes it feasible to transfer electrons with electroactive species in solution and the substrate; thus, CNF is used as a support material for catalysts, reinforcing fillers in polymer composites, hybrid fillings in plasticized carbon fibers, and for generating photocurrent in photochemical cells [48]. There are currently numerous synthetic methods for the preparation of carbon nanofibers. The carbon nanostructure, however, differs significantly in terms of diameter, crystallinity, crystal orientation, purity, and surface chemistry. These structural variations largely affect the intrinsic properties and processes and their behavior in the composite system [49]. The choice, therefore, largely depends on the nature of the material for the matrix, the processing technology, and the improvement of the desired properties. In the literature, various synthesis methods have been used for dispersing CNF in epoxy resin, thus leading to different levels of electrical conductivity of the composites [50].

CNF characteristics related to good electrical conductivity and large surface area make them suitable for the sustainable design of nanostructured carbon-based electrode materials for advanced electrooxidation of organic pollutants from water [33].

Graphene is a 2D carbon material that is one atom thick. It is made up of sp^2 hybridized carbon atoms arranged in a honeycomb shaped lattice and exhibits remarkable properties that could make it useful in a variety of applications [51].

Graphene is an ideal building block to study the fundamentals of carbon structures for 3D carbon forms, such as graphite, diamond, and amorphous carbon.

8.3.2 Dimensionally Stable Anodes

The conductive layer of dimensionally stable andoe (DSA) electrodes is generally formed by a mixture of active and inert metal oxides. The active oxides act as electrocatalysts, while the inert oxides modulate the electrochemical properties of the active components, providing high catalytic activity and higher lifetime [52].

DSAs form a class of electrodes that consist basically of a mixture of metallic support, such as titanium, which acts as an inert metal that controls the electrochemical properties of the active compounds, coated with an electrically active conductive oxide layer that acts as electrocatalyst. These electrodes have been widely used in electrochemical water treatment for the removal of organic pollutants due to their high catalytic activity, high stability to anodic corrosion, excellent mechanical stability, and high robustness [53]. These electrodes are classified as active electrodes and have strong interaction with the generated ·OH. In this situation, they have lower oxidizing power and promote selective oxidation (i.e. conversion) rather than mineralization of organic pollutants [54]. In these electrodes, oxidation can occur by direct electron exchange between the contaminant and the electrode surface, or by indirect in situ electrogeneration of catalytic species with high oxidizing power, such as H_2O_2, O_3, and Cl_2, which are able to promote contaminant oxidation [22, 55, 56].

DSA electrodes can be prepared by thermal decomposition of polymeric precursors, in what is known as the Pechini method. This method is based on the ability of certain hydroxycarboxylic acids, such as citric acid, to form chelates with various metallic cations [57]. Firstly, the complexation of the organic acid with the metallic cation takes place. These chelates undergo polyesterification when heated in the presence of a polyhydroxyl alcohol such as ethylene glycol [58]. After heating, a polymeric resin containing uniformly distributed metallic cations strongly bound to the carboxylic groups is obtained. This polymeric matrix can be deposited on a substrate and then thermally decomposed at elevated temperatures to form the desired oxide. The Pechini method offers advantages such as low cost, good composition homogeneity, and high purity; moreover, some works have reported that Ti/RuO_2 and Ti/IrO_2 electrodes prepared by this method have longer lifetime and higher electrochemically active area than those prepared by traditional methods [55].

DSAs are easy to fabricate and relatively inexpensive, but their main drawbacks, which are linked to the active nature of DSAs, are related to the incomplete degradation/mineralization of the pollutant compound and the formation of undesirable, partly degraded by-products. Such partial degradation of the target compound may be due to an indirect oxidation mechanism by the DSA and lower oxygen evolution overpotential of the metal oxides as well [59–61].

8.4 Performance Assessment Indicators and Operating Variables

The effort to improve the performance of electrooxidation processes is growing, mainly when related to the electrode composition, the reactor design, and the operating factors that affect the process performance. Besides the electrooxidation mechanism exploration, the study of the kinetics aspects completes the overall assessment of the electrooxidation performance. Considering the partial or total oxidation of the pollutants from water, the first performance indicators are *the degradation process efficiency* (η, %) and *the mineralization process efficiency* (η_{TOC}, %), calculated based on Eqs. (8.18) and (8.19) [62, 63]:

$$\eta(\%) = \frac{C_i - C_f}{C_i} \cdot 100 \tag{8.18}$$

where C_i and C_f are the initial and final organic concentration during the electrolysis period.

$$\eta_{TOC}(\%) = \frac{(TOC_i - TOC_f)}{TOC_i} \cdot 100 \tag{8.19}$$

where TOC_i and TOC_f are initial and final total organic concentration (mg C·dm^{-3}).

Futher, *the electrochemical degradation efficiency* (E, mg/C·cm^2) for organic pollutants degradation is expressed by Eq. (8.18'):

$$E = \frac{(C_i - C_f)}{C \cdot S} \cdot V \text{ mg}/C \cdot \text{cm}^2 \tag{8.18'}$$

where C is the electrical charge consumption during the electrolysis time (Coulombs), V is the sample volume (cm^3), and S is the area of the electrode surface (cm^2).

Also, *the electrochemical mineralization efficiency* for organic pollutants mineralization is determined based on Eq. (8.19'):

$$E_{TOC} = \frac{(TOC_i - TOC_f)}{C \cdot S} \cdot V \text{ mg}/C \cdot \text{cm}^2 \tag{8.19'}$$

The electrochemical degradation and mineralization efficiencies are considered for more accurate characterization of the electrode performance for a similar cell configuration, which is very useful for electrode design in the electrooxidation process.

*The mineralization current efficiency (*MCE*)* represents the percentage of current directed toward the mineralization of the substrate (molecule or TOC) and is calculated based on Eq. (20) [64, pp. 407–437, 65]:

$$\text{MCE} = \frac{n F V_S \Delta(\text{TOC})_{exp}}{4.32 \text{ x}10^7 \boldsymbol{mIt}} \cdot 100(\%) \tag{20}$$

where n is the number of electrons consumed in the mineralization process of 4-AP, F is the Faraday constant (= 96 487 C · mol^{-1}), V_s is the solution volume (dm^3), $\Delta(TOC)_{exp}$ is the experimental TOC decay (mg · dm^{-3}), 4.32 · 10^7 is a conversion factor for units homogenization (= 3600 s h^{-1} · 12 000 mg of carbon mol^{-1}), m is the number of carbon atoms in the organic pollutants, I is the applied current (A), and t is time (h). The number of electrons consumed is determined based on the overall mineralization reaction of organic pollutants to CO_2.

The specific energy consumption, W_{sp}, represents the economic aspect considered for the performance assessment, which can be related both to the amount of the organic pollutants mineralized Eq. (8.21), and to the volume of the electrolyzed water (Eq. 8.22):

$$W_{sp}(\text{kWh}/\text{g TOC removed}) = \frac{(I \Delta t V^{-1})U}{1000(TOC_i - TOC_f)} \tag{8.21}$$

$$W_{sp} = \frac{\left(I\Delta t V^{-1}\right)U}{1000} \text{ kWhdm}^{-3} \tag{8.22}$$

where U is the cell potential (V), I is the applied current (A), t is the electrolysis time (h), and V is the volume of the treated solution (dm^3).

An electrode reaction is controlled either by the electrical charge transfer or by the mass transfer of the electroactive species from the bulk of the solution to the electrode depending on the applied current density. When the applied current density is higher than the limiting current density (i_{lim}) the process is controlled by diffusion, and when the applied current density is lower than i_{lim} the process is controlled by electrical charge transfer. i_{lim} (A/m^2) is calculated according to Eq. (8.23) [66]:

$$i_{lim} = zF k_d C_R \tag{8.23}$$

where, i_{lim} is the limiting current density for organic pollutants mineralization (Am^{-2}), z is the number of electrons involved in the organic pollutants mineralization reaction, F is the Faraday constant 96 487 C mol^{-1}, k_d is the mass transfer coefficient ($m\cdot s^{-1}$) and C_R is the concentration of organic pollutants in the bulk solution ($mol\cdot m^{-3}$).

The mass transfer coefficient k_d was calculated from TOC analysis data according to Eq. (8.24):

$$\frac{\text{TOC}}{\text{TOC}_i} = \exp\left(-\frac{S}{V} k_d t\right) \tag{8.24}$$

where TOC_i and TOC are the initial concentration of total organic carbon and at time t, respectively, S is the anodic active surface area (m^2), V is the volume of the treated solution (m^3), and t is the eletrolysis time (s).

Considering the operating variables that are usually modified in the electrochemical oxidation processes to improve process performance, the current density (intensity per unit area of electrode) is one of the most important counted, because it controls the overall oxidation reaction rate. Even if it is expected that an increase in the current density enhances the oxidation efficiency or oxidation kinetics, however, its effect depends on the anode material composition and the characteristics of the effluent to be treated for a given anode material. Also, the usage of higher current densities implies an increase in energy consumption and, as implicit, higher operating costs.

Another important operating variable is the water/supporting electrolyte pH that influences the overall oxidation process in a different way from a thermodynamic point of view in comparison with kinetics aspects. It is well-known that acidic pH favors the reaction rate, while from the thermodynamic point of view, the alkaline pH means higher hydroxyl concentrations transformed into the hydroxyl radicals, which attack and degrade the pollutants. Also, during indirect oxidation through chlorine oxidant that is generated at the anode in accordance with Eq. (8.3), pH influences the chlorine speciation. At pH values lower than about 3.0, Cl_2 represents the primary active chloro species, while at higher pH values, by disproportionate reaction Cl_2 transported from the anode into water forms HClO at pH < 7.5 in according with Eq. (8.4) and ClO^- at pH > 7.5 (Eq. (8.5)). It is well-known that chlorine is the stronger oxidant in comparison with HClO, which suggests strong acidic conditions [67].

The matrix of the water/wastewater, in respect to the nature and concentration of electrolyte, the ranges of target pollutant concentrations, and pH value, also affects the electrochemical oxidation process. Based on the specificity of the electrochemical process to transport the electrical charge, the higher concentration of electrolyte, the higher conductivity and consequently, the lower cell voltage from reducing the ohmic drop, allows reducing energy coonsumption for a given current density. For this reason, electrochemical oxidation treatment is more convenient and cost-effective for water/wastewater characterized by high salinity.

8.5 Electrochemical Filtering Process: A Hybrid Process Based on Electrooxidation and Filtering

8.5.1 Introductory Aspects

The electrochemical filtering technology is a new trend in advanced water treatment that combines two advanced treatment processes with well-established steps within the advanced wastewater/water treatment technological train (see Figure 8.4), one that belongs to the separation category (sorption, catalysis) and the other, electro(catalytic)oxidation, to the AOP category.

The integration of the unitary filtering process with advanced electro(catalytic)oxidation processes creates a hybrid process that greatly improves global water quality with focus on the challenges for practical application given by all conventional and emerging indicator parameters of water quality [68]. This approach not only combines the advantages of the constituent treatment technologies but also eliminates the drawbacks. The concentrate from wastewater/water treated by filtration process has a relatively high concentration, which can reduce mass transfer limitations, but on the other hand, the sorbent concentration increases during the filtering process, which raises the conductivity of the whole system, leading to a significant reduction in energy consumption [69, 70].

It is well-known that the filtering process is one of the most effective methods for removing small molecules and low concentrations of organic pollutants from water and also from wastewater [71], and is often used as a tertiary treatment to produce high quality water that

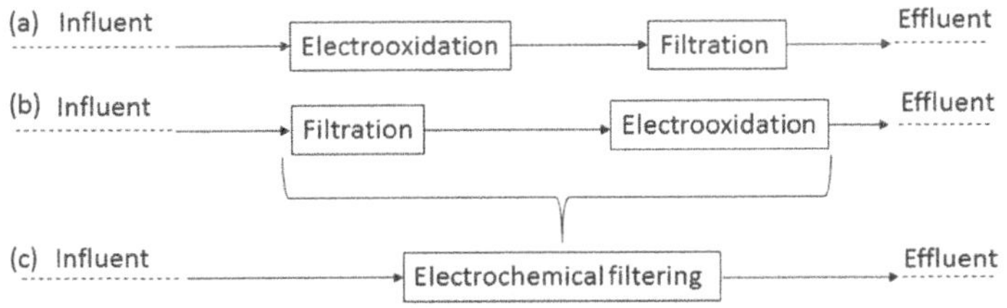

Figure 8.4 Simple schematic representation of the electrochemical filtering process derived from electrooxidation and filtering processes integrated in the advanced water/wastewater treatment technological flow.

is suitable for recycling or reclamation after removing organic contaminants, suspended solids, and other substrates by biological treatment [72–74].

The conventional filtering methods, however, exhibit drawbacks related to the sorbent/conventional filtering fouling and rejection during the course of filtration, due to the retention and accumulation of pollutants on the sorbent surface or inside the sorbent pores [75]. Also, large concentrate volumes generated during filtration are considered serious problems in water treatment/reclamation plants employing filtration processes [76, 77], which unavoidably debases sorbent/conventional filtering performance and life [78]. In fact, disposal of concentrates has become a serious problem in some places where no large surface water is available [76]. Although studies have suggested many approaches to minimize sorbent fouling and discharge options for the concentrate generated during treatment [76, 79], these problems still persist especially in drinking water treatment plants employing the conventional filtering/membranes/sorbents systems. Moreover, filtering is pointedly a physical process of retention and adsorption, which means that organic pollutants are not actually degraded. The need for pure drinking water and growing concerns about the presence of emerging contaminants in drinking water are serious challenges for practical application of the conventional filtering treatments in drinkable water treatment plants [80]. It is impossible, therefore, to achieve pure drinking water or qualitatively treated wastewater for reuse purpose only by filtration.

Advanced electro(catalytic)oxidation processes are a group of emerging technologies based on electrooxidation, which exhibit specific electrochemically favorable characteristics, such as no chemical reagents needed, easy process control, stable performance, and environmental friendliness, and they can decompose the organic pollutants to lower refractory products and even mineralize them to CO_2, H_2O, and other inorganic species [3, 20].

Although advanced electro(catalytic)oxidation processes are effective methods for removal of organic contaminants, several drawbacks remain for each individual electro(catalytic)oxidation process to scale up for large-scale industrial applications, e.g. the weak mass transfer of the pollutant molecules in the reactor, the energy consumption still at a relatively high level, and limited feasibility for treating large volumes of organic containing water.

Taking into account the development in water quality requirements related to emerging pollutants and corroborated with the main advantages and disadvantages of the individual filtering and electro(catalytic) oxidation processes, the hybrid process of electrofiltering should remove/degrade organic compounds from both conventional, priority, and emerging organic pollutants from any types of water (drinking water source, wastewater) [81, 139].

From the electrokinetic point of view, electrophoresis and electroosmosis are the two major phenomena that control the electrofiltering process [82]. In addition to these phenomena, various electrochemical reactions can also occur, which can influence negatively the efficiency of electrofiltering, due to an unwanted gas generation at the electrode surface [83, 84]. Also, the side reactions of water electrolysis are considered, which influence the pH of the concentrate and/or permeate during electrofiltering.

Although most of the experimental parameters that may affect the process are known, due to the presence of different electrokinetic effects, their probable and simultaneous interactions as well as their effects on permeate and concentrate characteristics, are still not fully understood for the electrofiltering process [85, 86].

8.5.2 Electrochemical Filtering System as a Unitary Process

The flexibility of these systems allows different combination strategies for distinct stages of electro(catalytic)oxidation processes and filtering configurations (see Figure 8.4):

a) Pretreatment stages, where the electro(catalytic)oxidation processes are applied before filtration, for reducing sorbent fouling by oxidizing organic pollutants that can be naturally or anthropically derived [87] and other pollutants that could foul or block the sorbent surface/pores, with the consequence of extending the operational time of the filtering process [88, 89].
b) Post-treatment stages where the electro(catalytic)oxidation processes are applied after the filtering (sorption) process, with the aim of mineralization of the sorbent concentrate before discharging into the environment, or treating permeate water to improve its quality for reuse [74, 80, 89]. The sorption process involves filtering to concentrate salts and organic pollutants in the sewage, with clean water exiting the system. Increased salinity improves conductivity of the sorbent concentrate, while high contaminant concentration levels reduce the diffusion limit, which ensures high degradation efficiency during the electro(catalytic)oxidation [90].

The hybrid process composed of the electro(catalytic)oxidation and filtration process are designed in the same unit, as simultaneous treatment where separation and degradation of pollutants occur concurrently.

The most important advantages of the integration of these two technologies relate to:

- The ability of the filtering/sorbent to retain compounds of different sizes with different physical and chemical properties.
- The improvement of the treatment efficiency and the filtering flow, due to the presence of an electrical field and the concentrate load of the sorbent/filter surface or the pores that can be more efficiently degraded under the electro(catalytic)oxidation, when the filtering media/bed is used as anode (named particulate) [91, 92].
- The ability to reduce the concentration load of the sorbent and slow the fouling rate.
- Self-cleaning sorbent capability or the presence of suspended catalyst and/or photocatalyst via in situ oxidative species generation through electrochemical process occurrence, leading to the recovery of water flow.
- Flexibility regarding the configuration of the cell/reactor and the geometry of the electrodes decreasing the construction expense due to its compact structure and small footprint [74, 80, 93].
- The high-volume rate of the integrated hybrid system that favors increase of the mass transfer coefficient in the liquid phase, as the flow of the feed solution drives the contaminants toward the filter electrode surface [141].
- The reduced energy consumed due to the large electroactive area that helps the process have a good performance at low current input [94].

8.5.2.1 Classification of the Electrochemical Filtering System

A main classification of the electrochemical filtering reactor/cell configuration is related to the integration or not of a supplementary so-called particulate electrode besides the anode and cathode that assures the electrical charge transportation (see Figure 8.5). If a

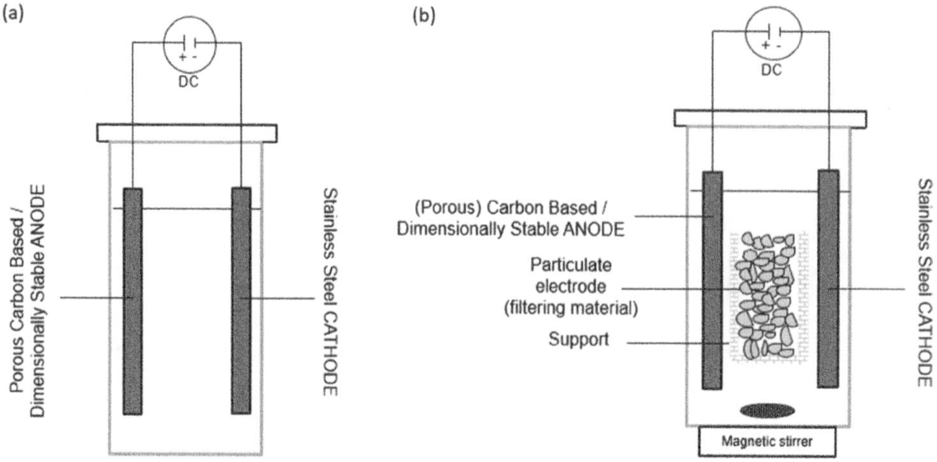

Figure 8.5 Simple schematic representation of the classification of electrochemical filtering process. (a) Porous anode (3D) (b) particulate electrode integration.

simple electrochemical cell configuration that consisted of the anode and the cathode is considered (see Figure 8.5a), it is very important the anode structure be 3D, which can be assured by the large porosity. In this situation, the anode has the role of pollutant sorption on the anode surface, which preconcentrates the local concentration of the pollutant at the electrode surface, improving the pollutant mass transport. The integration of the supplementary filtering system between anode and cathode with an applied electric field leads to a real combination of sorption/filtering and the electro(catalytic) oxidation process (see Figure 8.5b).

As for the improvement of any electrochemical process, the selection of the material of the electrode and the configuration of the reactor are very important.

For the choice of the electrode type, the electrode composition, geometry. and surface morphology are taken into account, since they determine the electrocatalytic activity, active electrode area, stability, service life, and mass transport characteristics. From this perspective, the electrochemical filtering system performance and the cost of the process are influenced by the electrode material. Also, the combination of these properties in a manner that improves all functional characteristics that dictate sorption and electrocatalytic properties toward organic pollutants removal and mineralization from water should constitute the key for the choice of 3D anode [95]. Nanostructured carbon-based electrodes, especially carbon-based nanotubes and carbon-based nanofiber material, can play the role of 3D anode, due to their porous 3D networks, which exhibit at the same time specific characteristics of both the adsorption process and the electrooxidation process, solving the technical challenge associated with the removal of water contaminants.

CNT as an emerging filter electrode material with the combined advantageous properties of stability, flexibility, chemical resistivity, and large specific area has been reported [143]. The CNT-based electrochemical filter combines separation and oxidation technology into one effective step. A CNT-based electrochemical filter has been reported to be effective for the adsorptive removal and oxidation of aqueous organic pollutants, salts [96], proteins [97], viruses [98], azo dyes [99], pharmaceuticals [100], and phenol [101], characterized by a typical removal efficiency higher than 90%. The fabrication of these filter systems, however, and application to water treatment is still at a premature stage, and numerous critical issues are yet to be effectively addressed. Commercial readiness, improving manufacturing scalability, and cutting down the CNTs' cost and potential toxic effects are the potential challenges to be overcome [102].

Although CNF shares a similar architectural skeleton to CNT, scientists have overlooked the potential of CNF for water treatment. For example, although CNF has been widely studied as a membrane for gas separation, only a few reports have been published on CNF-based water treatment applications. The main difference between CNF and CNT is diameter, and in the case of randomly oriented fibers, the size of the pores is larger. Larger CNF diameters will result in larger pores and a longer pore diffusion time. The overall cost of CNF is two orders of magnitude less than that of SWCNT as it is easier to produce and process. Development of a novel CNF-based water filtration system that would result in similar or even better performance than a CNT-based filter system is therefore important. Based on the limitations and drawbacks of the nanostructured carbon related to the toxicity to human health, it is important that these materials be supported on a substrate assuring the electrode stability. Their agglomeration should also be avoided to improve the surface area to favor the sorption process.

In the context of water treatment, graphene-based materials offer certain advances over CNTs [140]. Graphene-based electrochemical filters have been reported to exhibit excellent performance for physical sorption and electrochemical oxidization of organic contaminants, such as tetracycline, phenol, and oxalate, when used as the anode in the filter system for the electrochemical advanced oxidation process [103].

DSA can be considered as another type of 3D anode, if it possesses a porous matrix, which can improve the mass transfer at the electrode surface, influence the flow regime, and affect the pressure drop. The composition and surface morphology of 3D DSA electrodes should also improve the oxidation ability. DSAs consist of corrosion resistant material, such as titanium coated with a layer of metal oxides. Several types of DSAs have been tested (e.g. PbO_2, TiO_2, SnO_2, RuO_2, and IrO_2) due to high electrocatalytic activity, relatively low expense, and relatively long lifetime [104, 105, 142], but some of them have shown loss of activity due to surface fouling or limited service life [106–108].

Because the electrooxidation process is a main component of the electrochemical filtering system, active and nonactive electrodes are considered, as well. Active anodes (RuO_2) are characterized by a strong interaction of electrogenerated hydroxyl radicals and a low overpotential for O_2 evolution and show a reduced ability for conversion of organic contaminants. Nonactive electrodes (SnO_2) have an exquisite capacity to oxidize the organic contaminants, because they exhibit a lower electroactivity for O_2 evolution, which means a high overpotential of oxygen evolution reaction [109, 110].

Besides the porous carbon or DSA electrodes, a BDD electrode, a nonactive type, which interacts very weakly with ·OH radicals and acts mainly by direct oxidation of the organic pollutants [111], is considered for developing the electrochemical filtering system. BDD is an ideal electrode for water treatment, due to its properties, i.e. high stability, good mechanical strength, a large electrochemical window, good corrosion resistance, long life, high degradation/mineralization rate, and current efficiency [61, 112].

For scale-up of the electrodes, the manufacturing and process costs are the limitations for practical application [113]. In addition, the main limitations of anode materials are related to the selection of suitable substrate materials for BDD synthesis and its difficulty in maintaining stability during electrolysis [89, 114].

Integration of the fluidized bed filtering system between the anode and cathode, called a **particulate electrode (PE)**, overcomes some intrinsic drawbacks/limitations related to low mass transfer, small space–time yield, low area–volume ratio, and the temperature increase of 2D conventional electrochemical reactors that are generally used for electro(catalytic) oxidation processes, especially for water characterized by low conductivity. Also, by PE composition, which is basically granular, or fragmental material, the range of contaminants potentially removed could be extended and sorbent materials electrochemically regenerated. The selection of anodes and cathodes in a 3D electrode system is generally determined by the water/wastewater characteristics and the suitable treatment methods (electrooxidation, electro-Fenton, electrocoagulation) [112].

The composition of a PE plays an important role and deserves more efforts toward its development. Properties including large specific area for good sorption capacity, high electrocatalytic activity, and good conductivity, which can increase the current efficiency, are required for good particulate electrode materials. The catalyst present within the PE composition should promote the heterogeneous characteristics of the PE and should significantly contribute to the enhanced performance of a 3D process. PE characteristics are related to many complex processes that should occur (adsorption/desorption, oxidation/catalysis) including high specific surface area, good catalytic activity, and stability. A 3D electrode system, however, is more complicated than a 2D electrode system, and each part is closely linked in practical application. Small changes may induce differences in removal efficiency, such as the electrode position, aeration mode, and electrolyte concentration. A PE may also lose its sorption capacity and catalytic activity due to the accumulation of pollutants on the particle surface during continuous runs. Good results have been reported for organic pollutants and heavy metal removal from wastewater using various types of 3D electrode systems in relation with the anode material and PE composition [115–117].

In general, the activated carbon characterized by good sorption capacity can be integrated into a 2D system to construct a 3D electrode system. Due to its current cost of manufacturing, activated carbon will remain the most prominent carbon-based water filtration material [118]. An activated carbon particulate electrode was reported to improve a 3D system, being used as catalyst for the generation of hydrogen peroxide from the Fenton reaction by oxygen reduction to produce more hydroxyl radicals [119].

The degradation mechanism includes the direct oxidation on the anode surface and smart management of green engineering for the indirect oxidation through oxidants generated in the bulk solution. The mechanism for removal of pollutants in 3D electrochemical process, however, is very complex. The electrode material, the type of particulate electrode, and the various organic pollutants contained in water directly influence the degradation/mineralization mechanism. The electrochemical reactions on the electrodes, however, are the same as in a 2D system, except the role of the PE, which can behave as a sorbent, helping the ionic charge distribution, without a direct electric contact [120].

Granular activated carbon (GAC) is typically used as a PE, due to its good electrical conductivity and high surface area, enhancement of mass transfer, and sorption of the contaminants in the 3D electrochemical system. In a lot of studies on hybrid GAC–electrochemical systems, though, the GAC-PE was positioned between anode and cathode; the direct polarization of the PE at higher current and potential can cause its corrosion and worsen the treatment performance. In the presence of a large electric field, the GAC particles can be polarized without being in contact with the anode and cathode by shift of electric charge of these electrodes, converting each GAC particle into a charged microelectrode with one anodic and one cathodic side, inducing the electrosorption of pollutants and enabling the degradation of the adsorbed organic pollutants [116, 121–123].

The 3D reactors can be configured in two modes: the fixed bed electrode and fluidized bed electrode. In a fixed bed 3D reactor, the PE is positioned between the anode and cathode, overcoming some limitations frequently encountered in a 2D system (mass transfer limitation, small space–time yield, low area–volume ratio, and temperature increase). A drawback of this kind of system, however, is the need for periodic regeneration of the PE to keep its activity due to the pollutants always adsorbed or deposited on the electrode. The fluidized bed reactor was proposed to obtain a uniform temperature and a large interfacial area in the reactor. Fluidized bed with air injection is common for the electrochemical reactor to enhance mass transfer [112].

8.6 Integration of Electrooxidation-based Processes in Water/Wastewater Treatment Technological Flow

Even if the interest in the development of the electrooxidation process in water treatment has been growing, there are few examples of its application in full-scale water treatment technologies [110]. Nevertheless, the electrooxidation process is considered an emerging technology in the best available techniques (BATs) reference document for the treatment of wastewater in the chemical sector [124].

In addition, good efficiency of electrooxidation has been reported for industrial wastewater, for degrading several types of organic contaminants, e.g. pesticides [125], textile dyes [126], landfill leachate [127], pharmaceuticals [128], and explosive chemicals [129].

Electrooxidation can be applied either stand alone, because it can be developed as a compact module completely automated, or easily combined with other unitary processes to develop advanced technological flow for the treatment of wastewater. Considering the EO mechanism that governs the overall EO process through direct or indirect electrolysis, which defines the electrochemical conversion or incineration, the combination with a biological stage is commonly considered (see Figure 8.6). Applying before the biological stage results in the electrochemical conversion of the biorefractory organic pollutants into biodegradable subproducts, which can be further degraded until

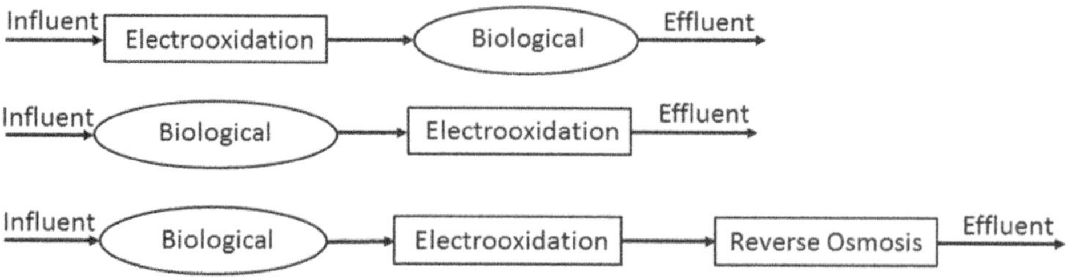

Figure 8.6 Integration of the electrooxidation process within the wastewater treatment technological flow.

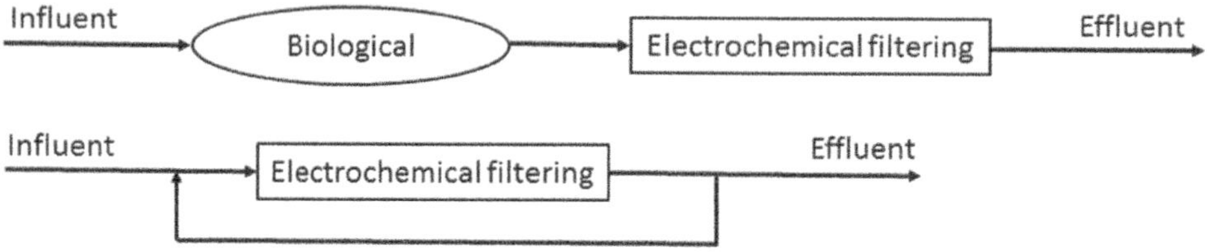

Figure 8.7 Integration of the electrochemical filtering process within the wastewater treatment technological flow.

their mineralization by the biological process, through the technological solution that can be optimized to achieve technical economic feasibility. EO placed after the biological stage should assure the finishing stage of the wastewater treatment flow from the point of view of the organic pollutants; furthermore, if the emission limit values are more restrictive with respect to wastewater reuse, further adding of another advanced unitary process (e.g. reverse osmosis) can be considered for removal of other types of contaminants which cannot degrade by EO (e.g. chloride or sulphate).

The limitation of mass transport in electrooxidation is one of the key obstacles for full-scale implementation of electrochemical oxidation. The electrochemical filtering process through 3D anodes and/or PEs can be employed in the flow-through mode to improve the treatment performance by mass transfer enhancement. Taking into account the combination of sorption/filtering with the electrooxidation process to develop the electrochemical filtering process, this process should be effective alone or combined with biological treatment in the tertiary advanced stage of wastewater treatment (Figure 8.7). Granular activated carbon, considered as a mature technology for water and wastewater treatment, has been studied for the development of the electrochemical filtering system for wastewater treatment, considering its suitability for retaining halogenated organic subproducts generated in the presence of chloride in the wastewater matrix [110].

Interest in the application of the electroxidation process in drinking water treatment has been growing especially toward the disinfection stage through chlorine generation at the anode, but also, for ammonium oxidation through electrochemical nitrification/denitrification and for natural organic matter degradation using boron-doped diamond electrodes [62, 130]. Considering surface and ground water as the main drinking water sources, with respect to their conventional treatment, electrooxidation can be integrated as a finishing stage as post-oxidation and coupled with reverse osmosis to get high quality treated drinking water, or as a pre-oxidation stage assuring the electrochemical conversion of the natural organic matter into byproducts, which are removed by the conventional processes of drinking water treatment (e.g. filtering, sorption) (see Figure 8.8). Taking into account the major limitations related to the slow mass transfer, the low conductivity of drinking water sources, and the necessity to couple electrooxidation with other conventional processes such as filtering, it is expected that electrochemical filtering can overcome these limitations and be appropriate either solely or coupled with reverse osmosis for drinking water treatment (see Figure 8.9).

More widespread applications of electrooxidation require advances in electrode materials and cell configuration to avoid electrode fouling during operation and to address the large spectrum of pollutants to be removed from water in any type of water body. Several examples related to the application of the electrocatalyst types as anodes in electrooxidation-based processes for removal of various organic pollutants from water are gathered in Table 8.1. It is obvious that from an technical economicidizd point of view, the process performance is linked directly to the electrode type, reactor configuration, and operating conditions.

All the discussed characteristics of the electrooxidation-based processes show them to be particularly well suited for decentralized water treatment, which is more and more accepted as the solution in sustainable water management. The design of the electrooxidation technology as a compact module that can be easily automated and remotely controlled, which make it suitable to be adjusted to variations in influent water composition, make it more favored than other conventional water treatment processes.

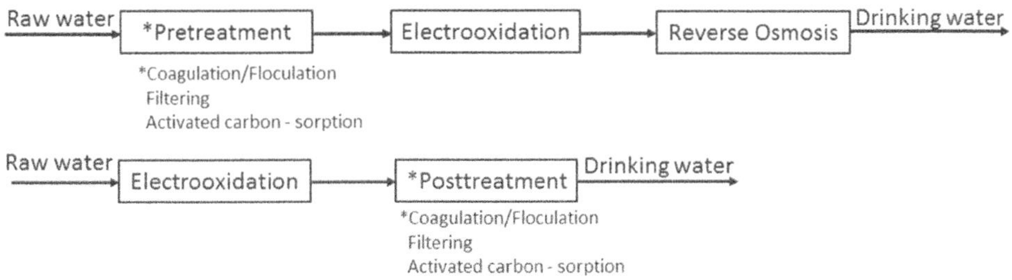

Figure 8.8 Integration of the electrooxidation process within the drinking water treatment technological flow.

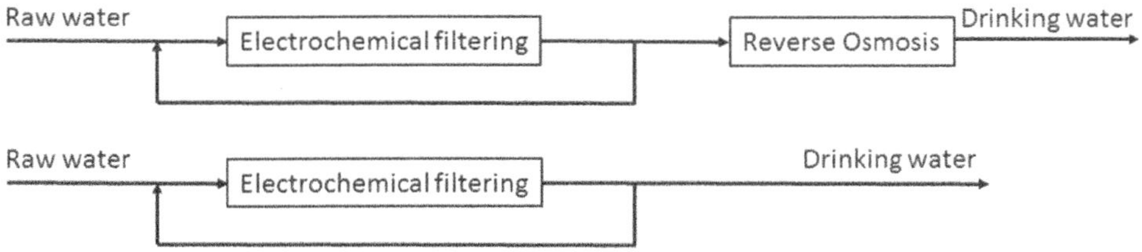

Figure 8.9 Integration of the electrochemical filtering process within the drinking water treatment technological flow.

Table 8.1 Electrocatalyst performances for removal of organic pollutants from water.

Anode	Pollutant	Process efficiency						Reference
		Mineralization efficiency, %	Current density, mA/cm^2	Time, min	Current efficiency, %	Power consumption	Cost of energy	
Graphite sheet (1104 cm^2)	Organic pollutants from textile effluents	93.3	4.5	150	12.3	293 kWh/kg COD	-	[131]
Graphite rod (918 cm^2)	Organic pollutants from textile effluents	90.9	4.5	150	11.9	268 kWh/kg COD	-	
BDD (70cm^2)	Aniline	91.36	33	240	-	1140 kWh/kg COD	32 euro/m^3	[132]
BDD (370 000 cm^2)	1.4-dioxane	40	-	-	-	154 kWh/kg COD	*5.2 Euro/m^3; **14 Euro/m^2	[133]
BDD	Organic pollutants from petrochemical industry	90	30	60	-	11 kWh/kg COD	-	[134]
Activated carbon powder PAC-3D electrode	Carbamazepine	89.8	9	10	15.65	70 kWh/kg TOC	-	[135]
Bismuth doped tin oxide coated CNT	Oxalate	93	5.7	50	-	25.7 kWh/kgCOD	-	[136]
Antimony doped tin oxide coated CNT	Oxalate	90	5.7	50	-	26.6 kWh/kg COD	-	

RuO2/IrO2/ TaO2-Ti (791 cm^2)	Organic pollutants from textile effluents	54.5	4.5	150	7.1	344 kWh/kg COD	-	[131]
Ti /RuO2 (107 cm^2)	Organic pollutants from petrochemical industry	100	30	***60	-	40.2 kWh/ kg COD	37.8 USS/kg COD	[137]
				****20	-	29.8 kWh / kg COD	28USS-/kg COD	
Ti/IrO_2-TaO_5	Organic pollutants from petrochemical industry	75	30	240	-	13 kWh/kg COD	-	[134]

-not reported
*price / EO unit
**price /BBD unit, for a wastewater plant capacity of 43 800m^3/y
***before flotation
****after flotation

Acknowledgements

This work was supported partially by a grant of the Romanian Ministry of Research and Innovation, CCDI-UEFISCDI, project code PN-III-P2-2.1-PED-2019-4492, contract number 441PED/2020 (3DSAPECYT), within PNCDI III, and partially by project "Program intern de stimulare si recompensare a activitatii didactice", contract number 10161/11.06.2021.

References

1 Wen, Y., Schoups, G., and van de Giesen, N. (2017). Organic pollution of rivers: combined threats of urbanization, livestock farming and global climate change. *Scientific Reports* 7: 43289. doi: 10.1038/srep43289.

2 Tak, S. and Vellanki, B.P. (2018). Natural organic matter as precursor to disinfection byproducts and its removal using conventional and advanced processes: state of the art review. *Journal of Water and Health* 16: 681–703. doi: 10.2166/wh.2018.032.

3 Moreira, F.C., Boaventura, R.A.R., Brillas, E. et al. (2017). Electrochemical advanced oxidation processes: a review on their application to synthetic and real wastewaters. *Applied Catalysis B: Environmental* 202: 217–261. doi: 10.1016/j.apcatb.2016.08.037.

4 Kapałka, A., Foti, G., and Comninellis, C. (2010). Basic principles of the electrochemical mineralization of organic pollutants for wastewater treatment. In: *Electrochemistry for the Environment* (ed. C. Comninellis and G. Chen), 1–23. New York: Springer. doi: 10.1007/978-0-387-68318-8_1.

5 Rajeshwar, K. and Ibanez, J.G. (1997). *Environmental Electrochemistry: Fundamentals and Applications in Pollution Sensors and Abatement*. San Diego: Academic Press.

6 Chen, G. (2004). Electrochemical technologies in wastewater treatment. *Separation and Purification Technology* 38: 11–41. doi: 10.1016/j.seppur.2003.10.006.

7 Rajeshwar, K., Ibanez, J.G., and Swain, G.M. (1994). Electrochemistry and the environment. *Journal of Applied Electrochemistry* 4: 1077–1091. doi: 10.1007/BF00241305.

8 Walsh, F.C. (2001). Electrochemical technology for environmental treatment and clean energy conversion. *Pure and Applied Chemistry* 73: 1819–1837. doi: 10.1351/pac200173121819.

9 Comninellis, C. (1994). Electrocatalysis in the electrochemical conversion/combustion of organic pollutants for waste water treatment. *Electrochimica Acta* 39: 1857–1862. doi: 10.1016/0013-4686(94)85175-1.

10 Yoshida, K., Yoshida, S., Seki, Y. et al. (2007). Basic study of electrochemical treatment of ammonium nitrogen containing wastewater using boron-doped diamond anode. *Environmental Research* 65: 71–73.

11 Feng, L., van Hullebusch, E.D., Manuel, A. et al. (2013). Removal of residual anti-inflammatory and analgesic pharmaceuticals from aqueous systems by electrochemical advanced oxidation processes. A review. *Chemical Engineering Journal* 22: 8944–8964. doi: 10.1016/j.cej.2013.05.061.

12 Panizza, M. and Cerisola, G. (2009). Direct and mediated anodic oxidation of organic pollutants. *Chemical Review* 109: 6541–6569. doi: 10.1021/cr9001319.

13 Miklos, D.B., Remy, C., Jekel, M. et al. (2018). Evaluation of advanced oxidation processes for water and wastewater treatment – critical review. *Water Research* 139: 118–131. doi: 10.1016/j.watres.2018.03.042.

14 Savariraj, A.D., Mangalaraja, R.V., Prabakar, K. et al. (2020). Electrochemical aspects for wastewater treatment. In: *Green Methods for Wastewater Treatment* (ed. M. Naushad, S. Rajendran, and E. Lichtfouse), 121–148. *Environmental Chemistry for a Sustainable World* 35. Springer, Cham. doi: 10.1007/978-3-030-16427-0_6.

15 Caliari, P.C., Pacheco, M.J., Ciríaco, L. et al. (2019). Tannery wastewater: organic load and sulfide removal dynamics by electrochemical oxidation at different anode materials. *Environmental Technology and Innovation* 14: 100345. doi: 10.1016/j.eti.2019.100345.

16 Chen, X., Chen, G., and Yue, P.L. (2001). Stable Ti/IrOx-Sb_2O_5-SnO_2 anode for O_2 evolution with low Ir content. *Journal of Physical Chemistry B* 105: 4623–4628. doi: 10.1021/jp010038d.

17 Makgae, M.E., Theron, C.C., Przybylowicz, W.J. et al. (2005). Preparation and surface characterization of Ti/SnO_2-RuO_2-IrO_2 thin films as electrode material for the oxidation of phenol. *Materials Chemistry and Physics* 92: 559–564. doi: 10.1016/j.matchemphys.2005.02.022.

18 Moradi, M., Vasseghian, Y., Khataee, A. et al. (2020). Service life and stability of electrodes applied in electrochemical advanced oxidation processes: a comprehensive review. *Journal of Industrial and Engineering Chemistry* 87: 18–39. doi: 10.1016/j.jiec.2020.03.038.

19 Zhao, W., Xing, J., Chen, D. et al. (2015). Study on the performance of an improved Ti/SnO_2-Sb_2O_3/PbO_2 based on porous titanium substrate compared with planar titanium substrate. *RSC Advances* 5: 26530–26539. doi: 10.1039/C4RA13492C.

20 Martinez-Huitle, C.A., Rodrigo, M.A., Sires, I. et al. (2015). Single and coupled electrochemical processes and reactors for the abatement of organic water pollutants: a critical review. *Chemical Review* 115: 13362–13407. doi: 10.1021/acs.chemrev.5b00361.

21 Sanchez-Montes, S., Perez, J.F., Saez, C. et al. (2020). Assessing the performance of electrochemical oxidation using DSA® and BDD anodes in the presence of UVC light. *Chemosphere* 238: 124575. doi: 10.1016/j.chemosphere.2019.124575.

22 Ganiyu, S. (2016). Electrochemical Advanced Oxidation Processes for removal of Pharmaceuticals from water: performance studies for sub-stoichiometric titanium oxide anode and hierarchical layered double hydroxide modified carbon felt cathode. *Geophysics [physics.geo-ph]*.

23 Wanekaya, A.K. (2011). Applications of nanoscale carbon-based materials in heavy metal sensing and detection. *Analyst* 136: 4383–4391. doi: 10.1039/C1AN15574A.

24 Bashir, M.J.K., Isa, M.H., Kutty, S.R.M. et al. (2009). Landfill leachate treatment by electrochemical oxidation. *Waste Management* 29: 2534–2541. doi: 10.1016/j.wasman.2009.05.004.

25 Dreyer, D.R., Park, S., Bielawski, C.W. et al. (2010). The chemistry of graphene oxide. *Chemical Society Reviews* 39: 228–240. doi: 10.1039/B917103G.

26 Rowley-Nealea, S.J., Randviir, E.P., Abo Dena, A.S. et al. (2018). An overview of recent applications of reduced graphene oxide as a basis of electroanalytical sensing platforms. *Applied Materials Today* 10: 218–226. doi: 10.1016/j.apmt.2017.11.010.

27 Biswas, C. and Lee, Y.H. (2011). Graphene versus carbon nanotubes in electronic devices. *Advanced Functional Materials* 21: 3806–3826. doi: 10.1002/adfm.201101241.

28 Li, F., Yu, Z., Han, X., and Lai, R.Y. (2019). Electrochemical aptamer-based sensors for food and water analysis: a review. *Analytica Chimica Acta* 1051: 1–23. doi: 10.1016/j.aca.2018.10.058.

29 Liu, Y., Zhou, J., Zhang, X. et al. (2009). Synthesis, characterization and optical limiting property of covalently oligothiophene-functionalized graphene material. *Carbon* 47: 3113–3121. doi: 10.1016/j.carbon.2009.07.027.

30 ElMekawy, A., Hegab, H.M., Losic, D. et al. (2017). Applications of graphene in microbial fuel cells: the gap between promise and reality. *Renewable and Sustainable Energy Reviews* 72: 1389–1403. doi: 10.1016/j.rser.2016.10.044.

31 Shestakova, M. and Sillanpää, M. (2017). Electrode materials used for electrochemical oxidation of organic compounds in wastewater. *Reviews in Environmental Science and Bio-technology* 16: 223–238. doi: 10.1007/s11157-017-9426-1.

32 Wiratini, N.M., Triyono, T., Trisunaryanti, W. et al. (2021). Graphite/NiO/Ni electrode for electro-oxidation of the remazol black 5 dye. *Bulletin of Chemical Reaction Engineering & Catalysis* 16: 847–856. doi: 10.9767/bcrec.16.4.11702.847-856.

33 Du, X., Oturan, M.A., Zhou, M. et al. (2021). Nanostructured electrodes for electrocatalytic advanced oxidation processes: from materials preparation to mechanisms understanding and wastewater treatment applications. *Applied Catalysis B: Environmental* 296: 120332. doi: 10.1016/j.apcatb.2021.120332.

34 Fudala-Ksiazek, S., Sobaszek, M., Luczkiewicz, A. et al. (2018). Influence of the boron doping level on the electrochemical oxidation of raw landfill leachates: advanced pre-treatment prior to the biological nitrogen removal. *Chemical Engineering Journal* 334: 1074–1084. doi: 10.1016/j.cej.2017.09.196.

35 Medeiros De Araujo, D., Canizares, P., Martinez-Huitle, C.A. et al. (2014). Electrochemical conversion/combustion of a model organic pollutant on BDD anode: role of sp^3/sp^2 ratio. *Electrochemistry Communications* 47: 37–40. doi: 10.1016/j.elecom.2014.07.017.

36 Pierpaoli, M., Jakobczyk, P., Sawczak, M. et al. (2021). Carbon nanoarchitectures as high-performance electrodes for the electrochemical oxidation of landfill leachate. *Journal of Hazardous Materials* 40: 123407. doi: 10.1016/j.jhazmat.2020.123407.

37 Goh, K., Karahan, H.E., Wei, L. et al. (2016). Carbon nanomaterials for advancing separation membranes: a strategic perspective. *Carbon* 109: 694–710. doi: 10.1016/j.carbon.2016.08.077.

38 Mauter, M.S. and Elimelech, M. (2008). Environmental applications of carbon-based nanomaterials. *Environmental Science & Technology* 42: 5843–5859. doi: 10.1021/es8006904.

39 Acquah, S.F.A., Penkova, A.V., Markelov, D.A. et al. (2017). Review—the beautiful molecule: 30 years of C_{60} and its derivatives. *ECS Journal of Solid-State Science and Technology* 6: M3155–M3162. doi: 10.1149/2.0271706jss.

40 Lim, E.K., Kim, T., Paik, S. et al. (2015). Nanomaterials for theragnostic: recent advances and future challenges. *Chemical Reviews* 115: 327–394. doi: 10.1021/cr300213b.

41 Kour, R., Arya, S., Young, S.J. et al. (2020). Review—recent advances in carbon nanomaterials as electrochemical biosensors. *Journal of the Electrochemical Society* 167: 037555. doi: 10.1149/1945-7111/ab6bc4.

42 Valcarcel, M., Cardenas, S., Simonet, B.M. et al. (2008). Carbon nanostructures as sorbent materials in analytical processes. *TrAC. Trends in Analytical Chemistry* 27: 34–43. doi: 10.1016/j.trac.2007.10.012.

43 Kang, I., Heung, Y.Y., Kim, J.H. et al. (2006). Introduction to carbon nanotube and nanofiber smart materials. *Composites: Part B* 37: 382–394. doi: 10.1016/j.compositesb.2006.02.011.

44 Li, R., Chang, X., Li, Z. et al. (2011). Multiwalled carbon nanotubes modified with 2-aminobenzothiazole modified for uniquely selective solid-phase extraction and determination of Pb (II) ion in water samples. *Microchimica Acta* 172: 269–276. doi: 10.1007/s00604-010-0488-9.

45 Ghasemi, M., Wan Daud, W.R., Hassan, S.H.A. et al. (2013). Nano-structured carbon as electrode material in microbial fuel cells: a comprehensive review. *Journal of Alloys and Compounds* 580: 245–255. doi: 10.1016/j.jallcom.2013.05.094.

46 Erkoç, Ş. (2000). Structural and electronic properties of carbon nanotubes. *International Journal of Modern Physics C* 11: 175–182. doi: 10.1021/jp993592k.

47 Kim, Y.A., Hayashi, T., Endo, M. et al. (2013). Carbon nanofibers. In: *Springer Handbook of Nanomaterials* (ed. R. Vajtai), 233–262. Springer, Berlin, Heidelberg: Springer Handbooks. doi: 10.1007/978-3-642-20595-8_7.

48 Qin, Y.H., Yang, H.H., Zhang, X.S. et al. (2010). Effect of carbon nanofibers microstructure on electrocatalytic activities of Pd electrocatalysts for ethanol oxidation in alkaline medium. *International Journal of Hydrogen Energy* 35: 7667–7674. doi: 10.1016/j.ijhydene.2010.05.034.

49 Choi, Y.K., Sugimoto, K., Song, S. et al. (2005). Mechanical and physical properties of epoxy composites reinforced by vapor grown carbon nanofibers. *Carbon* 43: 2199–2208. doi: 10.1016/j.carbon.2005.03.036.

50 Kotaki, M., Wang, K., Toh, M.L. et al. (2006). Electrically conductive epoxy/clay/vapor grown carbon fiber hybrids. *Macromolecules* 39: 908–911. doi: 10.1021/ma0522561.

51 Cao, S. and Yu, J. (2016). Carbon-based H_2-production photocatalytic materials. *Journal of Photochemistry and Photobiology C: Photochemistry Reviews* 27: 72–99. doi: 10.1016/j.jphotochemrev.2016.04.002.

52 Profeti, D., Lassali, T.A.F., and Olivi, P. (2006). Preparation of $Ir_{0.3}Sn_{(0.7-x)}$ Ti_xO_2 electrodes by the polymeric precursor method: characterization and lifetime study. *Journal of Applied Electrochemistry* 36: 883–888. doi: 10.1007/S10800-006-9149-4.

53 Castanho, M., Malpass, G.R.P., and Motheo, A.J. (2006). Photoelectrochemical treatment of the dye reactive red 198 using DSA® electrodes. *Applied Catalysis B: Environment* 62: 193–200. doi: 10.1016/j.apcatb.2005.07.011.

54 Radjenovic, J., Escher, B.I., and Rabaey, K. (2011). Electrochemical degradation of the β-blocker metoprolol by $Ti/Ru_{0.7}Ir_{0.3}O_2$ and Ti/SnO_2-Sb electrodes. *Water Research* 45: 3205–3214. doi: 10.1016/j.watres.2011.03.040.

55 Costa, C.R., Botta, C.M.R., Espindola, E.L.G. et al. (2008). Electrochemical treatment of tannery wastewater using DSA® electrodes. *Journal of Hazardous Materials* 153: 616–627. doi: 10.1016/j.jhazmat.2007.09.005.

56 Ramalho, A.M.Z., Martínez-Huitle, C.A., and da Silva, D.R. (2010). Application of electrochemical technology for removing petroleum hydrocarbons from produced water using a DSA-type anode at different flow rates. *Fuel* 89: 531–534. doi: 10.1016/j.fuel.2009.07.016.

57 Liu, W., Farrington, G.C., Chaput, F. et al. (1996). Synthesis and electrochemical studies of spinel phase $LiMn_2O_4$ cathode materials prepared by the Pechini process. *Journal of Electrochemical Society* 143: 879–884. doi: 10.1149/1.1836552.

58 Xu, Y., Yuan, X., Huang, G. et al. (2005). Polymeric precursor synthesis of $Ba_2Ti_9O_{20}$. *Materials Chemistry and Physics* 90: 333–338. doi: 10.1016/j.matchemphys.2004.10.022.

59 Chen, X., Chen, G., Gao, F. et al. (2003). High-performance Ti/BDD electrodes for pollutant oxidation. *Environmental Science & Technology* 37: 5021–5026. doi: 10.1021/es026443f.

60 Panizza, M. (2010). Importance of Electrode Material in the Electrochemical Treatment of Wastewater Containing Organic Pollutants. In: *Electrochemistry for the environment.* (ed. C. Comninellis and G. Chen), 25–54. New York: Springer.

61 Sonia, B.D., Patel, U.D., Agrawal, A. et al. (2017). Application of BDD and DSA electrodes for the removal of RB 5 in batch and continuous operation. *Journal of Water Process Engineering* 17: 11–21. doi: 10.1016/j.jwpe.2017.01.009.

62 Vlaicu, I., Pop, A., Manea, F. et al. (2011). Degradation of humic acid from water by advanced electrochemical oxidation method. *Water Science & Technology Water Supply* 11: 85–95. doi: 10.2166/ws.2011.013.

63 Wang, Y.H., Chan, K.Y., Li, X.Y. et al. (2006). Electrochemical degradation of 4-chlorophenol at nickel-antimony doped tin oxide electrode. *Chemosphere* 65: 1087–1093. doi: 10.1016/j.chemosphere.2006.04.061.

64 Brillas, E. (2011). Fenton-electrochemical treatment of wastewater for the oxidation of organic pollutants using BDD. In: *Synthetic Diamond Films, Preparation, Electrochemistry, Characterization, and Applications* (ed. E. Brillas and C.A. Martínez-Huitle), 407–437. Hoboken, NJ: John Wiley & Sons, Inc.

65 Guinea, E., Centellas, F., Garrido, J.A. et al. (2009). Solar photoassisted anodic oxidation of carboxylic acids in presence of Fe^{3+} using a boron-doped diamond electrode. *Applied Catalysis B: Environment* 89: 459–468. doi: 10.1016/j.apcatb.2009.01.004.

66 Chin, C.J.M., Chen, T.Y., Lee, M. et al. (2014). Effective anodic oxidation of naproxen by platinum nanoparticles coated FTO glass. *Journal of Hazardous Materials* 277: 110–119. doi: 10.1016/j.jhazmat.2014.02.034.

67 Anglada, A., Urtiaga, A., and Ortiz, I. (2009). Contributions of electrochemical oxidation to waste-water treatment: fundamentals and review of applications. *Journal of Chemical Technology & Biotechnology* 84: 1747–1755. doi: 10.1002/jctb.2214.

68 Wang, Y., Zhu, J., Huang, H. et al. (2015). Carbon nanotube composite membranes for microfiltration of pharmaceuticals and personal care products: capabilities and potential mechanisms. *Journal of Membrane Science* 479: 165–174. doi: 10.1016/j.memsci.2015.01.034.

69 Li, C., Zhang, M., Song, C. et al. (2018). Enhanced treatment ability of membrane technology by integrating an electric field for dye wastewater treatment: a review. *Journal of AOAC International* 101: 1341–1352. doi: 10.5740/jaoacint.18-0050.

70 Yang, Y., Li, J., Wang, H. et al. (2011). An electrocatalytic membrane reactor with self-cleaning function for industrial wastewater treatment. *Angewandte Chemie-International Edition in English* 123: 2196–2198. doi: 10.1002/anie.201005941.

71 Comerton, A.M., Andrews, R.C., and Bagley, D.M. (2009). The influence of natural organic matter and cations on the rejection of endocrine disrupting and pharmaceutically active compounds by nanofiltration. *Water Research* 43: 613–622. doi: 10.1016/j.watres.2008.11.003.

72 Filloux, E., Gernjak, W., Gallard, H. et al. (2016). Investigating the relative contribution of colloidal and soluble fractions of secondary effluent organic matter to the irreversible fouling of MF and UF hollow fiber membranes. *Separation and Purification Technology* 170: 109–115. doi: 10.1016/j.seppur.2016.06.034.

73 Urtiaga, A. (2021). Electrochemical technologies combined with membrane filtration. *Current Opinion in Electrochemistry* 27: 100691. doi: 10.1016/j.coelec.2021.100691.

74 Wei, K., Cui, T., Huang, F. et al. (2020). Membrane separation coupled with electrochemical advanced oxidation processes for organic wastewater treatment: a short review. *Membranes* 10: 337. doi: 10.3390/membranes10110337.

75 Miller, D.J., Dreyer, D.R., Bielawski, C.W. et al. (2017). Surface modification of water purification membranes. *Angewandte Chemie-International Edition in English* 56: 4662–4711. doi: 10.1002/anie.201601509.

76 Nederlof, M.M., Van Passan, J.A.M., and Jong, R. (2005). Nanofiltration concentrate disposal: experience in Netherlands. *Desalination* 178: 303–312. doi: 10.1016/j.desal.2004.11.041.

77 Van der Bruggen, B., Cornelis, G., Vandecasteele, C. et al. (2005). Fouling of nanofiltration and ultrafiltration membranes applied for wastewater regeneration in textile industry. *Desalination* 175: 111–119. doi: 10.1016/j.desal.2004.09.025.

78 Jhaveri, J.H. and Murthy, Z.V.P. (2016). A comprehensive review on anti-fouling nanocomposite membranes for pressure driven membrane separation processes. *Desalination* 379: 137–154. doi: 10.1016/j.desal.2015.11.009.

79 Van der Bruggen, B., Lejon, L., and Vandecasteele, C. (2003). Reuse, treatment and discharge of the concentrate of pressure driven membrane processes. *Environmental Science and Technology* 37: 3733–3738. doi: 10.1021/es0201754.

80 Ganiyu, S.O., van Hullebusch, E.D., Cretin, M. et al. (2015). Coupling of membrane filtration and advanced oxidation processes for removal of pharmaceutical residues: a critical review. *Separation and Purification Technology* 156 (891): 914. doi: 10.1016/j.seppur.2015.09.059.

81 Sirés, I., Brillas, E., Oturan, M.A. et al. (2014). Electrochemical advanced oxidation processes: today and tomorrow. A review. *Environmental Science and Pollution Research* 21: 8336–8367. doi: 10.1007/s11356-014-2783-1.

82 Enevoldsen, A.D., Hansen, E.B., and Jonsson, G. (2007). Electro-ultrafiltration of industrial enzyme solutions. *Journal of Membrane Science* 299: 28–37. doi: 10.1016/j.memsci.2007.04.021.

83 Jagannadh, S.N. and Muralidhara, H.S. (1996). Electrokinetic methods to control membrane fouling. *Industrial & Engineering Chemistry Research* 35: 1133–1140. doi: 10.1021/ie9503712.

84 Weber, K. and Stahl, W. (2002). Improvement of filtration kinetics by pressure electrofiltration. *Separation and Purification Technology* 26: 69–80. doi: 10.1016/S1383-5866(01)00118-6.

85 Brisson, G., Britten, M., and Pouliot, Y. (2007). Electrically-enhanced crossflow microfiltration for separation of lactoferrin from whey protein mixtures. *Journal of Membrane Science* 297: 206–216. doi: 10.1016/j.memsci.2007.03.046.

86 Hakimhashemi, M., Gebreyohannes, A.Y., Saveyn, H. et al. (2012). Combined effects of operational parameters on electro-ultrafiltration process characteristics. *Journal of Membrane Science* 403–404: 227–235. doi: 10.1016/j.memsci.2012.02.054.

87 Real, F.J., Benitez, F.J., Acero, J.L. et al. (2012). Combined chemical oxidation and membrane filtration techniques applied to the removal of some selected pharmaceuticals from water systems. *Journal of Environmental Science and Health Part A* 47: 522–533. doi: 10.1080/10934529.2012.650549.

88 Gonzalez-Olmos, R., Penadés, A., and Garcia, G. (2018). Electro-oxidation as efficient pretreatment to minimize the membrane fouling in water reuse processes. *Journal of Membrane Science* 552: 124–131. doi: 10.1016/j.memsci.2018.01.041.

89 Wang, X., Li, F., Hu, X. et al. (2021). Electrochemical advanced oxidation processes coupled with membrane filtration for degrading antibiotic residues: a review on its potential applications, advances, and challenges. *Science of the Total Environment* 784: 146912. doi: 10.1016/j.scitotenv.2021.146912.

90 Diogo, J.C., Morão, A., and Lopes, A. (2011). Persistent aromatic pollutants removal using a combined process of electrochemical treatment and reverse osmosis/nanofiltration. *Environmental Progress & Sustainable Energy* 30: 399–408. doi: 10.1002/ep.10497.

91 Khosravanipour, A., Zolfaghari, M., and Drogui, P. (2016). Electrofiltration technique for water and wastewater treatment and bio-products management: a review. *Journal of Water Process Engineering* 14: 28–40. doi: 10.1016/j.jwpe.2016.10.003.

92 Tsaia, Y.T., Weng, Y.H., Lin, A.Y.C. et al. (2011). Electro-microfiltration treatment of water containing natural organic matter and inorganic particles. *Desalination* 267: 133–138. doi: 10.1016/j.desal.2010.09.015.

93 Rosman, N., Salleh, W.N.W., Mohamed, M.A. et al. (2018). Hybrid membrane filtration-advanced oxidation processes for removal of pharmaceutical residue. *Journal of Colloid and Interface Science* 532: 236–260. doi: 10.1016/j.jcis.2018.07.118.

94 Mostafazadeh, A.K., Benguit, A.T., Carabin, A. et al. (2019). Development of combined membrane filtration, electrochemical technologies, and adsorption processes for treatment and reuse of laundry wastewater and removal of nonylphenol ethoxylates as surfactants. *Journal of Water Process Engineering* 28: 277–292. doi: 10.1016/j.jwpe.2019.02.014.

95 Rivera, F.F., de Leon, C.P., Nava, J.L. et al. (2015). The filter –press FM01-LC laboratory flow reactor and its applications. *Electochimica Acta* 163: 338–354. doi: 10.1016/j.electacta.2015.02.179.

96 De Lannoy, C.F., Jassby, D., Gloe, K. et al. (2013). Aquatic biofouling prevention by electrically charged nanocomposite polymer thin film membranes. *Environmental Science and Technology* 47: 2760–2768. doi: 10.1021/es3045168.

97 Sun, X., Wu, J., Chen, Z. et al. (2013). Fouling characteristics and electrochemical recovery of carbon nanotube membranes. *Advanced Functional Materials* 23: 1500–1506. doi: 10.1002/adfm.201201265.

98 Rahaman, M.S., Vecitis, C.D., and Elimelech, M. (2012). Electrochemical carbon-nanotube filter performance toward virus removal and inactivation in the presence of natural organic matter. *Environmental Science and Technology* 46: 1556–1564. doi: 10.1021/es203607d.

99 Vecitis, C.D., Gao, G., and Liu, H. (2011). Electrochemical carbon nanotube filter for adsorption, desorption, and oxidation of aqueous dyes and anions. *Journal of Physical Chemistry C* 115: 3621–3629. doi: 10.1021/jp111844j.

100 Li, H., Zhang, D., Han, X. et al. (2014). Adsorption of antibiotic ciprofloxacin on carbon nanotubes: pH dependence and thermodynamics. *Chemosphere* 95: 150–155. doi: 10.1016/j.chemosphere.2013.08.053.

101 Gao, G. and Vecitis, C.D. (2012). Doped carbon nanotube networks for electrochemical filtration of aqueous phenol: electrolyte precipitation and phenol polymerization. *ACS Applied Material and Interfaces* 4: 1478–1489. doi: 10.1021/am2017267.

102 Sadia, A.J. and Zhou, Z. (2016). Electrochemical carbon nanotube filters for water and wastewater treatment. *Nanotechnology Reviews* 5: 41–50. doi: 10.1515/ntrev-2015-0056.

103 Tonelli, F.M.P., Goulart, V.A.M., Gomez, K.N. et al. (2015). Graphene-based nanomaterials: biological and medical applications and toxicity. *Nanomedicine* 10: 2423–2450. doi: 10.2217/nnm.15.65.

104 Malpass, G.R.P., Miwa, D.W., Miwa, A.C.P. et al. (2007). Photo-assisted electrochemical oxidation of atrazine on a commercial $Ti/Ru_{0.3}Ti_{0.7}O_2$ DSA electrode. *Environmental Science and Technology* 41: 7120–7125. doi: 10.1021/es070798n.

105 Trasatti, S. (2000). Electrocatalysis: understanding the success of DSA. *Electrochimica Acta* 45: 2377–2385. doi: 10.1016/S0013-4686(00)00338-8.

106 Bagastyo, A.Y., Radjenovic, J., Mu, Y., Rozendal, R.A. et al. (2011). Electrochemical oxidation of reverse osmosis concentrate on mixed metal oxide (MMO) titanium coated electrodes. *Water Research* 45: 4951–4959. doi: 10.1016/j.watres.2011.06.039.

107 Perez, G., Fernandez-Alba, A.R., Urtiaga, A.M. et al. (2010). Electro-oxidation of reverse osmosis concentrates generated in tertiary water treatment. *Water Research* 44: 2763–2772. doi: 10.1016/j.watres.2010.02.017.

108 Pikaar, I., Rozendal, R.A., Yuan, Z. et al. (2011). Electrochemical sulfide oxidation from domestic wastewater using mixed metal-coated titanium electrodes. *Water Research* 45: 5381–5388. doi: 10.1016/j.watres.2011.07.033.

109 Krstić, V. and Pešovski, B. (2019). Reviews the research on some dimensionally stable anodes (DSA) based on titanium. *Hydrometallurgy* 185: 71–75. doi: 10.1016/j.hydromet.2019.01.018.

110 Radjenovic, J. and Sedlak, D.L. (2015). Challenges and opportunities for electrochemical processes as next- generation technologies for the treatment of contaminated water. *Environmental Science Technology* 49: 11292–11302. doi: 10.1021/acs.est.5b02414.

111 Ihos, M., Manea, F., Jitaru, M. et al. (2015). Diclofenac removal from aqueous solutions by electrooxidation at boron-doped diamond (BDD) electrode. *Environmental Engineering and Management Journal* 14: 1339–134. http://omicron.ch.tuiasi.ro/EEMJ.

112 Zhang, C., Jiang, Y., Li, Y. et al. (2013). Three-dimensional electrochemical process for wastewater treatment: a general review. *Chemical Engineering Journal* 228: 455–467. doi: 10.1016/J.CEJ.2013.05.033.

113 Panizza, M., Kapalka, A., and Comninellis, C. (2008). Oxidation of organic pollutant on BDD anodes using modulated current electrolysis. *Electrochimica Acta* 53: 2289–2295. doi: 10.1016/j.electacta.2007.09.044.

114 Yu, X., Zhou, M., Hu, Y. et al. (2014). Recent updates on electrochemical degradation of bio-refractory organic pollutants using BDD anode: a mini review. *Environmental Science and Pollution Research* 21: 8417–8431. doi: 10.1007/s11356-014-2820-0.

115 Xiong, Y., Strunk, P.J., Xia, H. et al. (2001). Treatment of dye wastewater containing acid orange II using a cell with three-phase three-dimensional electrode. *Water Research* 35: 4226–4230. 10.1016/s0043-1354(01) 00147-6.

116 Zhan, J., Li, Z., Yu, G. et al. (2019). Enhanced treatment of pharmaceutical wastewater by combining three-dimensional electrochemical process with ozonation to in situ regenerate granular activated carbon particle electrodes. *Separation and Purification Technology* 208: 12–18. doi: 10.1016/j.seppur.2018.06.030.

117 Zhou, M., Wang, W., and Chi, M. (2009). Enhancement on the simultaneous removal of nitrate and organic pollutants from groundwater by a three-dimensional bio-electrochemical reactor. *Bioresource Technology* 100: 4662–4668. doi: 10.1016/j.biortech.2009.05.002.

118 Ando, N., Matsui, Y., Kurotobi, R. et al. (2010). Comparison of natural organic matter adsorption capacities of super-powdered activated carbon and powdered activated Carbon. *Water Research* 44: 4127–4136. doi: 10.1016/j.watres.2010.05.029.

119 Can, W., Yao-Kun, H., Qing, Z. et al. (2014). Treatment of secondary effluent using a three-dimensional electrode system: COD removal, biotoxicity assessment, and disinfection effects. *Chemical Engineering Journal* 243: 1–6. doi: 10.1016/j.cej.2013.12.044.

120 Zhu, X., Ni, J., Xing, X. et al. (2011). Synergies between electrochemical oxidation and activated carbon adsorption in three-dimensional boron-doped diamond anode system. *Electrochimica Acta* 56: 1270–1274. doi: 10.1016/j.electacta.2010.10.073.

121 Gedam, N. and Neti, N.R. (2014). Carbon attrition during continuous electrolysis in carbon bed based three-phase three-dimensional electrode reactor: treatment of recalcitrant chemical industry wastewater. *Journal of Environmental Chemistry Engineering* 2: 1527–1532. doi: 10.1016/j.jece.2014.06.025.

122 McQuillan, R.V., Stevens, G.W., and Mumford, K.A. (2018). The electrochemical regeneration of granular activated carbons: a review. *Journal of Hazardous Materials* 355: 34–49. doi: 10.1016/j.jhazmat.2018.04.079.

123 Norra, G.F. and Radjenovic, J. (2021). Removal of persistent organic contaminants from wastewater using a hybrid electrochemical-granular activated carbon (GAC) system. *Journal of Hazardous Materials* 415: 125557. doi: 10.1016/j.jhazmat.2021.125557.

124 Brinkmann, T., Giner, G., Hande, S. et al. (2016). Best available techniques (BAT) reference document for common waste water and waste gas treatment/management systems in the chemical sector. EUR 28112 EN. doi: 10.2791/37535.

125 Cui, T., Zhang, Y., Han, W. et al. (2017). Advanced treatment of triazole fungicides discharged water in pilot scale by integrated system: enhanced electrochemical oxidation, up-flow biological aerated filter and electrodialysis. *Chemical Engineering Journal* 315: 335–344. doi: 10.1016/j.cej.2017.01.039.

126 Chatzisymeon, E., Xekoukoulotakis, N.P., Coz, A. et al. (2006). Electrochemical treatment of textile dyes and dye house effluents. *Journal of Hazardous Materials* 137: 998–1007. doi: 10.1016/j.jhazmat.2006.03.032.

127 Moreira, F.C., Soler, J., Fonseca, A. et al. (2015). Incorporation of electrochemical advanced oxidation processes in a multistage treatment system for sanitary landfill leachate. *Water Research* 81: 375–387. doi: 10.1016/j.apcatb.2015.09.014.

128 He, Y., Dong, Y., Huang, W. et al. (2015). Investigation of boron-doped diamond on porous Ti for electrochemical oxidation of acetaminophen pharmaceutical drug. *Journal of Electroanalytical Chemistry* 759: 167–173. doi: 10.1016/j.jelechem.2015.11.011.

129 Chen, Y., Hong, L., Han, W. et al. (2011). Treatment of high explosive production wastewater containing RDX by combined electrocatalytic reaction and anoxic–oxic biodegradation. *Chemical Engineering Journal* 168: 1256–1262. doi: 10.1016/j.cej.2011.02.032.

130 Manea, F., Bodor, K., Vlaicu, I. et al. (2021). Process for drinking water treatment. Romanian patent no. RO132097B1/30.09.2021, BOPI 9/2021.

131 Raju, G.B.M., Karuppiah, T., Latha, S.S. et al. (2009). Electrochemical pretreatment of textile effluents and effect of electrode materials on the removal of organics. *Desalination* 249: 167–174. doi: 10.1016/j.desal.2008.08.012.

132 Benito, A., Penades, A., Lliberia, J.L. et al. (2017). Degradation pathways of aniline in aqueous solutions during electrooxidation with BDD electrodes and UV/H2O2 treatment. *Chemeosphere* 166: 230–237. doi: 10.1016/j.chemosphere.2016.09.105.

133 Barndõk, H., Hermosilla, D., Negro, C. et al. (2018). Comparison and predesign cost assessment of different advanced oxidation processes for the treatment of 1,4-dioxane-containing wastewater from the chemical industry. *ACS Sustainable Chemistry & Engineering* 6: 5888–5894. doi: 10.1021/acssuschemeng.7b04234.

134 da Silva, A.J.C., dos Santos, E.V., Morais, C.C.O. et al. (2013). Electrochemical treatment of fresh, brine and saline produced water generated by petrochemical industry using Ti/IrO2–Ta2O5 and BDD in flow reactor. *Chemical Engineering Journal* 233: 47–55. doi: 10.1016/j.cej.2013.08.023.

135 Alighardashi, A., Aghta, R.S., and Ebrahimzadeh, H. (2018). Improvement of carbamazepine degradation by a three-dimensional electrochemical (3-EC) process. *International Journal of Environmental Research* 12: 451–458. doi: 10.1007/s41742-018-0102-2.

136 Liu, H. (2015). CNT-based electrochemical filter for water treatment: mechanisms and applications. Doctoral dissertation, Harvard University, Graduate School of Arts & Sciences. http://nrs.harvard.edu/urn-3:HUL.InstRepos:17467306.

137 Santos, I.D., Dezotti, M., and Dutra, A.J.B. (2013). Electrochemical treatment of effluents from petroleum industry using a Ti/RuO_2 anode. *Chemical Engineering Journal* 226: 293–299. doi: 10.1016/j.cej.2013.04.080.

138 Liu, H., Liu, J., Liu, Y. et al. (2014). Quantitative 2D electrooxidative carbon nanotube filter model: insight into reactive sites. *Carbon* 80: 651–664. doi: 10.1016/j.carbon.2014.09.009.

139 Liu, Z., Zhu, M., Zhao, L. et al. (2017). Aqueous tetracycline degradation by coal-based carbon electrocatalytic filtration membrane: effect of nano antimony-doped tin dioxide coating. *Chemical Engineering Journal* 314: 59–68. 10.1016/j.cej.2016.12.093.

140 Liu, Y., Dong, X., and Chen, P. (2012). Biological and chemical sensors based on graphene materials. *Chemical Society Reviews* 41: 2283–2307. doi: 10.1039/C1CS15270J.

141 Liu, S., Wang, Y., Zhou, X. et al. (2017). Improved degradation of the aqueous flutriafol using a nanostructure macroporous PbO_2 as reactive electrochemical membrane. *Electrochimica Acta* 253: 357–367. doi: 10.1016/j.electacta.2017.09.055.

142 Liu, L., He, H., Zhang, C. et al. (2012). Treatment of reverse osmosis concentrates using a three-dimensional electrode reactor. *Current Organic Chemistry* 16: 2091–2096. doi: 10.2174/138527212803532422.

143 Liu, Y., Lee, J.H.D., Xia, Q., Ma, Y., Yu, Y., Lanry, L.Y., Xie, Y.J., Ong, C.N., Vecitis, C.D., and Zhou, Z. (2014). A graphene-based electrochemical filter for water purification. *Journal of Materials Chemistry A* 2: 16554–16562. doi: 10.1039/C4TA04006F.

9

Electrocatalytic Removal of Heavy Metal Ions from Water

Prasenjit Bhunia[1,*] *and Kingshuk Dutta*[2]

[1] *Department of Chemistry, Silda Chandra Sekhar College, Jhargram, West Bengal, India*
[2] *Advanced Polymer Design and Development Research Laboratory (APDDRL), School for Advanced Research in Petrochemicals (SARP), Central Institute of Petrochemicals Engineering and Technology (CIPET), Bengaluru, Karnataka, India*
[*] *Corresponding author*

9.1 Introduction

Owing to the nondegradability and biological toxicity of heavy meal ions (HMs), and the fact that their contamination of aquatic environments threatens public health as well as ecosystems, they have recently become a source of growing global concern [1, 2]. HMs accumulate in the food chain, leading to a range of mild (such as tiredness and nausea) to serious (such as cancer) illnesses after consumption by humans [3]. The concentration of HMs has been continuously increasing with rapid growth in industrialization as well as agricultural activities. The major polluting sources include batteries, pesticides, petrochemicals, metal smelting, fluidized bed bioreactors, etc., plus activities from industries such as mining, plating and electroplating, tanning, textile, etc [4]. The maximum allowable concentration of HMs should essentially be below a few to a few tens of $\mu g\ L^{-1}$ according to national and international standards stipulated by the United States Environmental Protection Agency (USEPA) and the World Health Organization (WHO) [5, 6]. The most familiar contaminating HMs are zinc (Zn), lead (Pb), mercury (Hg), cadmium (Cd), nickel (Ni), copper (Cu), chromium (Cr), and arsenic (As). These HMs, along with some other HMs, viz. iron (Fe), silver (Ag), molybdenum (Mo), manganese (Mn), cobalt (Co), boron (B), calcium (Ca), antimony (Sb), etc., are usually found in contaminated water bodies and need to be efficiently expelled from wastewater [4]. The effects of traces of various HMs on human health are listed in Table 9.1 [4, 7]. Although various effective techniques have been devised in order to decontaminate HMs from water (such as chemical precipitation, adsorption, membrane separation, and ion exchange) [8–11], requirement of environmentally detrimental chemicals and excessive pretreatment steps limits their real life applications.

Photocatalysts and Electrocatalysts in Water Remediation: From Fundamentals to Full Scale Applications, First Edition. Edited by Prasenjit Bhunia, Kingshuk Dutta, and S. Vadivel.

Table 9.1 Various HMs that exist in wastewater along with their sources, health issues caused by the presence of inappropriate amounts, and the permitted quantities in drinking water following WHO recommendations. [4] / Springer Nature / CC BY 4.0.

Common heavy metal	Main sources	Main organ and system affected	Permitted amounts (μg)
Lead (Pb)	Lead-based batteries, solder, alloys, cable sheathing pigments, rust inhibitors, ammunition, glazes, and plastic stabilizers	Bones, liver, kidneys, brain, lungs, spleen, immunological system, hematological system, cardiovascular system, and reproductive system	10
Arsenic (As)	Electronics and glass production	Skin, lungs, brain, kidneys, metabolic system, cardiovascular system, immunological system, and endocrine	10
Copper (Cu)	Corroded plumbing systems, electronic and cables industry	Liver, brain, kidneys, cornea, gastrointestinal system, lungs, immunological system, and hematological system	2000
Zinc (Zn)	Brass coating, rubber products, some cosmetics, and aerosol deodorants	Stomach cramps, skin irritations, vomiting, nausea, anemia, and convulsions	3000
Chromium (Cr)	Steel and pulp mills and tanneries	Skin, lungs, kidneys, liver, brain, pancreas, testes, gastrointestinal system, and reproductive system	50
Cadmium (Cd)	Batteries, paints, steel industry, plastic industries, metal refineries, and corroded galvanized pipes	Bones, liver, kidneys, lungs, testes, brain, immunological system, and cardiovascular system	3
Mercury (Hg)	Electrolytic production of chlorine and caustic soda, runoff from landfills and agriculture, electrical appliances, industrial and control instruments, laboratory apparatus, and refineries	Brain, lungs, kidneys, liver, immunological system, cardiovascular system, endocrine, and reproductive system	6
Nickel (Ni)	Stainless steel and nickel alloy production	Lungs, kidneys, gastrointestinal distress, pulmonary fibrosis, and skin	70

Therefore, an efficient, fast, reliable, and accurate technique for the complete removal of HMs from water becomes essential for the sustainable development of the environment.

Electrochemistry, an interdisciplinary field of interfacial charge transfer, has been introduced for the decontamination of HMs. It has a contradictory duality, however, in the case of water treatment in particular, as it is a source of HMs in the first place as well as their remedy. In terms of remedy, electrocatalytic reduction/oxidation involves the conversion of HMs from their toxic to nontoxic form through the interaction of electrocatalysts and HM contaminants present in water; it therefore has attracted enormous attention in recent times. Particularly, due to the involvement of a very clean reagent, the electron, this technique, which closely resembles the electrocatalytic advanced oxidation process (EAOP), is considered to be environmentally benign. In addition, the safety arising out of its mild operating conditions, its versatility, its amenability toward automation, as well as its high energy efficiency are the factors that make this technique superior [12]. Consequently, several strategies involving this technique have been adopted, which include use of multistage systems and suitable electrocatalysts, preconcentration, promotion of turbulent regimes, etc [13]. The versatility of this technique has also been realized, from its ability to treat wastewater containing pollutants of varied natures, such as dyes, herbicides, pesticides, phenolic compounds, and personal care and pharmaceuticals products [14]. Therefore, the entire focus of this chapter will be on the fundamentals of electrocatalytic decontamination of HMs along with demonstration of an overview of the employed electrocatalysts in order to provide the readers an easy grasp and integral view of the field.

9.2 Fundamentals

Obviously, there are profound functions of electrochemistry in electrocatalytic reduction and oxidation of HMs present in wastewater. The efficiency of this technique depends on the appropriate modification of the electrodes as well as the electrocatalysts.

9.2.1 Structure of Electrocatalytic Reactors

Schematic diagrams of electrocatalytic reactors have been presented in Figure 9.1. Typically, a cell consists of a power source, a cathode, an anode, and the electrolyte (the effluent/wastewater under investigation). Also, electrochemical conversion/remediation of HMs from water is displayed in Figure 9.1 [15]. Within a cell, the effluent under investigation functions as the supporting electrolyte, into which the cathode–anode couple is immersed. In order to maintain a constant cell voltage or current between the cathode–anode couple, a DC power supply is employed. The most important challenge in the cell configuration is to maintain high mass transfer rates during the electrocatalytic reactions at the electrode surfaces. In this regard, the most common techniques that are employed to improve the mass transfer to the surface of the electrode are gas sparging, high fluid velocity, application of different types of turbulence promoters, and use of baffles.

In addition, sonoelectrochemical techniques may also be applied in order to enhance the rate of mass transfer [16].The cell should be configured, however, in such a manner

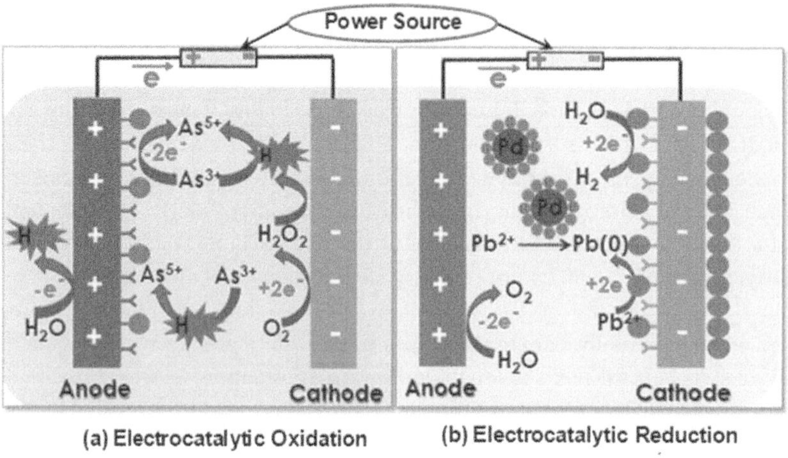

Figure 9.1 Schematic diagrams of electrocatalytic (a) oxidation and (b) reduction of HMs on the respective electrode surfaces [15] / with permission of Elsevier.

that all the units are conveniently exchangeable and accessible to the supporting electrolyte (i.e. effluent). In this connection, there are two types of electrodes that are known to exit, viz. two-dimensional and three-dimensional (high value of electrode surface-to-volume ratio) electrodes. The aforementioned electrodes can further be categorized into two classes, namely moving and static electrodes, out of which the former one leads to an improved mass transport coefficient because of turbulence promotion. More precisely, in recent years, static parallel and cylindrical electrodes have been designed and implemented in two-dimensional electrodes in reactors for the decontamination of wastewater through electrocatalytic oxidation. The parallel plate geometry in a filter press arrangement are widely employed in order to present larger electrode size through the use of either an increased number of stacked cells or electrode pairs for scaling up the process [13]. It should also be clarified here that the structure of the cell, i.e. either undivided or divided, must be taken into consideration during the design of the cell. In particular, an ion conducting membrane or a porous diaphragm must be used in order to separate the catholyte and the anolyte in case of divided cells. The deployment of divided cells, therefore, should be avoided as much as possible, since the separators are expensive and reduce the electrode distance [17]. In order to maintain the flow characteristics in the reactor, however, the limiting hydrodynamic behavior must be taken into account through the application of perfect mixing and plug flow.

9.2.2 Mechanisms of Electrocatalytic Oxidation/Reduction

9.2.2.1 Electrocatalytic Oxidation

In this process, the metal ions lose electrons and move from the lower to higher oxidation state either through a direct electrochemical process or through mediation of electrocatalysts. The electrochemical oxidation can therefore be categorized into two oxidation mechanisms, namely (a) direct oxidation, in which the electrons are directly transferred to the anode surfaces, and (b) indirect oxidation, in which the electrons are lost through assistance of catalysts along with generation of reactive radical species. For example, the electrochemical oxidation has been employed for the detoxification and removal of As^{3+} or Sb^{3+} from water [18, 19].

9.2.2.1.1 *Direct Oxidation*

In this process, the HMs are oxidized on the electrode surface followed by direct transfer of electrons to the anode surface (Figure 9.2). The efficiency of oxidation is highly dependent, however, on the electrode material, composition of the electrolyte, configuration of the electrolytic cell, flow rate, and current density. In this context, an electrode material comprised of a polymer–iridium oxide nanocomposite that can effectively oxidize As^{3+} into As^{5+} species at considerably low potentials of 0.3–0.7 V (vs. Ag/AgCl) has been investigated [20]. It has also been reported by Lacasa et al. [21] that the electrochemical conversion of As^{3+} to As^{5+} is a reversible process, and 100% conversion is possible at low current density in a divided cell. However, the conversion efficiency is only ranged from 70% to 90% in a nondivided cell even at high current density. In addition, a double potential step chronoamperometry approach has also been articulated with an oxidation efficiency of 91% for As^{3+} to As^{5+} conversion [22]. At higher pH (~10), the neutral As^{3+} species (H_3AsO_3) gets transformed into a negatively charged species ($H_2AsO_3^-$), followed by easy adsorption and oxidation at the anode surface. Comparatively higher electricity of 30 kWh mol^{-1}, however, is required for an initial As^{3+} concentration of 150 μg L^{-1}, which needs to be decreased in order to render the process feasible for practical implementation. In this regard, Faraday materials play a significant function in reducing the consumption of energy because of their high electron transfer efficiency as well as metal ion selectivity [18, 23, 24].

9.2.2.1.2 *Indirect Oxidation*

In indirect electrochemical oxidation (also known as mediated electro-oxidation), the low-valence metal ions, viz. As^{3+} and Sb^{3+}, are oxidized by the in situ electrogenerated oxidative species that include strong oxidizing agents (e.g. Cl_2 and H_2O_2) as well as reactive oxygenated free radicals (e.g. $OH^•$ and $O_2^{•-}$) (Figure 9.2) [25, 26]. The most commonly employed substance in indirect oxidation, however, is H_2O_2 due to its low redox potential,

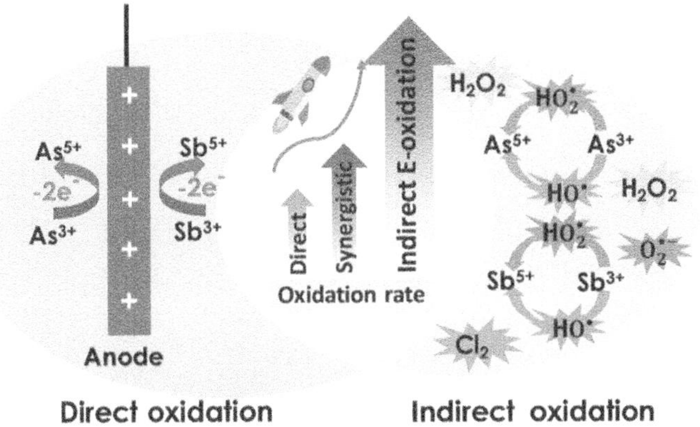

Figure 9.2 Schematic representation of the mechanisms involved in direct and indirect oxidation [15] / with permission of Elsevier.

as well as the fact that it can be continuously produced either by the anodic oxidization of H_2O (Eq. 9.1) or two-electron reduction of O_2 at the cathode (Eq. 9.2).

$$2H_2O - 2e^- \rightarrow H_2O_2 + 2H^+ \tag{9.1}$$

$$O_2 + 2H^+ + 2e^- \rightarrow H_2O_2 \tag{9.2}$$

Additionally, through the oxidation of free Cl^- ions in the electrolyte, Cl_2 can be produced at the anode surface (Eq. 9.3).

$$2Cl^- - 2e^- \rightarrow Cl_2 \tag{9.3}$$

It should be clarified that although the active radical species exist for a shorter period of time in the electrochemical oxidation system compared to the oxidizing agents, the efficiency of oxidation is far superior. In this regard, the $OH^•$ that can be formed through catalysis of Fe^{2+} (Eq. 9.4) (called the electro-Fenton method) and also through the reduction of H_2O (Eq. 9.5), is an important free radical [27].

$$Fe^{2+} + H_2O_2 + H^+ \rightarrow Fe^{3+} + H_2O + OH^• \tag{9.4}$$

$$H_2O - e^- \rightarrow OH^• + H^+ \tag{9.5}$$

On the other hand, $OH^•$ can also be produced by $O_2^{•-}$ induced through Fe^{2+} (Eqs. 9.6 and 9.7).

$$Fe^{2+} + O_2 \rightarrow Fe^{3+} + O_2^{•-} \tag{9.6}$$

$$H^+ + O_2^{•-} + e^- \rightarrow OH^• + HO_2^{•-} \tag{9.7}$$

In addition, the $HO_2^{•-}$ species is a precursor of H_2O_2 and can be generated by H_2O and O_2 reduction (Eq. 9.8).

$$H_2O + O^2 + 2e^- \rightarrow OH^- + HO_2^{•-} \tag{9.8}$$

The $E^0 = 2.8$ V vs. NHE for $OH^•$ and $E^0 = 1.0$ V for $HO_2^{•-}$ values indicate that these active species are strong oxidizing agents [18].

It must be mentioned here that the electrode materials trigger the formation of active species on the electrode surface. For instance, conductive diamond leads to more $OH^•$ formation on the anode surface, which is indicated by the highest oxygen evolution overpotential [16]. On the other hand, lead/lead dioxide electrodes improve the efficiency of electro-oxidation through the formation of large amounts of OCl^- [28]. It has also been articulated that the As^{3+} oxidation efficiency has been enhanced from 12.5% to 95.7% by

Au/Pd nanoparticles present on a CNT electrode through the generation of large amounts of $HO^•$ and $O_2^{•-}$ species [19]. In addition, the As^{3+} oxidation efficiency has been recorded as 99.1% within 40 min through the employment of MoS_2–carbon felt as an air-cathode. However, direct electro-oxidation without O_2 led to an oxidation efficiency of only 20.3% [29]. Additionaly, a multicycle galvanostatic charge–discharge process, employing hematite as the anode, has been reported to produce 98.6% removal of As^{3+} through the formation of large quantities of $OH^•$, ClO^-, and H_2O_2 [30].

Electron transfer between the metal ions and the anode surface is restricted by electrostatic repulsion between metal ions, competitive active sites, and diffusion resistance, hindering the oxidation efficiency in the direct electro-oxidation process. However, formation of large quantities of active oxidizing species in the electrolyte assist the oxidation of HMs in the shortest possible time; therefore, a close vicinity of the HMs and oxidizing agents enhances the oxidation efficiency. Based on the statistical analysis, the rate of oxidation is found to be the maximum for indirect electro-oxidation (29.70 mg L^{-1} min^{-1}), as compared to the direct process (2.32 mg L^{-1} min^{-1}). The increment of potential encourages the hydrogen evolution at the cathode surface, which results in the consumption or inhibition of the active substances (such as $OH^•$ and $HO_2^{•-}$). In contrast, as the other side reactions, such as conversion of H_2O_2 to O_2 ($E^0 = 0.49$ V vs. Ag/AgCl), reduction of O_2 to H_2O ($E^0 = 0.62$ V), or direct transformation of H_2O_2 to H_2O ($E^0 = 1.15$ V), begin to occur, a drop in H_2O_2 concentration at higher potential is observed [31].

9.2.2.2 Electrocatalytic Reduction

The electrocatalytic reduction techniques involve the conversion of high-valence HMs into lower or zero-valence states after receiving electrons. In this method, therefore, the active sites are freed through the conversion of HMs into their zero-valence state. This technique is also categorized into two types, viz. direct reduction, where HMs accept electrons directly from the cathode surface, and indirect reduction, where electrons are received through the assistance of reductants and through in situ generated active radicals.

9.2.2.2.1 Direct Reduction

As mentioned earlier, in this technique, the HMs first accept electrons directly from the cathode surface; this step is then followed by nucleation to produce metal oxides or elemental metals (Figure 9.3) [32]. For instance, uranium (U) has been decontaminated from ground water through this method, with 99.9% reduction efficiency at −0.6 V. It has been noticed that the transferred electrons were more or less double that required for the elimination of U^{6+} during the process, indicating complete utilization of electrons for the electroreduction of U^{6+} [33]. In this regard, for achieving high reduction efficiencies, two prerequisites are essential: (a) migration of the metal ions present in the electrolyte to the cathode surface, and (b) substantially higher rate of transfer of electrons from the surface of the cathode to the metal ions. Indeed, the charge transfer resistance, current density, and diffusion resistance have significant roles in the typical reduction technique. Nevertheless, it has been established that the diffusion resistance is the prime source of the electroreduction process taking Pd and Zn as the model HMs. Although the electroreduction is rapid, the slow migration rate of the metal ions to the electrode surface due to large diffusion resistance may hinder the entire electroreduction reaction [34].

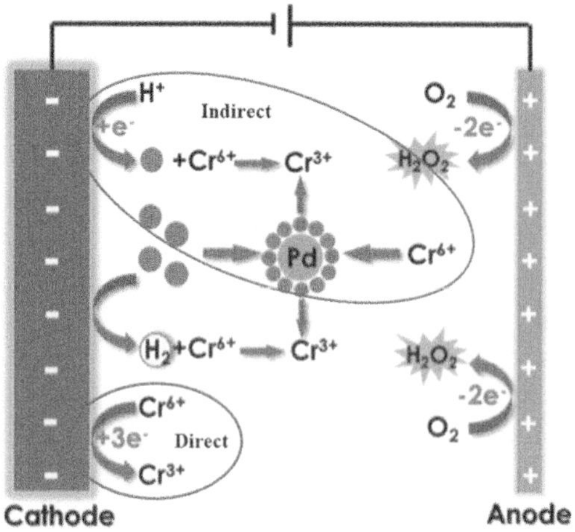

Figure 9.3 A schematic representation of the mechanisms involved in direct and indirect electroreduction of Cr^{6+} [15] / with permission of Elsevier.

9.2.2.2.2 Indirect Reduction

The reduction of HMs occurs through the in situ formation of active species, such as $H^{\bullet}$, e_{aq}^{-}, H_2O_2, etc., in the electrolyte in a typical indirect electroreduction method (Figure 9.3). Low-valence metal ion intermediate products (e.g. Fe^{2+}) and reductive polymers, such as polyaniline (PAni), polypyrrole (PPy), etc., are also used as the active substance for electroreduction. In particular, the reductive reactions are not dependent on the rate of migration, therefore leading to higher efficiency of reduction compared to that of the direct reduction process. It is worth mentioning that H_2O_2, produced through a two-electron redox reaction (Eq. 9.2), is considered to be a universal active substance [35].

It should also be noted here that atomic hydrogen ($H^{\bullet}$), formed from water electrolysis (Eq. 9.9), functions as a robust reducing species at acidic pH [36]; whereas, the hydrated electron (e_{aq}^{-}), which has a high quantum yield in the UV/sulfite process (Eq. 9.10), acts as a powerful reductant at alkaline pH [37].

$$H_2O + e^- \rightarrow OH^- + H^{\bullet} \tag{9.9}$$

$$SO_3^{2-} + h\upsilon \rightarrow SO_3^{\bullet -} + e_{aq}^{-} \tag{9.10}$$

Furthermore, reductive polymers, such as PPy and PAni, are able to shuttle back and forth reversibly between the neutral states (PPy^0, $PAni^0$) and oxidized states (PPy^+, $PAni^+$), and therefore function as spontaneous electron donors for the reduction of Cr^{6+} (Eqs. 9.11 and 9.12) [38, 39].

$$3PPy^0 + Cr^{6+} \rightarrow 3PPy^+ + Cr^{3+} \tag{9.11}$$

$$3PAni^0 + Cr^{6+} \rightarrow 3PAni^+ + Cr^{3+} \tag{9.12}$$

Additionally, the direct reduction of Cr^{6+} to Cr^{3+} through the mediation of Fe^{2+} ions, produced from the Fe anode surface (Eq. 9.13), has already been articulated [40].

$$Cr_2O_7{}^{2-} + 14H^+ + 6Fe^{2+} \rightarrow 2Cr^{3+} + Fe^{3+} + 7H_2O \tag{9.13}$$

The $H^\bullet$ species is the strongest reducing agent amongst the aforementioned active substances; therefore, the quantity of surface-adsorbed $H^\bullet$ triggers the rate of electroreduction [40]. In this connection, the Pd electrocatalysts have been reported to reduce molecular hydrogen to $H^\bullet$, which in turn causes reduction of Cr^{6+} within 30 min (with 95% conversion) [41]; however, the conversion is negligible in absence of Pd electrocatalysts. In fact, the chemisorbed $H^\bullet$ on the Pd surface converts the molecular H_2 and O_2 into H_2O_2, which has a significant role in the detoxification of Cr^{6+}. The indirect and direct electroreduction have been applied synergistically as they are complementary to each other. The Cr^{6+} reduction efficiency is reported to be enhanced significantly (from 37% to 100%) through the use of a reduced graphene oxide (RGO) cathode that leads to the formation of H_2O_2 when direct and indirect electroreduction functioned synergistically [43]. Similarly, a PPy-doped RGO aerogel electrode has been reported to promote the reduction efficiency toward Cr^{6+} [38].

9.3 Advantages and Disadvantages of the Electrocatalytic Approach

The electrocatalytic process has both merits and demerits. The most important advantage of this technique for water treatment is the avoidance of pollution complications: it is environmentally benign owing to the employment of a clean reagent (i.e. the electron). In addition, application of detrimental chemicals can be ignored in this technique, resulting in mild operation as well as involvement of straightforward equipment. Other major advantages include cost-effectiveness, versatility, remarkable efficiency, ease of automation, etc. Furthermore, this technique can be operated by the use of renewable energy. The scale-up of this process, however, is challenging due to operational limitations toward competent implementations. The advantages are:

a) Electrocatalytic reactions can be regulated (either restarted or terminated) by turning off the supply of power.
b) Discharge of unreacted wastes and volatile toxic components can be overridden, since this technique is operated at lower pressure and temperature than that used for non-electrochemical processes.
c) This process is capable of decontaminating a wide variety of industrial wastewater as well as domestic.
d) Electrical parameters applied in this process (e.g. current (I) and voltage (V)) are convenient for boosting the data process automation and data acquisition, and to regulate the control.

The foremost disadvantages of this process are:

a) Intrinsic: Various significant parameters, such as low space–time yield, low surface-to-volume ratio, mass transport limitations, and gradual temperature enhancement, have not been adequately elucidated as of now. Furthermore, the electrical conductivity of the effluent is a highly important parameter, but the conductivity of all waste streams are not enough for the applicability of the process.
b) Extrinsic: Low current efficiency, short life span, and/or expensiveness of some electrodes are important demerits. In addition, the influence on the efficiency of this process is not completely resolved for undesired reactions, effect of pH, type and concentration of contaminants, potential or current distribution, design of the reactors, and arrangement of electrodes. Moreover, severe electrode fouling can occur due to the adherence of reaction by-products on the electrode surface.

9.4 Summary

The electrocatalytic technique as an approach for the decontamination of HMs from water has been proven to be technologically as well as economically feasible. In-depth fundamentals of the technique have been provided in this chapter. Both the electrocatalytic oxidation and reduction processes have been elaborated along with the detailed mechanisms involved. It is clear from the discussion that the indirect electrocatalytic processes, both oxidation and reduction, occur through the reactive intermediates, and are more efficient than the direct processes. In an electrocatalytic decontamination system, the oxidation and reduction take place at the electrodes; therefore, it is important to develop appropriate electrodes/electrode materials with high selectivity as well as high density of active sites. Investigation on electrocatalytic HM decontamination techniques is limited, however, especially for large-scale applications. Finally, engineered electrocatalytic techniques for decontamination of HMs from water need to be devised, implemented, and optimized for real life applications.

References

1 Jansone-Popova, S., Moinel, A., Schott, J.A., Mahurin, S.M., Popovs, I., Veith, G.M., and Moyer, B.A. (2019). Guanidinium-based ionic covalent organic framework for rapid and selective removal of toxic Cr(VI) oxoanions from water. *Environmental Science & Technology* 53: 878–883. doi: 10.1021/acs.est.8b04215.

2 Zhang, Y., Xiong, Z., Yang, L., Ren, Z., Shao, P., Shi, H., Xiao, X., Pavlostathis, S.G., Fang, L., and Luo, X. (2019). Successful isolation of a tolerant co-flocculating microalgae toward highly efficient nitrogen removal in harsh rare earth element tailings (REEs) wastewater. *Water Research* 166: 115076. doi: 10.1016/j.watres.2019.115076.

3 Uzu, G., Sauvain, J.-J., Baeza-Squiban, A., Riediker, M., Hohl, M.S.S., Val, S., Tack, K., Denys, S., Pradère, P., and Dumat, C. (2011). In vitro assessment of the pulmonary toxicity and gastric availability of lead-rich particles from a lead recycling plant. *Environmental Science & Technology* 45: 7888–7895. doi: 10.1021/es200374c.

4 Qasem, N.A.A., Mohammed, R.H., and Lawal, D.U. (2021). Removal of heavy metal ions from wastewater: a comprehensive and critical review. *NPJ Clean Water* 4: 36. doi: 10.1038/s41545-021-00127-0.

5 USEPA Epa 822-f-18-001 (2018). *Edition of the drinking water standards and health advisories tables*. Washington, DC: Office of Water, U.S. Environmental ProtectionAgency.

6 WHO (2017). *World Health Organization Guidelines for Drinking Water Quality*, 4e. Geneva, Switzerland: World Health Organization.

7 United Nations (2015). *United Nations Guide to the Globally Harmonized System of Classification and Labeling of Chemicals (ghs)*. Geneva, Switzerland: United Nations.

8 Yang, S., Hu, J., Chen, C., Shao, D., and Wang, X. (2011). Mutual effects of Pb(II) and humic acid adsorption on multiwalled carbon nanotubes/polyacrylamide composites from aqueous solutions. *Environmental Science & Technology* 45: 3621–3627. doi: 10.1021/es104047d.

9 Zhang, Y., Zhang, S., and Chung, T.-S. (2015). Nanometric graphene oxide framework membranes with enhanced heavy metal removal via nanofiltration. *Environmental Science & Technology* 49: 10235–10242. doi: 10.1021/acs.est.5b02086.

10 Dai, C. and Hu, Y. (2015). Fe(III) hydroxide nucleation and growth on quartz in the presence of Cu(II), Pb(II), and Cr(III): metal hydrolysis and adsorption. *Environmental Science & Technology* 49: 292–300. doi: 10.1021/es504140k.

11 Chen, Z., Liang, Y., Jia, D., Chen, W., Cui, Z., and Wang, X. (2017). Layered silicate RUB-15 for efficient removal of UO_2^{2+} and heavy metal ions by ion-exchange. *Environmental Science: Nano* 4: 1851–1858. doi: 10.1039/C7EN00366H.

12 Rajeshwar, K., Ibanez, J.G., and Swain, G.M. (1994). Electrochemistry and the environment. *Journal of Applied Electrochemistry* 24: 1077–1091. doi: 10.1007/BF00241305.

13 Rajeshwar, K. and Ibanez, J.G. (1997). *Environmental Electrochemistry: Fundamentals and Applications in Pollution Abatement*. San Diego: Academic Press.

14 Wu, W., Huang, Z.-H., and Lim, T.-T. (2014). Recent development of mixed metal oxide anodes for electrochemical oxidation of organic pollutants in water. *Applied Catalysis A* 480: 58–78. doi: 10.1016/j.apcata.2014.04.035.

15 Yang, L., Hu, W., Chang, Z., Liu, T., Fang, D., Shao, P., Shi, H., and Luo, X. (2021). Electrochemical recovery and high value-added reutilization of heavy metal ions from wastewater: recent advances and future trends. *Environment International* 152: 106512. doi: 10.1016/j.envint.2021.106512.

16 Chen, G. (2004). Electrochemical technologies in wastewater treatment. *Separation and Purification Technology* 38: 11–41. doi: 10.1016/j.seppur.2003.10.006.

17 Wendt, H. and Kreysa, G. (1999). *Electrochemical Engineering: Science and Technology in Chemical and Other Industries*. Germany: Springer. doi: 10.1007/978-3-662-03851-2.

18 He, F., Hemmatifar, A., Bazant, M.Z., and Hatton, T.A. (2020). Selective adsorption of organic anions in a flow cell with asymmetric redox active electrodes. *Water Research* 182: 115963. doi: 10.1016/j.watres.2020.115963.

19 Sun, M., Zhang, G., Qin, Y., Cao, M., Liu, Y., Li, J., Qu, J., and Liu, H. (2015). Redox conversion of chromium(VI) and arsenic(III) with the intermediates of chromium(V) and arsenic (IV) via AuPd/CNTs electrocatalysis in acid aqueous solution. *Environmental Science & Technology* 49: 9289–9297. doi: 10.1021/acs.est.5b01759.

20 Rivera, J.F., Pignot-Paintrand, I., Pereira, E., Rivas, B.L., and Moutet, J.C. (2013). Electrosynthesized iridium oxide-polymer nanocomposite thin films for electrocatalytic oxidation of arsenic(III). *Electrochimica Acta* 110: 465–473. doi: 10.1016/j.electacta.2013.04.056.

21 Lacasa, E., Canizares, P., Rodrigo, M.A., and Fernández, F.J. (2012). Electro-oxidation of As (III) with dimensionally-stable and conductive-diamond anodes. *Journal of Hazardous Materials* 203–204: 22–28. doi: 10.1016/j.jhazmat.2011.11.059.

22 Song, Z., Garg, S., Ma, J., and Waite, T.D. (2019). Modified double potential step chronoamperometry (DPSC) method for As(III) electro-oxidation and concomitant As(V) adsorption from groundwaters. *Environmental Science & Technology* 53: 9715–9724. doi: 10.1021/acs.est.9b01762.

23 Su, X., Tan, K.-J., Elbert, J., Rüttiger, C., Gallei, M., Jamison, T.F., and Hatton, T.A. (2017). Asymmetric Faradaic systems for selective electrochemical separations. *Energy & Environmental Science* 10: 1272–1283. doi: 10.1039/C7EE00066A.

24 Song, Z., Garg, S., Ma, J., and Waite, T.D. (2020). Selective arsenic removal from groundwaters using redox-active polyvinylferrocene-functionalized electrodes: role of oxygen. *Environmental Science & Technology* 54: 12081–12091. doi: 10.1021/acs.est.0c03007.

25 Rajkumar, D. and Kim, J.G. (2006). Oxidation of various reactive dyes with in situ electro-generated active chlorine for textile dyeing industry wastewater treatment. *Journal of Hazardous Materials* 136: 203–212. doi: 10.1016/j.jhazmat.2005.11.096.

26 Chu, Y.Y., Qian, Y., Wang, W.J., and Deng, X.L. (2012). A dual-cathode electro-Fenton oxidation coupled with anodic oxidation system used for 4-nitrophenol degradation. *Journal of Hazardous Materials* 199–200: 179–185. doi: 10.1016/j.jhazmat.2011.10.079.

27 Liu, Y., Zhang, J., Liu, F., Shen, C., Li, F., Huang, M., Yang, B., Wang, Z., and Sand, W. (2020). Ultra-rapid detoxification of Sb(III) using a flow-through electro-fenton system. *Chemosphere* 245: 125604. doi: 10.1016/j.chemosphere.2019.125604.

28 El-Ashtoukhy, E.-S.Z., Amin, N.K., and Abdelwahab, O. (2009). Treatment of paper mill effluents in a batch-stirred electrochemical tank reactor. *Chemical Engineering Journal* 146: 205–210. doi: 10.1016/j.cej.2008.05.037.

29 Xue, Y., Zheng, S., Zhang, Y., and Reinforced, J.W. (2017). As(III) oxidation by the in-situ electro-generated hydrogen peroxide on MoS_2 ultrathin nanosheets modified carbon felt in alkaline media. *Electrochimica Acta* 252: 245–253. doi: 10.1016/j.electacta.2017.08.191.

30 Liu, L., Chen, H., Yang, X., Tan, W., Liu, C., Dang, Z., and Qiu, G. (2020). High-efficiency As (III) oxidation and electrocoagulation removal using hematite with a charge-discharge technique. *Science of the Total Environment* 703: 135678. doi: 10.1016/j.scitotenv.2019.135678.

31 Liu, Y., Zhang, L., Song, Q., and Xu, Z. (2020). Recovery of palladium as nanoparticles from waste multilayer ceramic capacitors by potential-controlled electrodeposition. *Journal of Cleaner Production* 257: 120370. doi: 10.1016/j.jclepro.2020.120370.

32 Kanani, N. (2004). Electrodeposition considered at the atomistic level. In: *Electroplating: Basic Principles, Processes and Practice*, 141–177. Amsterdam: Elsevier. doi: 10.1016/B978-185617451-0/50005-1.

33 Liu, T., Yuan, J., Zhang, B., Liu, W., Lin, L., Meng, Y., Yin, S., Liu, C., and Luan, F. (2019). Removal and recovery of uranium from groundwater using direct electrochemical reduction method: performance and implications. *Environmental Science & Technology* 53: 14612–14619. doi: 10.1021/acs.est.9b06790.

34 Zhang, L., Xu, Z., and He, Z. (2020). Selective recovery of lead and zinc through controlling cathodic potential in a bioelectrochemically-assisted electrodeposition system. *Journal of Hazardous Materials* 386: 121941. doi: 10.1016/j.jhazmat.2019.121941.

35 Scialdone, O., Galia, A., Gattuso, C., Sabatino, S., and Schiavo, B. (2015). Effect of air pressure on the electro-generation of H_2O_2 and the abatement of organic pollutants in water by electro-Fenton process. *Electrochimica Acta* 182: 775–780. doi: 10.1016/j.electacta.2015.09.109.

36 Zhou, Y., Zhang, G., Ji, Q., Zhang, W., Zhang, J., Liu, H., and Qu, J. (2019). Enhanced stabilization and effective utilization of atomic hydrogen on Pd-In nanoparticles in a flow-through electrode. *Environmental Science & Technology* 53: 11383–11390. doi: 10.1021/acs.est.9b03111.

37 Zhao, X., Cao, D., Su, P., and Guan, X. (2020). Efficient recovery of Sb(V) by hydrated electron reduction followed by cathodic deposition in a photoelectrochemical process. *Chemical Engineering Journal* 395: 124153. doi: 10.1016/j.cej.2020.124153.

38 Zhang, Y., Zhang, D., Zhou, L., Zhao, Y., Chen, J., Chen, Z., and Wang, F. (2018). Polypyrrole/reduced graphene oxide aerogel particle electrodes for high-efficiency electro-catalytic synergistic removal of Cr(VI) and bisphenol A. *Chemical Engineering Journal* 336: 690–700. doi: 10.1016/j.cej.2017.11.109.

39 Mo, X., Yang, Z.-h., Xu, H.-y., Zeng, G.-m., Huang, J., Yang, X., Song, P.-p., and Wang, L.-k. (2015). Combination of cathodic reduction with adsorption for accelerated removal of Cr(VI) through reticulated vitreous carbon electrodes modified with sulfuric acid–glycine co-doped polyaniline. *Journal of Hazardous Materials* 286, 493–502. doi: 10.1016/j.jhazmat.2015.01.002.

40 Zhou, L., Liu, D., Li, S., Yin, X., Zhang, C., Li, X., Zhang, C., Zhang, W., Cao, X., Wang, J., and Wang, Z.L. (2019). Effective removing of hexavalent chromium from wasted water by triboelectric nanogenerator driven self-powered electrochemical system – why pulsed DC is better than continuous DC? *Nano Energy* 64: 103915. doi: 10.1016/j.nanoen.2019.103915.

41 Mao, R., Zhao, X., Lan, H., Liu, H., and Qu, J. (2015). Graphene-modified Pd/C cathode and Pd/GAC particles for enhanced electrocatalytic removal of bromate in a continuous three-dimensional electrochemical reactor. *Water Research* 77: 1–12. doi: 10.1016/j.watres.2015.03.002.

42 Qian, A., Liao, P., Yuan, S., and Luo, M. (2014). Efficient reduction of Cr(VI) in groundwater by a hybrid electro-Pd process. *Water Research* 48: 326–334. doi: 10.1016/j.watres.2013.09.043.

43 Jin, W., Du, H., Yan, K., Zheng, S., and Zhang, Y. (2016). Improved electrochemical Cr(VI) detoxification by integrating the direct and indirect pathways. *Journal of Electroanalytical Chemistry* 775: 325–328. doi: 10.1016/j.jelechem.2016.06.030.

10

Combined Photoelectrocatalytic Techniques in Water Remediation

R. Suresh and Saravanan Rajendran*

Departamento de Ingeniería Mecánica, Facultad de Ingeniería, Universidad de Tarapacá, Avda. General Velásquez, Arica, Chile
**Corresponding author*

10.1 Introduction

In recent decades, advanced oxidation processes (AOPs) have been employed extensively in water treatment (industrial effluent, ground water, drinking water, etc.), soil remediation, and sludge treatment, as AOPs can remove a wider range of harmful chemical and biological pollutants, and improve odor and taste of water [1–3]. Basically, these processes involve in situ generation of hydroxyl ($^{\bullet}OH$, $E^{\circ} = 2.80$ V versus SHE, lifetime = a few 10^{-9} s) radicals, which rapidly react with organic contaminants, and may lead to degradation or complete mineralization (Eqs. (10.1) and (10.2)) [4].

$$R + {}^{\bullet}OH \rightarrow \text{Degradation by-products} \tag{10.1}$$

$$R + {}^{\bullet}OH \rightarrow CO_2 + H_2O \tag{10.2}$$

The in situ generated hydroxyl radicals may also reduce toxicity of inorganic species by redox reactions. For example, high toxicity of As(III) species can be decreased by oxidation through hydroxyl radicals [5].

Apart from hydroxyl radicals, in situ generated sulfate radicals ($SO_4^{\bullet -}$, $E^{\circ} = 2.6$ V versus SHE) can also be used for oxidative decomposition of organic pollutants [6]. Besides, electrochemical AOPs (EAOPs), a new kind of AOPs, were also developed for wastewater treatment. In the last few decades, the increasing attention of scientific society on EAOPs is reflected in the huge number of published journal and conference proceedings articles (Figure 10.1A). In EAOPs, the contaminated water is passed through an electrochemical cell that contains a cathode and anode along with a suitable electrolyte. The organic contaminants undergo oxidative decomposition at the anode. In addition, water undergoes oxidation at the anode to form hydroxyl radicals that oxidize organic molecules effectively

Photocatalysts and Electrocatalysts in Water Remediation: From Fundamentals to Full Scale Applications, First Edition. Edited by Prasenjit Bhunia, Kingshuk Dutta, and S. Vadivel.

(Figure 10.1B) [7]. EAOPs have many benefits, such as generation of clean reagent (electrons, $^{\bullet}OH$), easy procedure, and avoidance of secondary pollution. EAOPs were also coupled with the Fenton process [8], photocatalysis [9], and sonocatalysis [10] to accomplish efficient pollutant degradation/mineralization efficiency at a faster rate. In this chapter, application of photoelectrocatalysis (PEC), a widely studied EAOP for treatment of aquatic contaminants including chemical and biological agents will be discussed.

PEC is a hybrid EAOP, in which the electrochemical process (or electrocatalysis) is integrated with photocatalysis. Generally, in photocatalysis [11], when the semiconductor photocatalyst particles are illuminated by light with suitable energy (UV or visible), electrons in the valence band of the photocatalyst get excited to the conduction band, thereby leaving holes at the valence band. The photogenerated electrons (reducing agent) and holes

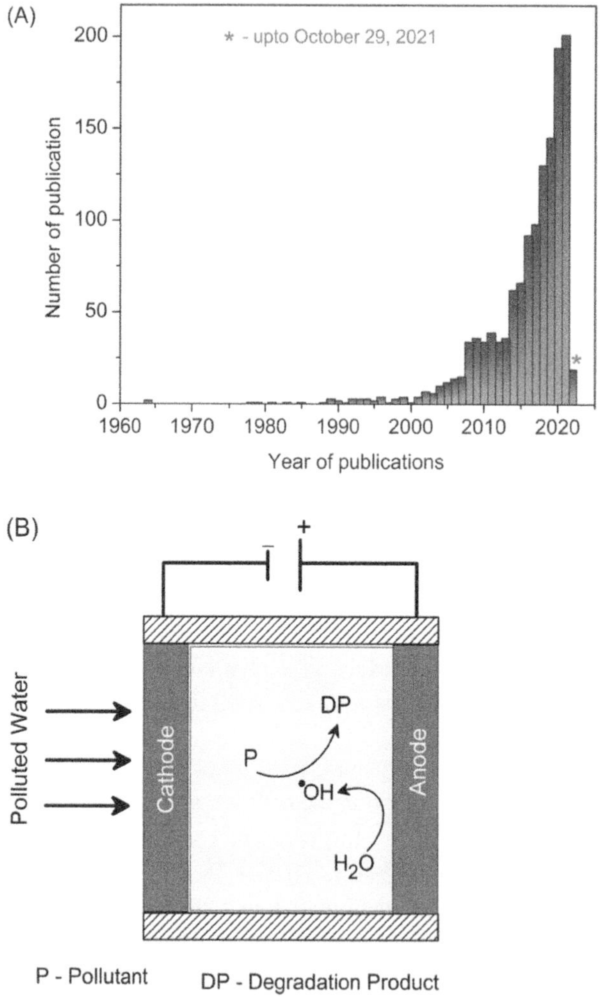

Figure 10.1 (A) Number of journal and conference publications (Scopus data, As on October 29, 2021) and (B) Schematic illustration of degradation of pollutants by electrochemical advanced oxidation process.

(oxidizing agent) subsequently participate in reductive and oxidative decomposition of pollutants, respectively [12]. Furthermore, the holes react with surface hydroxyl groups and/or water molecules to give hydroxyl radicals (Eq. 10.3). The photoexcited electrons react with dissolved oxygen molecules to give superoxide ($O_2^{\bullet-}$) radicals (Eq. 10.4).

$$h^+{}_{VB} + H_2O / OH^- \rightarrow {}^{\bullet}OH \tag{10.3}$$

$$e^-{}_{CB} + O_2 \rightarrow O_2^{\bullet-} \tag{10.4}$$

Both hydroxyl and superoxide radicals can cause decomposition of the pollutants. Lifetimes of the photoinduced electrons and holes are much shorter, however, and they have a strong electrostatic force of attraction between them [13]. Electrons and holes, therefore, recombine immediately, which strongly hinders catalytic activity of selected photocatalysts. When photocatalysis is coupled with an electrochemical process, under the applied bias potential, the photogenerated electrons get pulled out by cathode via an external circuit; thus, recombination of holes and electrons can be avoided efficiently [14]. Furthermore, distinct oxidation and reduction processes occur at the anode and cathode, respectively. It provides a way, therefore, for simultaneous oxidative and reductive removal of pollutants effectively. Meanwhile, accumulated holes in the anode generate hydroxyl radicals by reacting with water molecules. The accumulated electrons in the cathode react with oxygen molecules to give superoxide radicals. An external bias potential significantly affects the photoseparation efficiency of the photoelectrocatalysts.

The basic components of PEC are electrodes (two-electrode or three-electrode systems), electrochemical cell (e.g., quartz cell), electrolyte (e.g., Na_2SO_4), air pump, power source, and a light source (Xe lamp). In PEC with a two-electrode system [15], the photocatalyst-deposited/coated electrode acts as the photoanode, while a platinum or graphite rod acts as the cathode. Physical (sputtering and evaporation) and chemical (sol-gel process, successive ionic-layer adsorption) methods are available for the fabrication of the photoanode for the PEC process. In the case of PEC with a three-electrode configuration [15], photoelectrode (working electrode), a reference (Ag/AgCl, saturated calomel electrode), and a counter electrode are placed in the electrochemical cell along with a separator, and they are powered by a potentiostat. A schematic illustration of a PEC reactor with a three-electrode set-up is given in Figure 10.2. The photoexcited electrons generated at the photoelectrode move to the counter electrode, where H_2 or superoxide radicals may be formed (Eqs. (10.5) and (10.6)).

$$2H^+ + e^- \rightarrow H_2 \uparrow \tag{10.5}$$

$$O_2 + e^-(\text{electrode}) \rightarrow O_2^{\bullet-} \tag{10.6}$$

It was seen that the removal efficiency of pure photoelectrode is low, and its working ability is limited only under UV light. To achieve superior PEC performance under visible/solar light with high stability, photoelectrocatalysts were developed. Nanomaterials, such as metal oxides, metal sulfides, silver based materials, polymers, etc., were used as photoelectrocatalysts for removal of aquatic pollutants from water [14, 16].

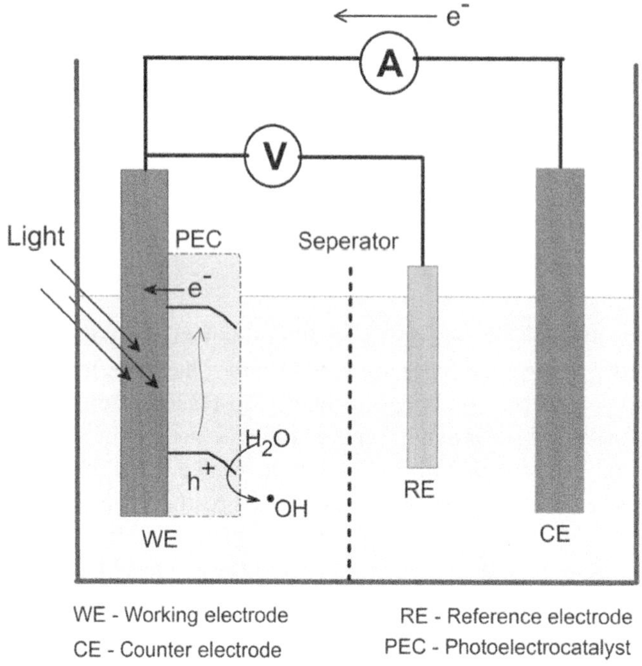

Figure 10.2 Illustration of a PEC reactor with three-electrode set-up and mechanism of production of free radicals for degradation of pollutants.

In this chapter, photoelectrocatalysts reported for water contaminants, including organic, inorganic, and pathogenic microbes, are discussed with mechanistic aspects. Improvement strategies like doping and composite formation in photoelectrocatalysts are discussed with appropriates studies.

10.2 Photoelectrocatalysts for Treatment of Water Contaminants

Photoelectrocatalysts are semiconducting materials with a light-harvesting tendency. Hence, band gap (0.1 to 4 eV) and the position of the conduction and valence band potentials of photoelectrocatalysts determine their PEC efficiency. Photoelectrocatalysts-deposited/immobilized conducting substrates were either used as anode or cathode in PEC pollutant removal reactions. The commonly used conducting substrates are indium tin oxide (ITO), fluorine doped tin oxide (FTO), titanium, stainless steel (SS), Ni foam, graphite, and boron-doped diamond (BDD). Section 10.2.1 describes applications of PEC for different types of water contaminants.

10.2.1 Organic Pollutants Removal

In general, oxidative decomposition of diverse organic pollutants, such as synthetic dyes, pharmaceutical and personal care products, pesticides, and phenolic compounds, occurs at the photoanode in PEC.

(a) Pure catalysts

Pristine TiO_2 nanopores and coated TiO_2 film photoelectrodes, constructed by anodization and chemical methods, respectively, were used in tetracycline degradation reactions [17]. The rate constant for PEC degradation of tetracycline at TiO_2 nanopores electrodes (electrochemical) is 0.560 ± 0.015 h^{-1} (R^2 = 0.991), which is much greater than that of pure electrocatalytic (0.012 ± 0.000 h^{-1}) and photocatalytic (0.084 ± 0.001 h^{-1}) processes. Further, compared to coated TiO_2 film (chemical), TiO_2 nanopore photoelectrodes exhibited better performance in PEC processes. Under UV illumination and an external potential of +0.5 V, tetracycline degradation efficiency of ~80% was achieved by a TiO_2 nanopore electrode within 3 h. The reason for enhanced PEC performance was identified as synergetic effect of electrocatalytic and photocatalytic coupling processes (decreased recombination of electron–hole pairs) and the unique nanoporous structure of TiO_2 (favoring the transport of electrons). The effect of the electrochemical reduction treatment of TiO_2 nanoporous photoelectrodes, fabricated by anodization methods, on pollutant degradation efficiency was further studied. For this purpose, a TiO_2 nanoporous/Ti plate working electrode was further subjected to electrochemical treatment in H_2SO_4 for 10 min and then used in PEC decomposition of acetaminophen and valacyclovir [18]. Importantly, pure photocatalytic and electrocatalytic processes didn't degrade acetaminophen and valacyclovir for up to three hours. However, PEC-comprised TiO_2 nanoporous electrodes rapidly removed acetaminophen (86.96%) and valacyclovir (53.12%), when compared with untreated TiO_2 nanoporous electrodes under UV–visible light irradiation (intensity = 130 mW cm^{-2}) and an external potential of 1.0 V (vs Ag/AgCl). The reason for superior activity of treated TiO_2 nanoporous electrodes was found to be occurrence of multiple oxidation states ($Ti^{3+/2+}$) and oxygen vacancies in TiO_2.

Apart from TiO_2, other metal oxides, such as WO_3 [19], Cu_2O [20], and ZnO [21], were also used as photoelectrodes in PEC for degradation of organics in water. For example, Sang et al. [20] have prepared cubic, cuboctahedra, truncated octahedra, and octahedral Cu_2O films on ITO substrates using electrochemical methods and used them as photocathodes in PEC degradation of o-chlorophenol under sunlight illumination and bias potential of −0.4 V. The octahedral Cu_2O photocathode with {111} facets showed higher PEC degradation efficiency (82.9%) than other morphologies, since it has greater catalytic active sites and lowest recombination rate of photocharges. Moreover, the suppression of electron–hole pairs mainly depends on morphology and facets of Cu_2O nanocrystals. By using electron spin resonance spectroscopy, superoxide radicals were found to be predominantly involved in the o-chlorophenol degradation reaction. The WO_3 nanoplatelets, prepared by anodization using NaF and H_2O_2, were utilized for PEC decomposition of methyl orange dye in water [22]. The aim of this study was to find the effect of supporting electrolytes used for the preparation of WO_3 electrodes on PEC assisted methyl orange degradation under bias potential of +1.0 V (light source = 1000 W Xe lamp). WO_3, prepared using H_2O_2, showed nearly 100% methyl orange decoloration efficiency within 60 min due to its distribution and small particle size. WO_3 photoelectrodes were also used in PEC degradation of febamiphos in water [23].

(b) Doped catalysts

Metal or nonmetal doping in a semiconductor usually increases its conductivity as well as light-harvesting properties. In this context, a Li^+ inserted TiO_2 nanotube electrode was reported for PEC degradation of phenol in water [24]. The insertion of Li ions was obtained by applying negative potentials to TiO_2 electrodes (Eq. 10.7) [24]:

$$> TiO_2 + xe^- + xLi^+ \rightarrow > Li_xTiO_2 \tag{10.7}$$

The inserted Li ions effectively separate photogenerated electron–hole pairs and favor enhanced production of hydroxyl radicals (Eqs. (10.8) and (10.9)) [24] under highly alkaline conditions.

$$h^+ + > Li^+ - Ti^{3+} - OH \rightarrow > Li^+ - Ti^{3+} - OH^{\bullet +} \tag{10.8}$$

$$> Li^+ - Ti^{3+} - OH^{\bullet +} + H_2O \rightarrow > Li^+ - Ti^{3+} - OH + OH^{\bullet} + H^+ \tag{10.9}$$

Dual elemental doping was also adopted to improve PEC efficiency of pristine photoelectrocatalysts. For instance, a nitrogen (N) and iron (Fe)-doped TiO_2 (Fe-N-TiO_2/Ti) electrode, fabricated through the sol-gel method using NH_4Cl and $Fe(NO_3)_3$ as a source of N and Fe respectively, was reported by Maulidiyah et al. [25]. This electrode showed a higher degradation rate constant 0.0580 min^{-1} toward thiamethoxam degradation under visible light illumination. Recently, an iodine (I) and phosphorous (P)-doped TiO_2 photoelectrode was also reported for PEC degradation of tetracycline [26]. Under the optimized condition, the highest PEC degradation rate constant of 4.20×10^{-2} min^{-1} was obtained for tetracycline degradation. Further, hydroxyl radicals and holes were responsible for the degradation of tetracycline. As expected, the doping process significantly decreases recombination of photocharges and enhances light absorption capacity of TiO_2. Based on the reports [24, 26], the general active radical's production pathway by doped photoelectrodes is shown in Figure 10.3A.

Recently, the effect of addition of oxidant on the PEC performance of doped photoelectrocatalysts was also determined. For instance, the influence of peroxydisulfate on PEC efficiency of self-doped TiO_2 toward bisphenol-A, 4-chlorophenol, sulfamethoxazole, and carbamazepine degradation reactions was studied by Son et al. [27]. The addition of peroxydisulfate greatly enhanced pollutant degradation efficiency due to enhanced production of hydroxyl radicals and selectivity of sulfate radicals. The order of increasing degradation rate was carbamazepine < sulfamethoxazole < 4-chlorophenol < bisphenol-A. It suggests that the PEC efficiency is dependent on the structure of pollutant molecules. Concentration of peroxydisulfate and applied potential bias have an effect on PEC/peroxydisulfate systems in degradation reactions. PEC/peroxydisulfate exhibits a three-fold higher degradation rate constant than that of a pure PEC system.

(c) Composite catalysts

Fabrication of composite is an important strategy for improving the property of pristine photoelectrocatalysts. Importantly, composite formation largely decreases photocorrosion of pure semiconductors and increases active surface area for interaction of organic pollutants. For example, PEC elimination of 2,4-dichlorophenol by an exfoliated graphite/CeO_2 (trimethoxymethylsilane) photoelectrode was demonstrated by Mafa et al. [28]. This electrode showed 98.7% of PEC degradation efficiency (pH 6.2) and 92.6% of total organic carbon removal efficiency toward degradation of 2,4-dichlorophenol within 180 min. In another study, g-C_3N_4/Ag_3PO_4/FTO was employed as a photoanode in PEC for Rhodamine B degradation [29]. Due to longer lifetime of photocharges, g-C_3N_4/Ag_3PO_4 exhibited a higher PEC efficiency for Rhodamine B degradation. The general mechanism of radical production by binary composite modified photoanodes is represented as Figure 10.3B. Some more examples of binary and ternary composite-based photoelectrodes and their performance in organic pollutant degradation have been listed in Table 10.1.

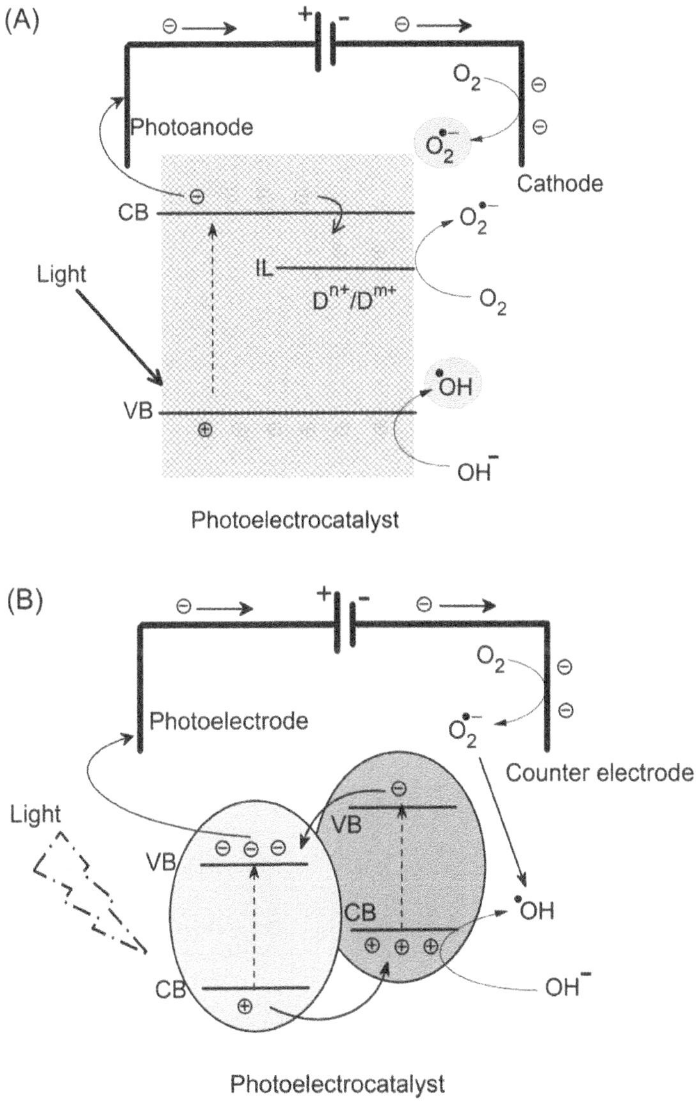

Figure 10.3 (A) Mechanism of free radical production by dual elemental doped PEC catalyst and (B) composite PEC catalyst. CB is conduction band, VB is valence band, D^{n+}/D^{m+} is dopant (metal ion).

Table 10.1 Examples of binary and ternary composite-based photoelectrodes and their PEC performance in organic pollutant degradation.

Photoelectrocatalyst (concentration)	Bias potential (V)	Light source	Pollutant (concentration)	Active species	PEC efficiency (%)	Reference
10-C_3N_4/α-Fe_2O_3	+1.5	500 W xenon lamp	p-nitrophenol, (10 mg L^{-1})	e-, h^+, $^{\bullet}OH$	87.1	[30]
Exfoliated graphite/CeO_2 (trimethoxymethyl-silane)	-	300 W xenon lamp	2,4-dichloro-phenol	h^+, $O_2^{\bullet-}$, $^{\bullet}OH$	98.7	[28]
WO_3/TiO_2	+0.2	UV lamp	Rhodamine B, methylene blue, orange G (6.0 mg L^{-1})	$^{\bullet}OH$	k = 0.9975 68.0 –	[31]
(0.04) Yb-$BiVO_4$-Bi_2O_3	+1.0	500 W xenon lamp	Methylene blue (10 mg L^{-1})	$O_2^{\bullet-}$, $^{\bullet}OH$	83.2	[32]
Poly (terthiophene)/ CuO_x	-1.0	300 W xenon lamp	Phenol	$O_2^{\bullet-}$, $^{\bullet}OH$	100	[33]
Pd/TiO_2	+3.0	35 W xenon lamp	Triclosan (10 mg L^{-1})	h^+, H_2O_2, $O_2^{\bullet-}$, $^{\bullet}OH$	99.7	[34]
CuS/TiO_2	+0.4	35 W xenon lamp	Penicillin G (5 mg L^{-1})	$^{\bullet}OH$	99.1	[35]
Sn_3O_4/Ni foam	+0.6	300 W xenon lamp	Polyacrylamide (20 mg L^{-1})	Reactive oxygen species	70.0	[36]
Ag/Ag_3PO_4(5% methyl viologen)/ exfoliated graphite	+1.5	300 W xenon lamp	Hydrochloro-thiazide (20 mg L^{-1})	h^+, $^{\bullet}OH$	87.1	[37]
ZnO/TiO_2/Ag_2Se	+1.0	36 W blue LED lamp	Oxytetracycline, (5 mg dm^{-3})	$^{\bullet}OH$	96.5	[38]
$Ag_3PO_4/MoS_2/TiO_2$	+0.3	300 W xenon arc lamp	Sulfadiazine (10 mg L^{-1})	$O_2^{\bullet-}$, $^{\bullet}OH$	70.0	[39]
Zn_2SnO_4/MoS_2/Ag/ AgCl	+1.0	300 W xenon lamp	Methylene blue, phenol	$^{\bullet}OH$, Cl^0	96.5 68.0	[40]

The inferences from Table 10.1 are:

a) Activity of photoelectrocatalysts mainly depends on nature and structure of organic molecules.
b) PEC efficiency is dependent on the applied external bias potential.
c) Composite optimization of constituents in binary and ternary composites is necessary to achieve high PEC efficiency in the targeted reaction.
d) Apart from hydroxyl radicals and holes, superoxide radicals also participated in the degradation reaction. In the case of chlorine based catalyst, Cl^0 also gets involved in the pollutant mineralization process.
e) Composites generally showed PEC activity under visible or solar light irradiation. It minimizes energy consumption of the PEC process.

10.2.2 Inorganic Pollutants

Water gets polluted by heavy metal ions, metalloids, and harmful anions. PEC was also actively applied to treat these inorganic contaminants. In principle, metal ions will be reduced at the cathode under the applied potential. The photoelectrocatalyst could improve reduction of metal ions by way of producing free radicals. On the other hand, in some cases like arsenic species, the oxidation process will decrease their toxicity. Hence, oxidative detoxification of inorganic species was also conducted using the PEC process. Some reports are discussed in this section.

(a) Pure catalysts

Zhao et al. [41] have constructed a tubular PEC reactor with TiO_2/Ti photoanode for recovery of Cu from a Cu-ethylenediamine tetraacetic acid (EDTA) complex. At the anode, TiO_2 produces hydroxyl radicals by absorbing UV light. The in situ formed hydroxyl radicals oxidize the Cu-EDTA complex and leave free Cu^{2+} ions in solution (Eq. 10.10). At the cathode, Cu^{2+} ions get reduced to form Cu^0 particles (Eq. 10.11) [41].

$$\mathrm{Cu{-}EDTA} + {}^{\bullet}\mathrm{OH} \rightarrow \mathrm{Cu}^{2+} + \text{oxidized products} + \mathrm{ne}^- \tag{10.10}$$

$$\mathrm{Cu}^{2+} + 2\mathrm{e}^- \rightarrow \mathrm{Cu}^0 \tag{10.11}$$

The experimental parameters, such as current density, pH, and initial complex concentration, are important factors that affect Cu-EDTA decomposition followed by the Cu recovery process. Under the optimum conditions (current density = 1.13 A m^{-2}, time = 60 min, pH = 3.5), the efficiency of Cu recovery and Cu-EDTA degradation were found as 85% and 80%, respectively. A list of some toxic metal ions and their PEC removal by pristine metal oxide based photoelectrodes is given in Table 10.2.

From Table 10.2, it can be inferred that (a) oxidative/reductive conversion efficiencies of photoelectrocatalysts vary with the nature of the metal ions, and (b) external bias potential should vary based on the inorganic pollutant.

To enhance further the PEC performance in the inorganic pollutant removal process, chloride-containing electrolyte was tested. For example, WO_3 photoanode with chloride-containing electrolyte was used in PEC decomposition of Cu–cyanide complexes, followed

Table 10.2 Toxic inorganic ion removal by the PEC method.

Inorganic pollutant (concentration)	Photoelectrocatalyst	Bias potential (V)	Light source	Efficiency (%)/Rate constant	Time (min)	Reference
Cr(VI) ($2.0\ g\ mL^{-1}$)	TiO_2 spheres	+2.0	500 W high-pressure-mercury lamp	100.0	180	[42]
Cu(II)	WO_3	+1.5	300 W xenon lamp	95.6	120	[43]
Sb(III)	TiO_2 nanotubes	+1.5	UV LED lamps	96.7	30	[44]
As(III) ($500\ \mu g\ L^{-1}$)	α-Fe_2O_3	+0.8	Sun 2000 Solar Simulator	$0.0176\ min^{-1}$	450	[45]
Ni(II) (500 mM)	TiO_2/Ni-Sb-SnO_2	+3.0	5 W low pressure mercury lamp	78	24 h	[46]

by the Cu recovery process [43]. It was found that higher chloride ion concentration (>20 mM) improves decomposition of Cu–cyanide complexes due to the formation of free chlorine. The total nitrogen removal efficiency was also high (95.6%) in this PEC process. It is due to production of chlorine radicals. The Cu^{2+} is reduced at the cathode while CNO^- is converted into N_2 at the anode.

Interestingly, the effect of rotating speed of the cathode on metal–EDTA complexes and the metal recovery process was examined. Chen et al. [47] have used a PEC reactor with a Ti rotating cathode and a TiO_2/Ti anode for the destruction of M–EDTA (M = Cu^{2+}, Co^{2+}, Ni^{2+}, and Zn^{2+}) complexes, followed by recovery of metal ions. By varying the rotating speed of the cathode, the efficiency of M–EDTA complex decomposition and the metal recovery process was changed intensively. Particularly, at rotation speed = 100 rpm, pH = 3.18 and current density = $0.5\ mA\ cm^{-2}$, the Cu recovery efficiency was found to be 75.54%. Under the optimized cathode rotating speed, the order of increasing recovery efficiency was $Cu^{2+} > Co^{2+} > Ni^{2+} > Zn^{2+}$. From these discussed reports, the detoxification or complete recovery of metal ions by a photoelectrocatalyst is schematically illustrated in Figure 10.4.

The PEC process was also applied exclusively to eliminate cyanide (CN^-) ions from wastewater. A TiO_2 anode and $CuFe_2O_4$/graphite felt cathode were employed in this PEC reactor [48]. Addition of persulfate improved CN^- removal efficiency from 31.10% to 98.21% within 120 min. Under an optimized bias potential of +1.0 V, the hydroxyl and sulfate radicals, produced by a Fenton-like process were responsible for CN^- removal.

(b) Doped catalysts

A carbon (2%, 5% and 10%)-doped TiO_2 electrode was used for PEC decomposition of sodium oxalate in aqueous solution under UV irradiation [49]. At an applied bias potential of +1.2 V and pH 7, a 5% carbon-doped TiO_2 electrode exhibited greater oxalate mineralization efficiency than other doped TiO_2 electrodes. The fluorine (F) and Ti-doped hydroxyapatite/ITO photoanode was applied to detoxify (reduce) Cr(VI) species [50]. For 5.6 wt% F and Ti-doped sample (pH = 3.4), 84% of PEC Cr(VI) conversion efficiency was obtained under exposure of

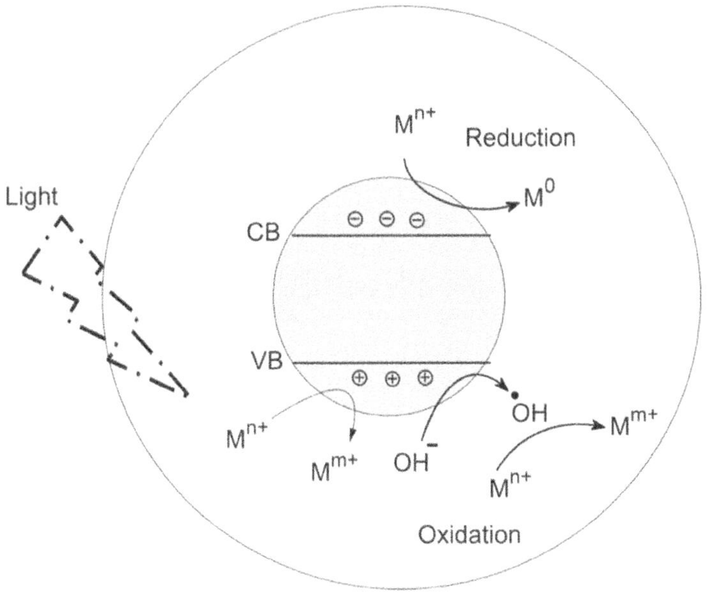

Figure 10.4 PEC reduction of metal ions at a photoelectrocatalyst. CB is conduction band, VB is valence band, M^{n+}/M^{m+} are metal ions.

UV irradiation for 90 min. Interestingly, from the experimental results it was found that the Cr(VI) reduction occurred at the surface of the doped hydroxyapatite photoanode.

(c) Composite catalysts

Besides the doping process, composites were also tested as photoelectrodes in PEC assisted elimination of inorganic pollutants. For instance, Mo/W polyoxometalates/CdS binary composites were synthesized and studied for their PEC performances toward oxidation of As(III) to As(V) [51]. It should be mentioned that As(V) species have less toxicity than As(III) species [52]. The PEC As(III) efficiency of binary composites (85%) was superior to that of pristine CdS.

10.2.3 Biological Pollutants

Studies on PEC inactivation of pathogens, such as bacteria, fungi and viruses, have been performed worldwide [53].

(a) Pure catalysts

A pure TiO_2-coated ITO electrode-placed PEC reactor was assembled for deactivation of MS-2 bacteriophage in the presence of blacklight blue (4 W) irradiation [54]. An anodic potential leads greatly to bacteriophage inactivation, while a cathodic potential presents a complete obstacle towards inactivation. This is due to the fact that under the positive potential, negatively charged viral capsid gets attracted to catalyst surfaces due to the electrostatic force of attraction, and also causes greater production of hydroxyl radicals. Furthermore, *Escherichia coli*, *Staphylococcus aureus*, *Klebsiella pneumoniae*, *Bacillus subtilis* spores, and *Cryptosporidium parvum* oocyst were destroyed by this photoelectrode.

The general process of degradation of pathogenic microorganisms by the PEC process is shown in Figure 10.5.

Halogen radicals are well known antibacterial agents [55] and can be easily generated through a PEC system. However, in PEC with a TiO_2 electrode reactor, only chloride (Cl^-) and bromide (Br^-) ions are possible to use for inactivation of microorganisms, since the potentials of $Cl^•/Cl^-$ and $Br^•/Br^-$ are +2.41 V and +1.96 V, respectively [56]. Furthermore, these halides can produce stable di-halide radical anions (X_2^-, X = Cl or Br). These radicals will destroy protein capsid and attack pathogens to further damage intracellular macromolecules. However, iodide ($E_0[I^•/I^-]$ = +1.33 V) [57] and fluoride ($E_0[F^•/F^-]$ = +3.6 V) [58] ions cannot be used, as their potential values are not suitable to form radicals by TiO_2. In this context, a PEC system using halides (bromide, chloride) was used for disinfection of *Escherichia coli* [59]. A continuous flow-through PEC cell, with TiO_2 photoanode, Ag/AgCl and Pt disc as working reference, and counter electrode, respectively, was used for this study. It was found that when compared to chloride (Cl^-) ions, even a low concentration (50 μM) of bromide (Br^-) ions is capable of inactivating 100% of *E. coli* (9.0×10^6 CFU mL^{-1}) within 1.57 s under UV-LED array illumination. The damaged *E. coli* cells were also clearly identified by using scanning electron microscopic images. In a PEC-Br^- system (potential bias = +0.3 V), di/mono bromide radical anions ($Br^{•-}/Br_2^{•-}$) are generated, and they are responsible for cell rupture of E. coli (Eqs. (10.12)–(10.14)) [59].

$$TiO_2 + X^- \rightarrow X^• \tag{10.12}$$

$$X^• + X^- \rightarrow X_2^{•-} \tag{10.13}$$

$$X^• / X_2^{•-} + \text{Bacteria} \rightarrow X^- + \text{Oxidized bacteria} \tag{10.14}$$

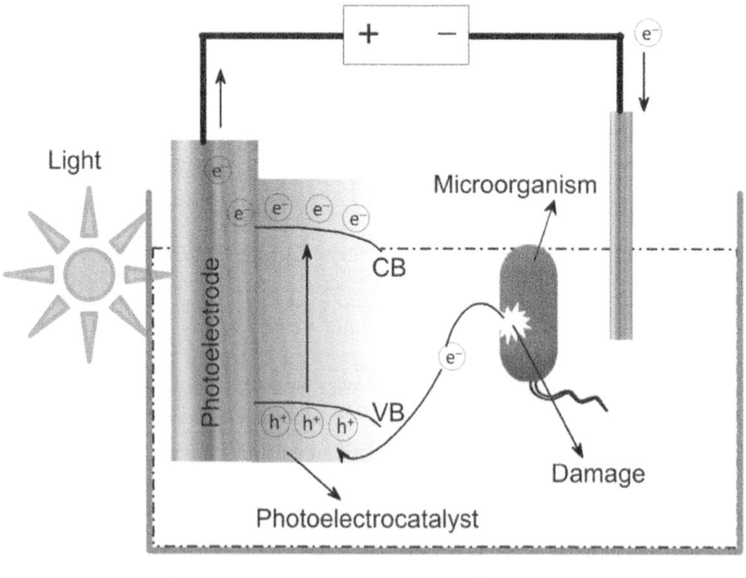

Figure 10.5 PEC inactivation of microorganisms. CB is the onduction band, VB is the valence band.

In another investigation, PEC-X (Br^- and Cl^-) systems with a UV-LED array light source were employed for degradation of replication-deficient recombinant adenovirus [56]. PEC, consisting of a TiO_2 electrode along with 1.0 mM Br^- ions, showed complete degradation of the replication-deficient recombinant adenovirus population within 31.7 s. The effective virucidal performance is ascribed to enhanced generation of active oxygen species, holes, and halogen radicals.

(b) Composite catalysts

Silver-containing composites were effectively used in PEC inactivation of pathogenic bacteria. For example, the disinfectant property of Ag (0–16%)/TiO_2 nanotube array binary composites was evaluated using the PEC process [60]. Under the applied potential of 1.5 V, an Ag (0–16%)/TiO_2 photoelectrode displayed 100% and 99.6% *M. smegmatis* disinfection efficiencies within 3 min and 30 min of UV and visible light irradiations, respectively. In other research, a PEC process consisting of an Ag/TiO_2 binary composite modified ITO photoanode, stainless steel cathode and Ag/AgCl reference electrode was used for inactivation of *Pseudomonas aeruginosa* (pH = 5.9) under UVA irradiation [61]. The process time, bias potential, and electrolyte were 5 min, 1.70 V, and 25 mM Na_2SO_4 aqueous solution, respectively. The in situ generated hydroxyl radicals attacked the outer cell wall of the bacteria effectively. The optimum content of silver nanoparticles in Ag/TiO_2 composite was determined as 4%. Besides, *Bacillus atrophaeus* was more slowly (15 min) deactivated owing to its different cell wall structure. The CdS-Pt/TiO_2 ternary composite was fabricated and used as working electrode material in PEC with a three-electrode system [62]. In presence of simulated solar light irradiation (300 W Xe lamp, intensity = 74 $mWcm^{-2}$), a CdS-Pt/TiO_2 electrode exhibited superior *Escherichia coli* disinfection efficiency than that of Pt–TiO_2 or pure TiO_2 electrodes. The disinfectant efficiencies of 99.2, 85.1, and 51.4% were obtained for bias potentials of 0.6, 0.3, and 0 V, respectively. The active free radicals generated through Eqs. (10.15)–(10.21) [62] were responsible for rapid bacterial damage.

$$CdS + h\nu \rightarrow CdS(e^{?}) + CdS(h^+) \quad (10.15)$$

$$CdS(h^+) + H_2O \rightarrow H^+ + OH^{\bullet} \quad (10.16)$$

$$CdS(e^-) + TiO_2 \rightarrow CdS + TiO_2(e^-) \quad (10.17)$$

$$TiO_2(e^-) + Pt \rightarrow TiO_2 + Pt(e^-) \quad (10.18)$$

$$2Pt(e^-) + 2H^+ + O_2 \rightarrow H_2O_2 \quad (10.19)$$

$$Pt(e^-) + O_2 \rightarrow O_2^{\bullet-} \quad (10.20)$$

$$H_2O_2 + O_2^{\bullet-} \rightarrow OH^- + OH^{\bullet} + O_2 \quad (10.21)$$

Solid state PEC with a Pt/activated carbon fiber/TiO_2 photoanode can also destroy *Penicillium expansum* spores [63]. Based on the X-ray photoelectron spectroscopic analysis, oxidation states of Pt in this composite were 0, +2, and +4, i.e. Pt existed as Pt^0, PtO, and PtO_2, respectively. The existence of PtO and PtO_2 on activated carbon fiber/TiO_2 improved the absorption of ultraviolet light and also increased charge separation. Therefore, PEC performance of this ternary composite photoanode was good towards *Penicillium expansum* disinfection.

10.3 Simultaneous Removal of Organic and Inorganic Pollutants

Usually, wastewater consists of recalcitrant organic contaminants along with heavy metals and/or metalloids [64]. The PEC technique can be used for treatment of organic and inorganic contaminants due to the following reasons. Both oxidation and reduction reactions occur separately and simultaneously in PEC reactors (Figure 10.6). The light induced excited electrons present in the conduction band of the photoelectrode will migrate to the cathode, where they reduce positive charged metal ions (Cu^{2+} and Ni^{2+}). Besides, due to the potential difference, negative charged metal species like $Cr_2O_7^{2-}$ will diffuse to the anode rather than the cathode [65]. Meanwhile, at the photoanode, holes effectively participate in oxidation of organic pollutant molecules. For instance, Liu et al. [65] have studied simultaneous reduction of Cr(VI) species and degradation of 2,4-dichlorophenoxyacetic acid using the PEC process. Both concentration of chromium and pH strongly influence the reduction of Cr(VI) ion at the cathode's surface. To achieve effective PEC efficiencies,

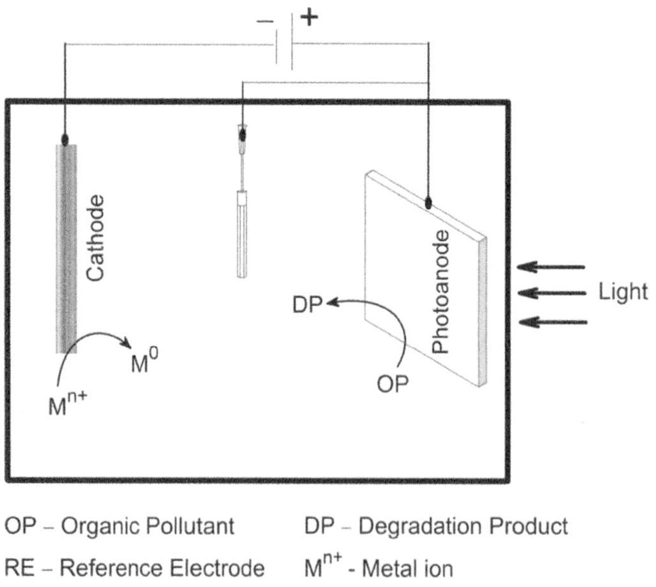

Figure 10.6 Simultaneous removal of organic and inorganic pollutants by the PEC process.

PEC with an amino (polyethylenimine) functionalized carbon cloth cathode and a reduced graphene oxide/TiO_2 photoanode was constructed. Both two-electrode and three-electrode systems were used along with solar light illumination (500 W xenon arc lamp, λ = 280–2000 nm, and flux = 100 mW cm^{-2}). Under acidic conditions, amino groups undergo protonation to give quaternary ammonium cations, which can adsorb $Cr_2O_7^{2-}$ ions via ionic interaction. The reduced graphene oxide in the fabricated photoanode momentously adsorbs 2,4-dichlorophenoxyacetic acid, and thus facilitates pollutant mineralization. The removal rate of Cr(VI) and 2,4-dichlorophenoxyacetic acid was faster even at lower voltage, and the electric energy consumption decreased by 50%. A PEC reactor containing a g-C_3N_4/TiO_2 working electrode was also used for rapid degradation of phenol/benzyl alcohol and reduction of Cr(VI) species in a Cr(VI)/phenol or benzyl alcohol mixture solution under UV–visible light illumination [66]. It was determined that hydroxyl and superoxide radicals were responsible for the elimination of organic and Cr(VI) species.

In another study [67], a two-electrode-configured (counter electrode = Pt foil, potential = 1 V, light source = 500 W xenon lamp, and support electrolyte = 0.1 M K_2SO_4 aqueous solution) PEC reactor with a visible light active TiO_2/Cu (active area = 1.8 cm^2) working electrode was employed for degradation of Rhodamine B and Cr(VI) species simultaneously. The TiO_2/Cu working electrode was fabricated by hydrothermal deposition of Cu nanoparticles on TiO_2 nanotubes using variable glucose concentration (0, 0.0625, and 0.125 mmol). Importantly, PEC removal of Rhodamine B and Cr(VI) species was dependent on glucose content during the preparation of the working electrode. About 95.7% Rhodamine B degradation efficiency was obtained for TiO_2/Cu prepared with 0.0625 mmol of glucose (Irradiation time = 120 min). The visible light performance of TiO_2/Cu electrode is due to dispersion of Cu nanoparticles. The generation of Schottky barriers at the interface of Cu and TiO_2 effectively avoids photocharges recombination, which leads to enhanced degradation efficiencies. The Ag_3PO_4/TiO_2 composite, prepared by the successive ionic layer adsorption and reaction approach was also applied as a solar-light responsive photoelectrode in a PEC reactor for removal of methyl orange dye pollutant and Cr(VI) species [68]. Only the photocatalytic studies showed that Ag_3PO_4/TiO_2 composite exhibited removal efficiencies of 42.9% and 48.8% toward methyl orange and Cr(VI), respectively (light exposure time = 180 minutes). But, PEC removal efficiency of Ag_3PO_4/TiO_2 for methyl orange (81.4%) and Cr(VI) species (79.6%) was higher than the pure photocatalytic process. Generation of active free radicals by this composite follows the Z-scheme mechanism. Moreover, the ratio between silver and phosphorous determined the PEC activity of Ag_3PO_4/TiO_2 (Ag/P = 3.36).

Ag_2S-Bi_2S_3/TiO_2 nanotube arrays were fabricated by anodization followed by a successive ionic layer adsorption and reaction route [69]. This ternary composite photoelectrode displayed excellent photocurrent response (photocurrent density = 3.71 mA cm^{-2}) under visible light irradiation due to the enhanced charge transfer via heterojunction surface and decreased recombination rate of electron–hole pairs. The deposition cycles number of Ag_2S and Bi_2S_3 nanoparticles on TiO_2 nanotube arrays has a vital effect on PEC activity. Especially, Ag_2S(2)-Bi_2S_3(4)/TiO_2 composite exhibited better PEC activity towards methylene orange, and Cr(VI) species, while Ag_2S(3)-Bi_2S_3(3)/TiO_2 showed the best PEC efficiency towards Rhodamine B degradation. The PEC removal efficiencies of methylene orange, Rhodamine B and Cr(VI) species were 81.7, 84.3, and 92.4%, respectively.

Wastewater also contains both chemical and pathogenic microorganisms. Therefore, simultaneous removal of these pollutants is an essential requirement to obtain clean water. For this need, Ag/β-SiC composites with varying silver content (1, 3, and 5 wt% Ag) were prepared and evaluated for PEC decomposition (voltage = 1, 3, and 5 V) of orange G, and *E. coli* under irradiation of UV light [70]. The optimum 3 wt% Ag/β-SiC composite with surface area of 103 $m^2 g^{-1}$ showed 98% (time = 120 min) and 100% (time = 100 min) of PEC degradation efficiency for orange G and *E. coli*, respectively (potential = 3 V). Formation of a Schottky barrier in the interface between Ag and β-SiC nanoparticles enhanced charge separation, and consequently degradation efficiency increased. He et al. [71] have also successfully fabricated a Sb-SnO_2-Ag/TiO_2 photoelectrode and utilized it for simultaneous elimination of 17α-ethinylestradiol (68%) and *E. coli* (>5-log) through the PEC method within sixty minutes.

10.4 Conclusions and Perspective

A wide range of aquatic pollutants, including organic, inorganic, and pathogenic, have been treated by using the PEC process effectively. Semiconductor photoelectrodes, especially TiO_2, were extensively used in PEC reactors. Metal and/or nonmetal doping and composite formation improved PEC performance of pristine photoelectrodes, by way of enhancing light-harvesting capacity, decreasing photocharge recombination, and absorbing pollutants. There are several experimental factors which affect PEC removal efficiencies of photoelectrodes. They are (a) method of photoelectrode fabrication, (b) nature, morphology, and facet of photoelectrocatalysts, (c) additional electrochemical reduction treatment of fabricated electrode, (d) nature of supporting electrolyte used during preparation of photoelectrode, (e) nature and concentration of pollutants, (f) additional oxidants like peroxydisulfate, and (g) applied bias potential. Simultaneous elimination of selected organic and inorganic contaminants was also found to be possible by the PEC method. Further research is still needed, however, as many of these reports are based on laboratory experiments using simulated contaminated waters, rather than real polluted waters. In terms of economic and environmental safety concerns, a focus on development of solar (including near-infrared) active metal-free semiconducting photoelectrocatalysts should be given. Optimized experimental conditions for PEC treatment of real contaminated waters should be explored. It can be expected that further improvements in PEC technology would lead this technology to be effective in water treatment plants.

References

1 Antonopoulou, M., Evgenidou, E., Lambropoulou, D., and Konstantinou, I. (2014). A review on advanced oxidation processes for the removal of taste and odor compounds from aqueous media. *Water Research* 53: 215–234. doi: 10.1016/j.watres.2014.01.028.

2 Maniakova, G., Salmerón, I., Inmaculada, M., López, P., Oller, I., Rizzo, L., and Malato, S. (2021). Simultaneous removal of contaminants of emerging concern and pathogens from

urban wastewater by homogeneous solar driven advanced oxidation processes. *Science of the Total Environment* 766: 144320. doi: 10.1016/j.scitotenv.2020.144320.

3 Chen, Y.D., Duan, X., Zhou, X., Wang, R., Wang, S., Ren, N.Q., and Ho, S.H. (2021). Advanced oxidation processes for water disinfection: features, mechanisms and prospects. *Chemical Engineering Journal* 409: 128207. doi: 10.1016/j.cej.2020.128207.

4 Sirés, I., Brillas, E., Oturan, M.A., Rodrigo, M.A., and Panizza, M. (2014). Electrochemical advanced oxidation processes: today and tomorrow. A review. *Environmental Science and Pollution Research* 21: 8336–8367. doi: 10.1007/s11356-014-2783-1.

5 Dutta, P.K., Pehkonen, S.O., Sharma, V.K., and Ray, A.K. (2005). Photocatalytic oxidation of arsenic(III): evidence of hydroxyl radicals. *Environmental Science & Technology* 39: 1827–1834. doi: 10.1021/es0489238.

6 Giannakis, S., Lin, K.Y.A., and Ghanbari, F. (2021). A review of the recent advances on the treatment of industrial wastewaters by Sulfate Radical-based Advanced Oxidation Processes (SR-AOPs). *Chemical Engineering Journal* 406: 127083. doi: 10.1016/j.cej.2020.127083.

7 Titchou, F.E., Zazou, H., Afanga, H., Gaayda, J.E., Akbour, R.A., Nidheesh, P.V., and Hamdani, M. (2021). An overview on the elimination of organic contaminants from aqueous systems using electrochemical advanced oxidation processes. *Journal of Water Process Engineering* 41: 102040. doi: 10.1016/j.jwpe.2021.102040.

8 Lofrano, G., Faiella, M., Carotenuto, M., Murgolo, S., Mascolo, G., Pucci, L., and Rizzo, L. (2021). Thirty contaminants of emerging concern identified in secondary treated hospital wastewater and their removal by solar Fenton (like) and sulphate radicals-based advanced oxidation processes. *Journal of Environmental Chemical Engineering* 9: 106614. doi: 10.1016/j.jece.2021.106614.

9 Morales, U., Escudero, C.J., Rivero, M.J., Ortiz, I., Rocha, J.M., and Hernández, J.M.P. (2018). Coupling of the electrochemical oxidation (EO-BDD)/photocatalysis (TiO2-Fe-N) processes for degradation of acid blue BR dye. *Journal of Electroanalytical Chemistry* 808: 180–188. doi: 10.1016/j.jelechem.2017.12.014.

10 Rayaroth, M.P., Aravindakumar, C.T., Shah, N.S., and Boczkaj, G. (2021). Advanced oxidation processes (AOPs) based wastewater treatment – unexpected nitration side reactions – a serious environmental issue: a review. *Chemical Engineering Journal* doi: 10.1016/j.cej.2021.133002.

11 Zhang, C., Li, Y., Li, M., Shuai, D., Zhou, X., Xiong, X., Wang, C., and Hu, Q. (2021). Continuous photocatalysis via photo-charging and dark-discharging for sustainable environmental remediation: performance, mechanism, and influencing factors. *Journal of Hazardous Materials* 420: 126607. doi: 10.1016/j.jhazmat.2021.126607.

12 Kumar, R., Sejdic, J.T., and Padhye, L.P. (2020). Conducting polymers-based photocatalysis for treatment of organic contaminants in water. *Chemical Engineering Journal Advances* 4: 100047. doi: 10.1016/j.ceja.2020.100047.

13 Xu, Q., Zhang, L., Cheng, B., Fan, J., and Yu, J. (2020). S-scheme heterojunction photocatalyst. *Chem* 6: 1543–1559.

14 Ibanez, P.F., McMichael, S., Cabanillas, A.R., Alkharabsheh, S., Moranchel, A.T., and Byrne, J.A. (2019). New trends on photoelectrocatalysis (PEC): nanomaterials, wastewater treatment and hydrogen generation. *Current Opinion in Chemical Engineering* 1: 100725. doi: 10.1016/j.coche.2021.100725.

15 Pullupaxi, P.A., Montero, P.J.E., Pallo, C.S., Vargas, R., Fernande, L., Hernandez, J.M.P., and Paz, J.L. (2021). Fundamentals and applications of photoelectrocatalysis as an efficient process to remove pollutants from water: a review. *Chemosphere* 281: 130821. doi: 10.1016/j.chemosphere.2021.130821.

16 Xiao, F.X., Miao, J., Tao, H.B., Hung, S.F., Wang, H.Y., Yang, H.B., Chen, J., Chen, R., and Liu, B. (2015). One-dimensional hybrid nanostructures for heterogeneous photocatalysis and photoelectrocatalysis. *Small* 11: 2115–2131. doi: 10.1002/smll.201402420.

17 Liu, Y., Gan, X., Zhou, B., Xiong, B., Li, J., Dong, C., Bai, J., and Cai, W. (2009). Photoelectrocatalytic degradation of tetracycline by highly effective TiO_2 nanopore arrays electrode. *Journal of Hazardous Materials* 171: 678–683. doi: 10.1016/j.jhazmat.2009.06.054.

18 Xie, G., Chang, X., Adhikari, B.R., Thind, S.S., and Chen, A. (2016). Photoelectrochemical degradation of acetaminophen and valacyclovir using nanoporous titanium dioxide. *Chinese Journal of Catalysis* 37: 1062–1069. doi: 10.1016/S1872-2067(15)61101-9.

19 Peleyeju, M.G. and Viljoen, E.L. (2021). WO_3-based catalysts for photocatalytic and photoelectrocatalytic removal of organic pollutants from water – a review. *Journal of Water Process Engineering* 40: 101930. doi: 10.1016/j.jwpe.2021.101930.

20 Sang, W., Zhang, G., Lan, H., Ana, X., and Liu, H. (2017). The effect of different exposed facets on the photoelectrocatalytic degradation of o-chlorophenol using p-type Cu_2O crystals. *Electrochimica Acta* 231: 429–436. doi: 10.1016/j.electacta.2017.02.073.

21 Goulart, L.A., Santos, G.O.S., Eguiluz, K.I.B., Banda, G.R.S., Lanza, M.R.V., Saez, C., and Rodrigo, M.A. (2021). Towards a higher photostability of ZnO photo-electrocatalysts in the degradation of organics by using MMO substrates. *Chemosphere* 271: 129451. doi: 10.1016/j.chemosphere.2020.129451.

22 Domene, R.M.F., Tovar, R.S., Granados, B.L., Zamora, C.S.G., and Antón, J.G. (2018). Customized WO_3 nanoplatelets as visible-light photoelectrocatalyst for the degradation of a recalcitrant model organic compound (methyl orange). *Journal of Photochemistry and Photobiology A: Chemistry* 356: 46–56. doi: 10.1016/j.jphotochem.2017.12.010.

23 Marquez, G.R., Domene, R.M.F., Tovar, R.S., and Anton, J.G. (2020). Photoelectrocatalyzed degradation of organophosphorus pesticide fenamiphos using WO_3 nanorods as photoanode. *Chemosphere* 246: 125677. doi: 10.1016/j.chemosphere.2019.125677.

24 Kang, U. and Park, H. (2013). Lithium ion-inserted TiO_2 nanotube array photoelectrocatalysts. *Applied Catalysis B: Environmental* 140–141: 233–240. doi: 10.1016/j.apcatb.2013.04.003.

25 Maulidiyah, T.A., Nurwahidah, A.T., Wibowo, D., and Nurdin, M. (2017). Photoelectrocatalyst of Fe co-doped N-TiO_2/Ti nanotubes: pesticide degradation of thiamethoxam under UV–visible lights. *Environmental Nanotechnology, Monitoring & Management* 8: 103–111. doi: 10.1016/j.enmm.2017.06.002.

26 Liu, D., Li, H., Gao, R., Zhao, Q., Yang, Z., Gao, X., Wang, Z., Zhang, F., and Wu, W. (2021). Enhanced visible light photoelectrocatalytic degradation of tetracycline hydrochloride by I and P co-doped TiO_2 photoelectrode. *Journal of Hazardous Materials* 406: 124309. doi: 10.1016/j.jhazmat.2020.124309.

27 Son, A., Lee, J., Lee, C., Cho, K., Lee, J., and Hong, S.W. (2021). Persulfate enhanced photoelectrochemical oxidation of organic pollutants using self-doped TiO_2 nanotube arrays: effect of operating parameters and water matrix. *Water Research* 191: 116803. doi: 10.1016/j.watres.2021.116803.

28 Mafa, P.J., Mamba, B.B., and Kuvarega, A.T. (2020). Photoelectrocatalytic evaluation of EG-CeO_2 photoanode on degradation of 2,4-dichlorophenol. *Solar Energy Materials & Solar Cells* 208: 110416. doi: 10.1016/j.solmat.2020.110416.

29 Amedlous, A., Majdoub, M., Amaterz, E., Anfar, Z., and Benlhachemi, A. (2021). Synergistic effect of g-C_3N_4 nanosheets/Ag_3PO_4 microcubes as efficient n-p-type heterostructure based photoanode for photoelectrocatalytic dye degradation. *Journal of Photochemistry & Photobiology, A: Chemistry* 409: 113127. doi: 10.1016/j.jphotochem.2020.113127.

30 Wang, F., Ou, R., Yu, H., Lu, Y., Qu, J., Zhu, S., Zhang, L., and Huo, M. (2021). Photoelectrocatalytic PNP removal using C_3N_4 nanosheets/α-Fe_2O_3 nanoarrays photoanode: performance, mechanism and degradation pathways. *Applied Surface Science.* doi: 10.1016/j.apsusc.2021.150597.

31 Davaslıoglu, I.Ç., Özdokur, K.V., Koçak, S., Çırak, Ç., Çaglar, B., Çırak, B.B., and Ertas, F.N. (2021). WO_3 decorated TiO_2 nanotube array electrode: preparation, characterization and superior photoelectrochemical performance for Rhodamine B dye degradation. *Journal of Molecular Structure* 1241: 130673. doi: 10.1016/j.molstruc.2021.130673.

32 Xue, J., Li, J., Bi, Q., Tang, C., Zhang, L., and Leng, Z. (2020). Yb-substitution triggered $BiVO_4$-Bi_2O_3 heterojunction electrode for photoelectrocatalytic degradation of organics. *Colloids and Surfaces A* 593: 124640. doi: 10.1016/j.colsurfa.2020.124640.

33 Mo, Z., Wang, K., Yang, H., Ou, Z., Tong, Y., Yu, T., Wang, Y., Tsiakaras, P., and Song, S. (2020). Heterojunction architecture of pTTh nanoflowers with CuO_x nanoparticles hybridized for efficient photoelectrocatalytic degradation of organic pollutants. *Applied Catalysis B: Environmental* 277: 119249. doi: 10.1016/j.apcatb.2020.119249.

34 Zhang, X., Nengzi, L.C., Li, B., Liu, L., and Cheng, X. (2020). Design and construction of a highly efficient photoelectrocatalytic system based on dual-Pd/TNAs photoelectrodes for elimination of triclosan. *Separation and Purification Technology* 235: 116232. doi: 10.1016/j.seppur.2019.116232.

35 Ma, Q., Zhang, H., Guo, R., Li, B., Zhang, X., Cheng, X., Xie, M., and Cheng, Q. (2018). Construction of CuS/TiO_2 nano-tube arrays photoelectrode and its enhanced visible light photoelectrocatalytic decomposition and mechanism of penicillin G. *Electrochimica Acta* 283: 1154–1162. doi: 10.1016/j.electacta.2018.07.026.

36 Yang, R., Ji, Y., Zhang, J., Zhang, R., Liu, F., Chen, Y., Liang, L., Han, S., Yu, X., and Liu, H. (2019). Efficiently degradation of polyacrylamide pollution using a full spectrum Sn_3O_4 nanosheet/Ni foam heterostructure photoelectrocatalyst. *Catalysis Today* 335: 520–526. doi: 10.1016/j.cattod.2019.02.019.

37 Mafa, P.J., Patala, R., Mamba, B.B., Liu, D., Gui, J., and Kuvarega, A.T. (2020). Plasmonic Ag_3PO_4/EG photoanode for visible light-driven photoelectrocatalytic degradation of diuretic drug. *Chemical Engineering Journal* 393: 124804. doi: 10.1016/j.cej.2020.124804.

38 Changanaqui, K., Brillas, E., Alarcon, H., and Sires, I. (2020). ZnO/TiO_2/Ag_2Se nanostructures as photoelectrocatalysts for the degradation of oxytetracycline in water. *Electrochimica Acta* 331: 135194. doi: 10.1016/j.electacta.2019.135194.

39 Teng, W., Xu, J., Cui, Y., and Yu, J. (2020). Photoelectrocatalytic degradation of sulfadiazine by Ag_3PO_4/MoS_2/TiO_2 nanotube array electrode under visible light irradiation. *Journal of Electroanalytical Chemistry* 868: 114178. doi: 10.1016/j.jelechem.2020.114178.

40 Xue, J., Li, S., Lei, D., Bi, Q., Tang, C., and Xu, N. (2021). Multi-component Zn_2SnO_4/MoS_2/Ag/AgCl for enhancing solar-driven photoelectrocatalytic activity. *Applied Surface Science* 544: 148922. doi: 10.1016/j.apsusc.2020.148922.

41 Zhao, X., Guo, L., and Qu, J. (2014). Photoelectrocatalytic oxidation of Cu-EDTA complex and electrodeposition recovery of Cu in a continuous tubular photoelectrochemical reactor. *Chemical Engineering Journal* 239: 53–59. doi: 10.1016/j.cej.2013.10.088.

42 Zhao, Y., Chang, W., Huang, Z., Feng, X., Ma, L., Qi, X., and Li, Z. (2017). Enhanced removal of toxic Cr(VI) in tannery wastewater by photoelectrocatalysis with synthetic TiO_2 hollow spheres. *Applied Surface Science* 405: 102–110. doi: 10.1016/j.apsusc.2017.01.306.

43 Xiao, K., Zhou, B., Chen, S., Yang, B., Zhang, J., and Zhu, C. (2019). Enhanced photoelectrocatalytic breakdown of Cu-cyanide complexes and copper recovery using photoelectrogenerated free chlorine. *Electrochemistry Communications* 100: 34–38. doi: 10.1016/j.elecom.2019.01.018.

44 Jiang, W., Liu, Y., Liu, F., Li, F., Shen, C., Yang, B., Huang, M., Liu, J., Wang, Z., and Sand, W. (2020). Ultra-fast detoxification of Sb(III) using a flow-through TiO_2-nanotubesarray-mesh based photoelectrochemical system. *Chemical Engineering Journal* 387: 124155. doi: 10.1016/j.cej.2020.124155.

45 Spanu, D., Santo, V.D., Malara, F., Naldoni, A., Turolla, A., Antonelli, M., Dossi, C., Marelli, M., Altomare, M., Schmuki, P., and Recchia, S. (2018). Photoelectrocatalytic oxidation of As(III) over hematite photoanodes: a sensible indicator of the presence of highly reactive surface sites. *Electrochimica Acta* 292: 828–837. doi: 10.1016/j.electacta.2018.10.003.

46 Zhang, J., Djellabi, R., Zhao, S., Qiao, M., Jiang, F., Yan, M., and Zhao, X. (2020). Recovery of phosphorus and metallic nickel along with HCl production from electroless nickel plating effluents: the key role of three-compartment photoelectrocatalytic cell system. *Journal of Hazardous Materials* 394: 122559. doi: 10.1016/j.jhazmat.2020.122559.

47 Chen, Y., Zhao, X., Guan, W., Cao, D., Guo, T., Zhang, X., and Wang, Y. (2017). Photoelectrocatalytic oxidation of metal-EDTA and recovery of metals by electrodeposition with a rotating cathode. *Chemical Engineering Journal* 324: 74–82. doi: 10.1016/j.cej.2017.05.031.

48 Guo, T., Dang, C., Tian, S., Wang, Y., Cao, D., Gong, Y., Zhao, S., Mao, R., Yang, B., and Zhao, X. (2018). Persulfate enhanced photoelectrocatalytic degradation of cyanide using a $CuFe_2O_4$ modified graphite felt cathode. *Chemical Engineering Journal* 347: 535–542. doi: 10.1016/j.cej.2018.04.143.

49 Egerton, T.A., Janus, M., and Morawski, A.W. (2006). New TiO_2/C sol–gel electrodes for photoelectrocatalytic degradation of sodium oxalate. *Chemosphere* 63: 1203–1208. doi: 10.1016/j.chemosphere.2005.08.074.

50 Feng, X., Shang, J., and Chen, J. (2017). Photoelectrocatalytic reduction of hexavalent chromium by Ti-doped hydroxyapatite thin film. *Journal of Molecular Catalysis A: Chemical* 427: 11–17. doi: 10.1016/j.molcata.2016.09.031.

51 Song, Y., Fang, W., Liu, C., Sun, Z., Li, F., and Xu, L. (2020). Effect of mixed Mo/W polyoxometalate modification on photoelectrocatalytic activity of CdS nanocrystals for arsenic(III) oxidation. *Journal of Physics and Chemistry of Solids* 141: 109395. doi: 10.1016/j.jpcs.2020.109395.

52 Rahman, M.A., Hogan, B., Duncan, E., Doyle, C., Krassoi, R., Rahman, M.M., Naidu, R., Lim, R.P., Maher, W., and Hassler, C. (2014). Toxicity of arsenic species to three fresh water organisms and biotransformation of inorganic arsenic by fresh water phytoplankton (Chlorella sp. CE-35). *Ecotoxicology and Environmental Safety* 106: 126–135. doi: 10.1016/j.ecoenv.2014.03.004.

53 Segura, S.G., Arotiba, O.A., and Brillas, E. (2021). The pathway towards photoelectrocatalytic water disinfection: review and prospects of a powerful sustainable tool. *Catalysts* 11: 921. doi: 10.3390/catal11080921.

54 Cho, M., Cates, E.L., and Kim, J.H. (2011). Inactivation and surface interactions of MS-2 bacteriophage in a TiO_2 photoelectrocatalytic reactor. *Water Research* 45: 2104–2110. doi: 10.1016/j.watres.2010.12.017.

55 Huang, H.H., Anand, A., Lin, C.J., Lin, H.J., Lin, Y.W., Harroun, S.G., and Huang, C.C. (2021). LED irradiation of halogen/nitrogen-doped polymeric graphene quantum dots triggers the photodynamic inactivation of bacteria in infected wounds. *Carbon* 174: 710–722. doi: 10.1016/j.carbon.2020.11.092.

56 Li, G., Liu, X., Zhang, H., Wong, P.K., An, T., Zhou, W., Li, B., and Zhao, H. (2014). Adenovirus inactivation by in situ photocatalytically and photoelectrocatalytically generated halogen viricides. *Chemical Engineering Journal* 253: 538–543. doi: 10.1016/j.cej.2014.05.059.

57 Rogers, J.E., Abraham, B., Rostkowski, A., and Kelly, L.A. (2001). Mechanisms of photoinitiated cleavage of DNA by 1,8-naphthalimide derivatives. *Photochemistry and Photobiology* 74: 521–531. doi: 10.1562/0031-8655(2001)0740521MOPCOD2.0.CO2.

58 Lv, K. and Xu, Y.M. (2006). Effects of polyoxometalate and fluoride on adsorption and photocatalytic degradation of organic dye X3B on TiO2: the difference in the production of reactive species. *Journal of Physical Chemistry B* 110: 6204–6212.

59 Li, G., Liu, X., Zhang, H., Wong, P.K., An, T., and Zhao, H. (2013). Comparative studies of photocatalytic and photoelectrocatalytic inactivation of E. coli in presence of halides. *Applied Catalysis B: Environmental* 140–141: 225–232. doi: 10.1016/j.apcatb.2013.04.004.

60 Brugnera, M.F., Miyata, M., Zocolo, G.J., Leite, C.Q.F., and Zanoni, M.V.B. (2012). Inactivation and disposal of by-products from Mycobacterium smegmatis by photoelectrocatalytic oxidation using Ti/TiO_2-Ag nanotube electrodes. *Electrochimica Acta* 85: 33–41. doi: 10.1016/j.electacta.2012.08.116.

61 Espíndola, R.B.D., Casamada, C.B., Martínez, S.S., Araujo, R.M., Brillas, E., and Sirés, I. (2019). Photoelectrocatalytic inactivation of Pseudomonas aeruginosa using an Ag decorated TiO_2 photoanode. *Separation and Purification Technology* 208: 83–91. doi: 10.1016/j.seppur.2018.05.005.

62 Kang, Q., Lu, Q.Z., Liu, S.H., Yang, L.X., Wen, L.F., Luo, S.L., and Cai, Q.Y. (2010). A ternary hybrid CdS/Pt–TiO_2 nanotube structure for photoelectrocatalytic bactericidal effects on Escherichia Coli. *Biomaterials* 31: 3317–3326. doi: 10.1016/j.biomaterials.2010.01.047.

63 Su, S.F., Ye, L.M., Tian, Q.M., Situ, W.B., Song, X.L., and Ye, S.Y. (2020). Photoelectrocatalytic inactivation of Penicillium expansum spores on a Pt decorated TiO_2/activated carbon fiber photoelectrode in an all-solid-state photoelectrochemical cell. *Applied Surface Science* 515: 145964. doi: 10.1016/j.apsusc.2020.145964.

64 Borjac, J., Joumaa, M.E., Kawach, R., Youssef, L., and Blake, D.A. (2019). Heavy metals and organic compounds contamination in leachates collected from Deir Kanoun Ras El Ain dump and its adjacent canal in South Lebanon. *Heliyon* 5: e02212. doi: 10.1016/j.heliyon.2019.e02212.

65 Liu, C., Ding, Y., Wu, W., and Teng, Y. (2016). A simple and effective strategy to fast remove chromium (VI) and organic pollutant in photoelectrocatalytic process at low voltage. *Chemical Engineering Journal* 306: 22–30. doi: 10.1016/j.cej.2016.07.043.

66 Zhang, Y., Wang, Q., Lu, J., Wang, Q., and Cong, Y. (2016). Synergistic photoelectrochemical reduction of Cr(VI) and oxidation of organic pollutants by g-C_3N_4/TiO_2-NTs electrodes. *Chemosphere* 162: 55–63. doi: 10.1016/j.chemosphere.2016.07.064.

67 Zhong, J.S., Wang, Q.Y., Zhou, J., Chen, D.Q., and Ji, Z.G. (2016). Highly efficient photoelectrocatalytic removal of RhB and Cr(VI) by Cu nanoparticles sensitized TiO_2 nanotube arrays. *Applied Surface Science* 367: 342–346. doi: 10.1016/j.apsusc.2016.01.189.

68 Wang, Q., Zheng, Q., Jin, R., Gao, S., Yuan, Q., Rong, W., and Wang, R. (2017). Photoelectrocatalytic removal of organic dyes and Cr(VI) ions using Ag_3PO_4 nanoparticles sensitized TiO_2 nanotube arrays. *Materials Chemistry and Physics* 199: 209–215. doi: 10.1016/j.matchemphys.2017.06.051.

69 Wang, Q., Jin, R., Yin, C., Wang, M., Wang, J., and Gao, S. (2017). Photoelectrocatalytic removal of dye and Cr(VI) pollutants with Ag_2S and Bi_2S_3 co-sensitized TiO_2 nanotube arrays under solar irradiation. *Separation and Purification Technology* 172: 303–309. doi: 10.1016/j.seppur.2016.08.028.

70 Adhikari, S., Eswar, N.K., Sangita, S., Sarkar, D., and Madras, G. (2018). Investigation of nano Ag-decorated SiC particles for photoelectrocatalytic dye degradation and bacterial inactivation. *Journal of Photochemistry and Photobiology A: Chemistry* 357: 118–131. doi: 10.1016/j.jphotochem.2018.02.017.

71 He, H., Sun, S., Gao, J., Huang, B., Zhao, T., Deng, H., Wang, X., and Pan, X. (2020). Photoelectrocatalytic simultaneous removal of 17α-ethinylestradiol and E. coli using the anode of Ag and SnO_2-Sb 3D-loaded TiO_2 nanotube arrays. *Journal of Hazardous Materials* 398: 122805. doi: 10.1016/j.jhazmat.2020.122805.

Index

Note: Page numbers followed by "*f*" and "*t*" refers to figures and tables respectively

Photocatalysts and Electrocatalysts in Water Remediation: From Fundamentals to Full Scale Applications, First Edition. Edited by Prasenjit Bhunia, Kingshuk Dutta, and S. Vadivel.